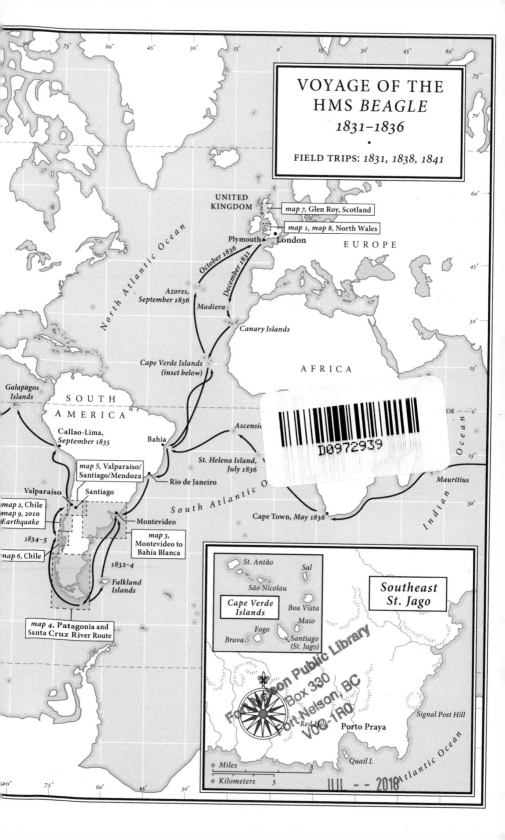

VOYAGE OF THE HMS *BEAGLE*
1831–1836
•
FIELD TRIPS: *1831, 1838, 1841*

UNITED KINGDOM

map 7, Glen Roy, Scotland

map 1, map 8, North Wales

Plymouth • London

EUROPE

North Atlantic Ocean

October 1836

December 1831

Azores, *September 1836*

Madiera

Canary Islands

Cape Verde Islands (inset below)

AFRICA

Galapagos Islands

SOUTH AMERICA

Callao-Lima, *September 1835*

Bahia

Ascension

St. Helena Island, *July 1836*

South Atlantic O

Mauritius

Indian Ocean

map 5, Valparaíso/ Santiago/Mendoza

Rio de Janeiro

Cape Town, *May 1836*

Valparaíso   Santiago

map 2, Chile
map 9, 2010 Earthquake

Montevideo

1834–5

map 3, Montevideo to Bahía Blanca

map 6, Chile

1832–4

Falkland Islands

map 4, **Patagonia and Santa Cruz River Route**

Southeast St. Jago

St. Antão

Sal

São Nicolau

Cape Verde Islands

Boa Vista

Maio

Fogo

Brava

Santiago (St. Jago)

Red Hill

Porto Praya

Signal Post Hill

Quail I.

Atlantic Ocean

o Miles   3

o Kilometers   3

DARWIN'S
FIRST
THEORY

DARWIN'S FIRST THEORY

EXPLORING DARWIN'S QUEST
TO FIND A THEORY OF THE EARTH

ROB WESSON

PEGASUS BOOKS

NEW YORK LONDON

DARWIN'S FIRST THEORY

Pegasus Books Ltd.
148 W 37th Street, 13th Floor
New York, NY 10018

First Pegasus Books edition April 2017

Interior design by Maria Fernandez

Library of Congress Cataloging-in-Publication Data is available.

ISBN: 978-1-68177-316-2

10 9 8 7 6 5 4 3 2 1

Printed in the United States of America
Distributed by W. W. Norton & Company

For all those who have suffered
from the paroxysms of Planet Earth . . .

And for those curious ones who render
its behavior ever less mysterious.

CONTENTS

ILLUSTRATIONS LIST

A Note on Quotations, Spelling, and Names

I n the nearly two centuries since the voyage of the HMS *Beagle*, styles, spelling, and conventions in both English and Spanish, as well as geographic names, have changed, leading to consequences ranging from quaint amusement to downright confusion.

In the quotations from letters, journals, articles, and books from the period, I have tried to maintain the original spelling and punctuation. Spelling was not as standardized then as it is now, and Darwin himself was not a particularly good speller. His confusion with compass directions offers a hint of dyslexia. The punctuation and construction in his notes and letters reflect the style of the times, but are also to some extent idiosyncratic. Among Darwin's correspondents, Charles Douglas, an Englishman, residing on the Chilean island of Chiloé, may be the worst speller ever to allude to Milton's *Paradise Lost*, although Douglas likely didn't have the opportunity for an education to match his native intelligence. To capture the spirit and tenor of the times I have tried to preserve these mannerisms, quirks, and foibles—warts and all—only interfering when the meaning is obscured.

Geographic names and their spellings present another set of problems. The country that we now know as Chile, was sometimes referred to as Chili. The city in Chile and the island in the Cape Verde Islands, both now known universally as Santiago, were referred to in English as St. Jago. In 1834, shortly after the departure of the *Beagle*, the name of the city on Chiloé Island, previously known as San Carlos de Chiloé, was formally changed to Ancud. The standardization of the spelling of some geographic names derived from Mapudungun and its relatives, the languages spoken by the native peoples of south-central Chile and Argentina, remains a work in progress. The translation of the sounds *hua* and *hui* are particular problems. What was called Huafo Island by Robert FitzRoy and Darwin, is now Guafo Island. Huamblin Island is now Guamblin Island. Darwin and FitzRoy also used English names for several places for which Spanish names are commonly used today. In quotations and where the meaning is clear, I have used the names in the original sources. Where confusion might arise, I have used the modern name.

In the U.K., Welsh and Scottish place names present their own special charm to those of us speakers of English who have not grown up with them. For me—a product of the State of Washington—Seattle and Suiattle, Snohomish and Skykomish, all roll off my tongue, but Betws-y-Coed catches me up short.

Another possible source of confusion is that there are several rivers named Río Negro in South America. The two that I mention herein, are first, the most important river of Uruguay, and the second, a river farther south in Argentina, in northern Patagonia.

I have tried to provide a path for the reader through this linguistic and nomenclatural obstacle course that is, at the same time, true to the original, yet also intelligible to the modern reader, especially to one who is familiar with the modern names or who may wish to consult a map or atlas. I beg the reader's forgiveness for any shortcomings or lapses that may remain.

Last but not least, and as unlikely as it might seem, the two men that I met at the potato field in Lipimávida, Chile—one the owner, the other the overseer and laborer who was caught in the tsunami of 2010—as well as a school administrator in Tirúa were all named Don José Luis. To avoid confusion, and with apologies to both, I refer to two of these gentlemen by their last names, the land owner in Lipimávida as Señor Ruíz, and the school administrator in Tirúa as Señor Montero.

A Peculiar Obsession

His eyes told the story. Dark eyes, rimmed with red. Sad, weary, haunted eyes. They didn't tell the whole story, not by a long shot. But one person's story, his story.

Behind us across the dunes I could hear the muffled crash of breakers on the beach. We stood in the jumble of his yard, strewn with scattered piles of boards and odd bits of furniture. Don José Luis, a slender, lanky man with a slight stoop, wore a broad, flat-brimmed straw hat. It shaded those eyes from the sun. He was not so much old as worn. His slow gait, his roughened hands, his deeply lined face patched with gray stubble, spoke of work in the potato field a hundred yards down the road. "A man of the country-side," Marco later said. His small, pink one-story house lay before us, twisted off its foundation, reminiscent of Auntie Em's house from *The Wizard of Oz*, windows boarded, one wall patched with fiberboard. But the eyes of Don José Luis—mournful, downcast, remembering eyes—those eyes told his story of the tsunami.

Marco pulled out his notebook and began the interview. Don José Luis spoke quietly and slowly, answering Marco's questions. I looked on, struggling to follow the colloquial Chilean Spanish. Marco read his notes aloud as he wrote, checking them with Don José Luis for accuracy. On the night of February 27, 2010, José Luis Díaz Farías was sleeping in his house located only a few yards from the sea. Between three and four in the morning the ground began to shake. It was a large earthquake that lasted about two minutes. "It moved the whole house, making it jump," he said. After the shaking stopped, he checked on his neighbor next door. Then he and his sister, Mercedes, began to prepare some coffee to calm their nerves.

"Why didn't you go up the hill when you first felt the earthquake?" Marco asked. This after all is the first tenet of tsunami education and understanding tsunamis. Helping the people of Chile to prepare for them is Marco's passion.

"Nobody told us," Don José Luis said. Nobody told them. Marco and I shook our heads later as we repeated his words.

As they fixed the coffee, suddenly, through the window his sister saw the sea was rising.

"I was surprised to see her move so fast," he said. He tried to follow her, as she ran across the road toward the hill, but his legs would not move as quickly. He couldn't keep up. The first wave of the tsunami caught him as he was crossing the road, the water rising to his knees. Then the water began to recede, tugging at his legs, dragging him out to sea. He grabbed the branches of a large bush and held on. Held on for dear life.

Finally escaping the grip of the receding water he made his way up the hill, joining his sister. Together, they watched the second wave approach, seeing it clearly thanks to the full moon. This wave rose much higher than the first, already carrying debris. A horrific din rumbled through the night, the roar of breaking wood and glass, as the wave crushed and shattered the houses along the road.

We walked around the remains of Don José Luis's house lying akimbo off its foundation, one of the few nearby that hadn't been completely washed away, and continued a few yards down the road. A hedge of bushes rose above our heads. Don José Luis stopped at a gap in the hedge. Here, he said, showing us the branch. The branch that saved his life. He grasped the branch, and looked to the sky, reenacting his salvation.

Marco Cisternas and I had come to the tiny settlement of Lipimávida along the Chilean coast to study the effects of this devastating tsunami. It was neither our first visit here, nor to the potato field where Don José Luis earned his living. But it was our first since the giant earthquake and tsunami. On visits before the earthquake we had found a layer of black beach sand buried four feet down in the potato field, the trace, we suspected, of a historic tsunami. But we had harbored doubts. Was the buried sand layer evidence of a tsunami, or of something else? We puzzled through all the possibilities that we could think of. Now a fresh layer of sand covered the potato field. Slabs of asphalt, ripped from the road, lay tossed about like dry leaves scattered by an autumn wind. Nearby beach houses lay in shambles, or just gone, completely washed from their foundations. Our doubts ebbed. Our suspicions about that buried layer of black sand were almost certainly correct. It had been deposited by a tsunami.

And tsunamis at Lipimávida now had a human face—the sad, red-rimmed eyes of Don José Luis.

◆

One hundred and seventy five years before, two other inquiring men began their study of a similar earthquake and tsunami, possibly the very tsunami responsible for the layer of sand buried in the potato field at Lipimávida.

That day, February 20, 1835, dawn broke peacefully enough. The small but stout sailing ship HMS *Beagle* bobbed gently at anchor near Valdivia, Chile, about 350 miles south of Lipimávida. Her sails were folded and stowed below. Her three straight masts, jutting bowsprit, and bare angular rigging contrasted sharply with the soft green tangle of rain forest lining the banks of the estuary. The tide flowed out as the sun rose. Ripples lapped at the *Beagle's* sides as she tugged at the cables holding her, creaking softly. The pungent, organic smell of sea life and the squawking of gulls hung in the still, austral summer air. Onshore a few hundred yards away rose the massive stone block walls of Corral Fort. The fort, almost in ruins, its cannons more a hazard than a threat, belonged to a string of decaying fortresses built by the Spanish to defend the town of Valdivia, ten miles up the Calle-Calle River. But this morning, so serene at its outset, would end rather differently, and for none more so than for the young man who would become the *Beagle* expedition's most famous member, its unofficial geologist, Charles Darwin.

Arriving ten days before, the *Beagle* crew had set up an observation point on the shore near the fort. Based in tents, the officers carried out a familiar suite of measurements: determining angles to the sun and stars, the direction and intensity of the magnetic field, the time and height of the tide, all to the exacting standards of their captain, Robert FitzRoy. In their whaleboats and cutter, the crew plied the surrounding waters, making soundings with pole and lead line to measure the water's depth, and mapping the details of the coast, noting especially rocks and shoals that could present a hazard to an unwary ship. Their bosses at the British Admiralty envisioned that these charts and observations would facilitate British merchant trade with the newly independent Republic of Chile.

The officers ashore carefully gauged the altitude of the sun to determine the precise moment of noon. They compared this moment with the time back at the Royal Observatory in Greenwich, read from an array of carefully tended, wind-up chronometers on board the *Beagle*—chronometers suspended in gimbals and packed in sawdust to insulate them from shipboard jarring and changing temperature—painstakingly determining the longitude at the observation point. A reference for their local map, this longitude would also become a link in a chain of observations of longitude circling the globe. Their results near Corral Fort differ by only about three miles from what would be determined in a few seconds by a GPS (or more formally, Global Positioining System) receiver today.

At 6:00 A.M. the ship's meteorological log reported an air temperature of 59°F (it would rise to 67°F by noon) and a water temperature of 59.5°F under a blue sky with drifting clouds. Later FitzRoy and some of the officers ran errands in town along its wooden-paved streets. A sketch made by one shows a few residents of the sleepy town ambling through the small plaza beneath the twin bell towers of the modest church, an old man with a broad-brimmed hat and a cane, a boy chasing a stray dog with a stick, a woman carrying a bowl on her head, an abandoned two-wheel cart parked nearby.

In late morning, Darwin lay not far away taking a rest from his explorations in the woods. Perhaps he was just relaxing, or maybe suffering from his disappointment at the obstacle to studying the geology of the region: the dense, green, broad-leafed rain forest covering the countryside and obscuring the underlying rocks.

But at 11:40 A.M., the unnatural swaying of the trees of the forest, the creaking of buildings in town, the rumbling of a giant earthquake, jolted them all to attention.

Darwin came to quickly, recording later in his journal, "It came on suddenly, and lasted two minutes; but the time appeared much longer. The rocking of the ground was most sensible . . ." He was perhaps two hundred miles from the source of the earthquake, so the shaking where he sat in the forest was much less intense than at points closer to the center of destruction.

In town, FitzRoy also felt the shock, "At Valdivia the shock began gently, increased gradually during two minutes, was at its strongest about one minute, and then diminished." He too had no difficulty standing, "but the houses waved and cracked . . . All the dwelling-houses being strongly built of wood, withstood the shock."

The light wooden buildings in Valdivia withstood the shaking fairly well. Later a man and a woman, apparently trying to take advantage of the unusually low water level to gather shellfish across the bay from the Corral Fort, drowned when the waters of the small tsunami rose more quickly than they had anticipated. But as Darwin and FitzRoy would learn in the days ahead, the towns of Concepción and Talcahuano, much closer to the source, were not so fortunate. The earthquake shattered the brick and adobe buildings of Concepción, and the waves of the tsunami dashed what was left of Talcahuano after the shaking calmed. When the *Beagle* arrived twelve days later, both towns lay in ruins.

Through the weeks that followed, Darwin and FitzRoy recorded every detail they could find about these peculiar and terrifying geologic events. Thanks to Darwin and FitzRoy this earthquake would eventually become the pivotal evidence for the nineteenth-century notion of how continents rose from the sea, of how even the most dramatic of the earth's features, including the mighty Andes, were formed; the uplift caused by this earthquake one more step in their skyward ascent, the *vera causa* for this great range, *in vivo*.

◆

The earthquake on February 27, 2010, struck the same area, had similar consequences, and was a near repeat of the earthquake in 1835. I learned of the earthquake from CNN as I ate powdered sugar doughnuts in the breakfast room of a Holiday Inn Express in southern Arizona. I was on a bicycle tour with my wife and some friends, to escape the winter weather of Colorado, and to pay back my wife who'd suffered winter in Colorado as I'd spent January enjoying summer in Chile.

I'd spent that January as I'd spent the two previous winters—and would spend the six succeeding winters—following the geologic footsteps of Darwin and FitzRoy in southern South America. I roamed Chile, Argentina, and Uruguay, and tramped across the Andes. My aim was to explore the geologic legacy of Darwin and FitzRoy, to try to understand how these geologic pioneers had inter-preted what they saw—then still in the early days of the science, to fast-forward to the present to consider how the same phenomena are interpreted today, and to reflect on what this kind of science has meant, and will continue to mean for the people living on our planet. Now Darwin and FitzRoy's earthquake had happened again.

But I am getting way ahead of the story. Let's get a couple of things straight.

First, I am a geologist. I've spent a career studying earthquakes. I've earned my living puzzling about them. They fascinate me. I'm an earthquake guy. But it's not just earthquakes. It's mountains and glaciers and volcanoes and floods and sea level and climate. I am cap-tivated by the way the earth changes. How it's changed in the past and how it will change in the future. And, in truth, it's not only geology. It's how Homo sapiens deals with these changes in particular.

Second, until middle age I knew almost nothing about Darwin, FitzRoy, or the *Beagle* expedition. I didn't know that Darwin had joined the expedition to serve as its geologist, or what the two men had achieved in their investigations of the earth.

A decade ago, indulging my passion for wild places, I went on vacation to Patagonia. For reading material I took along Darwin's *The Voyage of the Beagle*. I hiked among the jagged teeth of the Torres del Paine in southern Chile. I climbed to the glaciers at the base of the sheer granite monolith named for FitzRoy. I traversed the dry plains of Argentine Patagonia. I cruised the waters of the Beagle Channel. I was in a geologist's heaven. On buses and during evenings I read Darwin's groundbreaking descriptions of the wondrous geology he encountered; from the immense to the microscopic, of the landscape of Patagonia, of fossil shells from a shallow sea now at nearly 14,000 feet on the crest of the Andes, and of tiny bubbles frozen in a volcanic bomb—a lump of lava that exploded during the eruption of a volcano on Ascension Island in the Atlantic. When I came home to Colorado, I kept reading.

While I was amazed at Darwin's powers of observation, I was even more impressed by his ability to synthesize and explain what he saw. I also became intrigued by FitzRoy, his fellow explorer, who, although not strictly speaking a geologist, was fascinated by science, even though he later reverted to Biblical explanations for the origin of the earth. The two men seemed to embody the sense of inquiry and adventure that had drawn me to geology in the first place, but had been missing in my nearly forty years of sitting in front of a computer screen, attending tedious meetings, and supporting other geologists who experienced the challenges and rewards of working in the field instead. Had I taken the wimp's way out, focusing on quantitative analysis and management, instead of confronting geology face to face? I needed to find out if I could do it too. Even though my own fascination with the earth had been kindled in the field, I had drifted away.

The seeds of this passion initially began to sprout in the summer before beginning ninth grade. I had the chance to join a ten-day hiking trip with a group of Explorer Scouts into the wilderness of the North Cascade Mountains near my home in Seattle. One day, leaving the evergreen forest, we climbed above the tree line through meadows of blooming pink heather, finally reaching bare rocks at the foot of a glacier. We camped among stark boulders that seemed to shout that they had only recently escaped the clutches of ice. The next day, beginning before dawn, we trekked across the glacier, with crampons and ropes, to climb Glacier Peak, one of Washington's less-known volcanoes. Approaching the summit we climbed above the glacier onto a ridge of sandy ash and loose, rotten pieces of lava, riddled with small pits and holes, like chunks of sponge frozen in stone. Even to a fourteen-year-old kid, it seemed pretty obvious that these rocks had cooled from a froth during an eruption. I grabbed a few, threw them in my pack and carried them down, even as the day turned into one of the longest and most arduous of my young life. We returned to our camp near dark, exhausted.

During the following school year, in a class on the geography and history of Washington State, the teacher listed the state's volcanoes. He did not include Glacier Peak. I raised my hand to object. Mr. Davis, a short, chubby, well-liked teacher with a horseshoe of white fringe surrounding his shiny pate—his real claim to fame was that he played music on a saw at school assemblies—dismissed my arguments summarily. Unbowed, in the notebook that each of us had to prepare as individual class projects, I Scotch-taped a tiny chip of the lava, evidence to support the position of the young smarty pants. Today volcanologists view the resumption of explosive eruptions at Glacier Peak as a serious possibility.

Later in high school, intrigued by the mechanics of glaciers, I even tried to make a working model of a glacier out of a big slab of raspberry Jell-O resting on a tilted sheet of plywood. The experiment proved a dismal failure and the subject of great hilarity to the teenage friend who witnessed the Jell-O slide down and off the plywood with a disappointing plop. But from there—trying to understand how a glacier moves—it really wasn't all that big a step to puzzling about earthquakes.

As a university student, I was seduced by the exactitude of mathematical models away from the critical analysis and synthesis of observations. Now five decades later, I traded my chair in front of a computer for a sleeping bag and a trenching shovel to get back to the field. Here it makes a difference whether the sun is shining or the rain is pouring down, the air still or the wind howling. And here I could learn by experiencing and seeing the geology—the coasts, the mountains, the rocks, the rivers, the dirt—for myself, directly, rather than by taking someone else's word for it. I became obsessed without really being able to explain why, or what I hoped to learn, but I was hooked once again. I wanted to try my hand at parsing the landscape, to understand its history and development as Darwin had done.

This path would lead me across southern South America, and also to Snowdonia in Wales and Glen Roy in Scotland; to mountains and beaches; to caves where both ancient Patagonian people and now-extinct giant ground sloths lived only several thousand years ago; to dimly lit archives filled with scholars silently leafing through ancient documents; to the church in Wales where Darwin's college girlfriend lies buried; and to the *teatro municipal* in the Chilean village of Maullín filled with *chilenos* seeking answers to questions about their own experiences with an abruptly changing Planet Earth, their lives once again jolted by a giant earthquake; to my meeting with Don José Luis, and to his haunting eyes, looking skyward, seeking answers; and to reigniting my own passion for understanding the earth and Homo sapiens's place upon it.

PART ONE

DARWIN IN THE FIELD

CHAPTER ONE

The Lieutenant and the Beetle Collector

*If ever I left England again on a similar expedition,
I would endeavour to carry out a person qualified
to examine the land; while the officers, and myself,
would attend to hydrography.*

—Robert FitzRoy

*Geology is a capital science to begin, as it requires
nothing but a little reading, thinking & hammering.*

—Charles Darwin

For one week each December, members of a rather strange subspecies of Homo sapiens stride the morning streets of San Francisco. Thousands of them. Some might be hard to distinguish from the general population, maybe a touch more in need of a haircut, a few more beards among the men, the older ones perhaps a bit tweedier. Many, perhaps most, are easier to pick out. These wear Gore-Tex or fleece jackets from The North Face, Mammut, or Millet, and hiking shoes. Eyes rimmed by glasses, many bear backpacks or messenger bags. Their ages run from young to old, almost as

3

many women as men. Some speak German to those hurrying beside them, others Italian, French, Japanese, Spanish, and many, Chinese. But even the English speakers seem to communicate in a patois of familiar and unfamiliar words. They all scurry purposefully along the sidewalks from the BART, the Caltrain, the buses, and from the less expensive hotels of the city, all in the direction of the Moscone Center.

The dead giveaway is the tube.

Carried under their arms, held in their hands, or slung from a strap over their shoulders, these three-foot-long tubes—brown, orange, or white cardboard, black or gray plastic—provide a sufficient, if not necessary, condition for a positive identification. Rolled in each of these tubes is a form of scientific communication unknown in the time of Charles Darwin—the poster.

The members of this subspecies are all headed for a kind of convention, an unusual convention for San Francisco. Taxi drivers report that the hookers take vacation for this week each year, the annual fall meeting of the American Geophysical Union (AGU). What truly defines this subspecies is a passion for understanding the earth, its insides and out, its surroundings in space, how it got to be the way it is, where it's going in the future, and even, increasingly, the creatures and vegetation that grow upon it. These "geo-ists" (as in geochemist, geomorphologist, geophysicist, geobotanist . . .) come to listen to and to give talks, to view and present these posters, and, of course, to network.

These are scientific descendants of Charles Darwin and his captain, Robert FitzRoy of the *Beagle* expedition. It would be nearly impossible to find a single one of the presentations of the more than ten thousand at the meeting this year, that could not be traced (perhaps in some cases a little tortuously) to something that Darwin or FitzRoy wrote.

The problem is that so many have so much to say; the posters offer a partial solution. In contrast to a proper lecture, familiar to Darwin and FitzRoy, with tea, or perhaps after dinner, the presentation of these posters seems more like a combination of window shopping, a science fair, and speed dating. Rows of poster boards divide an immense exhibit hall at the Moscone Center into long aisles. Each presenter tacks his or her poster onto a preassigned space measuring four feet high by six feet wide. The poster typically contains too much text to read, but through a mix of words, graphs, maps, photos, and equations, describes some incremental advance of science. As the presenters stand in front, ready to explain, expand, and extol, the attendees stream past, stopping to read, chat, and sometimes challenge. Crowds gather at popular posters, blocking the aisles. The din of thousands of simultaneous conversations makes verbal communication difficult.

I've been coming to these meetings for four decades, to listen, to talk, to get new ideas, and to meet old friends, but this year I have a different mission. I must reach outside my comfortable circle of colleagues to try something that I haven't done before. This morning, as I join my fellow attendees on the crowded streets, I am headed for Moscone West, a massive metal and glass monument that is proof—despite the insinuations of twenty-first-century technology to the contrary—that human beings actually do want to meet face-to-face. My mission is to inveigle an invitation, an invitation that could spirit me away from this myriad of subspecialties and posters, and back to geology in the raw.

The AGU attendees represent different tribes, each with its own idiom. A certain kind of etiquette guides this intertribal event. Each attendee is most comfortable powwowing with the other members of his or her own tribe; there the jargon is familiar, the common assumptions already agreed upon, the agreements and disagreements well-established, and knowledge of the tribal hierarchy, its history, and its foibles taken for granted. Cross-tribal interactions remain a sought-after, but elusive goal.

My own tribe, the seismologists, a large and domineering bunch, have planned dozens of arcane sessions with titles like "Rethinking Seismicity Declustering" and "Earthquake Source Inversion Under Scrutiny: Validation, Resolution, Robustness." Instrumental, rather than personal, on-the-spot observations are our thing. Other tribes, such as the paleoseismologists and tectonic geologists, are more field-oriented geologists who rely more on their own senses, rather than fancy gadgets. They sometimes refer to us as "black-boxers," to them a pejorative term.

But if I hope to get my feet into the footsteps of Darwin and FitzRoy—to truly bolt from my computer screen and reconnect with the essence of my profession—I must hop over these tribal boundaries. I want to join a real geologist, someone who is working in the field in South America on a problem related to one that intrigued Darwin and FitzRoy. I have been thinking for weeks and asking around. Now I have a lead. I am seeking out an acquaintance, Brian Atwater, a field geologist, and a very good one, who has worked on earthquakes and tsunamis around the world, but for me most importantly, in Chile. I've been acquainted with Brian for many years, but I don't know him well. I could have emailed him, but I wanted to do this in person. I've heard that he is planning a return trip to Chile. I need a few quiet moments to make my case and pop my question: Is there any chance that he might let me come along?

I slip into the back of a cavernous lecture room to listen to a series of talks about giant earthquakes of the kind that affected Sumatra in 2004 and southern Chile in 1960. As one speaker concludes, I notice out of the corner of my eye, my prey leaving through a side door. This is my chance, I think, as I race after him. Exiting the dark lecture room into the brightly lit corridor, I call after him, "Hi Brian, do you have a minute?"

Brian stops and turns, smiling politely, his eyes crinkling. "What's up?"

The gray creeping into his short curly hair and closely cropped beard only hint at Brian's high status within his own subtribe of field geologists, the paleoseismologists, not to mention his inclusion in *Time* magazine's list of the one hundred most influential people of 2005, a list that also included a then-little-known senator from Illinois, a future president of the United States.

"I hear that you are going to Chile," I blurt out, "and I was wondering if there is any chance that I might tag along?" stammering some of the reasons for my interest.

"¿Hablas español?" he asks.

"Un poquito [a little bit]," I reply, stretching the truth.

"Chileans will like that use of the diminutive," he chuckles, explaining that the trip is not yet for sure, but that it might be possible for me to come. He'll know in a few weeks. He will be in touch. In the meantime, he is meeting for lunch with some colleagues to discuss planning a future meeting in Chile to mark the fiftieth anniversary of the largest earthquake since the birth of seismology, the great Chilean earthquake of 1960. He suggests that I might want to come along. I instantly agree.

Of course it's not for sure, but I feel like I'm already on my way. I've taken the first, and possibly most difficult step.

◆

The 1820s—as Darwin and FitzRoy were coming of age—saw the British Empire settling into its "Imperial Century," an unrivaled global power. Its Royal Navy ruled the seas. The notion of the "white man's burden" flourished as a component of Britain's image of itself, even if Rudyard Kipling wouldn't write his poem for another seventy years. Back then geology amounted to little more than a new arrival, a captivating toddler, among the sciences. Physics had its Newton; chemistry its Lavoisier; but perhaps geology's best candidate up until then, James Hutton—who credited natural processes with the structure and composition of the earth's crust—wrote such cryptic prose that his ideas were not yet fully appreciated more than two decades after his death. Now Adam Sedgwick, Charles Lyell, and a vibrant band of English geologists were gathering momentum. But none of these come to modern lips as easily as Newton.

It's not that geology landed like a meteorite from the cosmos and led the canal builder William Smith to make the geologic map of

England and Wales that he finished in 1815, the triumph of practical science that Simon Winchester describes in *The Map That Changed the World*. No, as with almost everything in the history of science, one can find the fingerprints, albeit a little smudged by the passage of time, of both the Greeks led by Xenophanes and Aristotle, and the Persian, one-man MIT of the eleventh century, Avicenna. Two factors came together to get the ball of modern geology rolling in the nineteenth century, however. First, the Industrial Revolution brought practical needs to dig up useful minerals from mines—coal, iron, and copper—and to make big holes in the ground, like the construction of canals that William Smith supervised and provided the stimulus for his map. And second, the Age of Enlightenment allowed the curious to freely inquire about what the rocks and sediments had to say about how the modern landscape formed . . . and to question the Biblical version of events.

One central Biblical issue was the Deluge, the flood that tossed Noah and his ark around in the book of Genesis. Were the deposits of rocks and gravel scattered all around the countryside dumped at once, the result of Noah's great flood; or accumulated from many floods, large and small? (Some even argued that the complexity was just God's way of throwing the proto-geologists off track, a strange idea that has resurfaced in the arguments of the antievolutionists.) Was the earth shaped by apocalyptic cataclysms, or by processes going on all the time, processes that continue to the present day? These two points of view received fancy names: catastrophism and uniformitarianism. Ultimately FitzRoy would come down on one side, Darwin on the other.

In December 1831, the HMS *Beagle* set sail from Plymouth, England, on a voyage around the world. Only one of numerous expeditions launched by the British Admiralty during this era, it would take almost five years. Notwithstanding its impact on science, the voyage of the *Beagle* had a very practical purpose: to make nautical charts. Up until this point, the British lagged behind the Portuguese, Dutch, Spanish, and French in navigation and map-making. Britain had neither the incentive nor the technology to produce her own. Sir Francis Drake sailed around the world using not only captured Portuguese and Spanish charts, but also unfortunate Portuguese and Spanish pilots whom he kidnapped, then abandoned along the way. As late as the mid–seventeenth century, English seafarers suffered the indignity of being forced to use Dutch charts for navigating their own coasts and harbors.

The art and science of navigation involves finding the way from one place to another. Generally this requires knowledge of where you are, and a chart showing the location of your destination, the best ways to get there, and the places that might pose dangers along the way. Today we all learn in grade school about latitude, the angle measured north and south of the equator, and longitude, the angle now measured 180 degrees around the globe east and west from Greenwich, England. We learn that we can read these arcane coordinates off a map. But how does one learn the latitude and longitude of a point on a featureless plain, in a dense forest, or from the deck of a rolling ship far beyond the sight of land? With today's technology it's easy, but it wasn't until the end of the fifteenth century that determining latitude both at land and at sea—using an astrolabe or one of its descendants to measure the altitude of the sun or a star together with an almanac—became almost a piece of cake. And then by the late eighteenth century, the problem of determining longitude was conquered through accurate timekeeping. By comparing the local time from observing the sun or stars, with the time on a precise clock, called a chronometer, kept to show the time at a place of known longitude, the relative longitude could be fixed. In those days, while the English used Greenwich as the reference, or zero point, the French used Paris, and the Spanish, Cádiz. Greenwich was finally adopted as the international standard for the zero meridian in 1884.

Every hour of difference between the local time and the time on the chronometer represents fifteen degrees of longitude. As an example, say you were at the center of the Plaza de Armas, the main square in Chile's capital city of Santiago. By observing the sun, you determined the exact moment of noon. Checking your trusty chronometer at that instant, you would see that the time in Greenwich was 4:42:36 P.M., and doing the arithmetic you would find a west longitude of 70°39'1.7".

More accurate tools were developed through the 1700s for measuring the angles from the horizon to the sun, moon, and stars (the sextant), and between land-based points (the theodolite). A three-armed protractor, called a "station pointer," was invented to facilitate the plotting of a position on a chart from two angles measured between three known points. These new tools supplemented the previous old standbys of mapmaking: the plane table for plotting positions on a piece of paper, the chain for measuring distance, the compass for determining the direction of magnetic north, and the pole and lead-line for measuring, or sounding, water depth. By the 1820s Britain had the tools, the skills, and the imperative to make her own nautical charts.

In the first decades of the nineteenth century, as the former Spanish and Portuguese colonies along the southern coasts of South America gained independence—becoming Argentina, Bolivia, Chile, Peru, and Uruguay—their markets became objects of intense desire for British merchants. The passage around the southern tip of South America was critical to reaching the resources and markets along the west coast of the continent. The need for accurate charts of these coasts (at this time Bolivia still had one) became acute. Spain guarded her nautical charts as state secrets; they were available neither to Britain's Royal Navy nor to its merchant fleet. In 1823 the British Admiralty broke from tradition and allowed distribution of its own Admiralty charts to merchant vessels. Then in 1826 the British Admiralty targeted southern South America for new surveys and chart making, and launched an expedition under the overall command of Captain Phillip Parker King aboard the HMS *Adventure*. The expedition also included the HMS *Beagle*, then under the command of Captain Pringle Stokes.

From a base at dourly named Port Famine (now Puerto de Hambre), about halfway through the Straits of Magellan at the southern extremity of mainland South America (near what is today

Punta Arenas, Chile), Stokes and his men sailed west toward the Pacific Ocean, mapping the rocky and treacherous coasts. Stokes's journal tells the story of sunken rocks, dreary landscapes, heavy clouds, violent gales, mountainous seas, incessant rain, constant danger, and a sick crew. "Around us, and some of them distant no more than two-thirds of a cable-length [about four hundred feet], were rocky islets, lashed by tremendous surf; and, as if to complete the dreariness and utter desolation of the scene, even the birds seemed to shun its neighborhood." And then the punch line, "The weather was that in which (as Thompson [sic] emphatically says) 'the soul of a man dies in him.'"

Sitting one day in sunny Colorado, I sought out a copy of *The Seasons*, by James Thomson, the eighteenth-century Scottish poet, from which Stokes quotes this last line. As Stokes read the "Winter" section of the poem, with its descriptions of "the fierce-conflicting brine," "a thousand raging waves," "winds across the howling waste," it must have rung all too true to his surroundings. The whole section of the poem, including the line that Stokes quoted in his journal likely epitomized both his dismal environment and his deteriorating state of mind, and seems to portend the means he used to resolve his plight.

> . . . Thus Winter falls,
> A heavy gloom oppressive o'er the world,
> Through Nature shedding influence malign,
> And rouses up the seeds of dark disease.
> The soul of man dies in him, loathing life,
> And black with more than melancholy views.

Even through all the years and miles, I could glimpse his despair.

Stokes found the work exhausting, the weather abysmal, the loneliness of command unbearable. In August 1828, caught in the vise of depression, and alone in his cabin, Pringle Stokes shot himself in the head. He lingered for eleven days, a pistol ball embedded in his brain. When the end came at last, his crew buried him nearby at Port Famine, where his grave may be found today.

King, the expedition commander, spent much of Stokes's last days by his side, trying to comfort him, and, during his periods of lucidity, extracting what he could about Stokes's progress on the mission. He learned that the responsibility for much of what had been accomplished lay with Lieutenant William Skyring. He named Skyring acting captain of the *Beagle*. But upon return to Rio de Janeiro, Admiral Robert Otway, commander of the Royal Navy fleet in South American waters, overruled King, and named his own golden boy, his Flag Lieutenant Robert FitzRoy, to the job. In Otway's consideration, FitzRoy scored off the charts in two key categories: merit and connections. A brilliant, hardworking young officer, he also had serious pull through his upper-crust family connections in military and political circles, what the young naval officers of the day called "interest" or "a handle to his name."

Members of FitzRoy's family were, and indeed still are, aristocrats. From a puritan American perspective, it seems a bit odd that a family that traces its roots to the licentious romping of a bawdy king would be blue-blooded aristocrats. But that's the way it was, and is. One needs look no farther than the late Princess Diana to find a modern aristocrat who shared FitzRoy's lineage from the same wild seed. Of their mutual ancestor, Charles II, a friend wrote:

> Restless he rolls from whore to whore
> A merry monarch, scandalous and poor.

Despite these indelicate roots, and the name to prove it (FitzRoy after all means "son of the king"), through the intervening generations, FitzRoy's family trumpeted a tradition of public service. Robert's grandfather, a man sympathetic to the concerns of the pesky American colonists, served briefly as prime minister. Unfortunately he lost out to the hard-liners and the War of Independence followed shortly thereafter.

Young Robert was raised in this somber tradition of service and duty. His father, an army officer himself, shipped twelve-year-old

Robert off to the Royal Naval College to begin a career as an officer in the Royal Navy. A letter from the schoolboy to his father provides a glimpse into his adolescent frame of mind. "We had another Examination last Monday & I took three more places which were all that I could take for I am now at the head of the part of the College in which I stay, so I could not have taken any more, I think that I have got through pretty well as yet for I have not been reported either in or out of School & I have got a very good Character."

He excelled at school. His on-the-job training—literally learning the ropes high in the rigging of a ship—went as well. Displaying intelligence, leadership, and a bit of derring-do, as when he raced in a whaleboat filled with sailors to rescue some of their fellows shanghaied by a Danish captain in the harbor at Montevideo, he climbed the ladder rapidly. Thus buoyed by both his merit and his family connections, in December 1828, FitzRoy was appointed to replace poor Stokes as captain of the *Beagle*. He earned the trust of his new crew, excelled at meeting the challenges of the elements, and delighted in the discipline and science of chart making.

◆

As a boy Charles Darwin wandered the woods and fields of Shropshire, working with his father in the greenhouse and with the gardener, taking notes on the germination and blooming of their plants. He learned to hunt grouse and rabbits, and conducted chemistry experiments with his big brother Erasmus, fusing minerals and producing oxygen from rusting iron.

Later as a halfhearted medical student in Edinburgh he gathered specimens of sponges and other creatures from the Firth of Forth, the estuary spreading below the city. He learned how to dissect sea creatures from the young biologist Robert Grant, then an extramural lecturer to the medical students in Edinburgh and later a professor of comparative anatomy at University College London. Darwin gained a taste for the camaraderie of science from his membership in a group of students called the Plinian Society, which presented talks to one another and debated scientific topics. He even learned about the harsh, competitive side of science when Grant, his erstwhile friend and mentor, took credit for what Darwin considered his own, albeit minor, discovery. He also attended lectures in geology, but—at that time—found the subject dull.

Ultimately the terror of operations without anesthetics drove Darwin away from medicine. He abandoned his medical studies at Edinburgh in the spring of 1827, leaving the footsteps of his grandfather, father, and older brother, and disappointing his father profoundly. At a loss for what else to do with his unfocused son, Dr. Robert Darwin concluded that there was no hope but for Charles to become a country parson. Not very interested in the doctrines of the church or in proselytizing souls, Charles nonetheless saw an opportunity to indulge his curiosity for the natural world and a comfortable life in this role. Many clergymen of the day were also naturalists or hobby geologists, following their scientific curiosity as a means to understand God's handiwork. Dr. Darwin sent young Charles off to Cambridge to prepare for ordination.

J.B.S. Haldane, a glib twentieth-century British biologist, in response to a question about what could be said about the Creator from the analysis of what he had created, replied, perhaps apocryphally, "an inordinate fondness for beetles." Not an odd answer considering that beetles make up about one fourth of all the species of animals. But Haldane might just as well have been describing Darwin. In the fall of 1828, the competition for Darwin's passions lay between his girlfriend, Fanny Owen, and collecting beetles, an avocation he developed as a student. By 1831, the beetles had won.

Fanny did have her attractions. Darwin described her, "Fanny, as all the world knows, is the prettiest, plumpest, Charming personage that Shropshire possesses, ay & Birmingham too." Over the summer and fall of 1828, Darwin spent considerable time at her father's estate, relaxing, going to musical events, hunting, riding, gathering and eating wild strawberries, and even collecting a few beetles with Fanny. She did tease him about the beetles, writing as she was fading out of the picture in January 1830, "Why did you not come home this Xmas? I fully expected to have seen you–but I suppose some *dear little Beetles*, in Cambridge or London kept you away—I know when *a Beetle is in the case* every other *paltry* object *gives way*—if I could have sent to tell you I had found a SCROFULUM MORTURORUM perhaps you might have been induced to come down!–how does the *mania* go on, are you as constant *as ever?*"

In fairness, perhaps it wasn't beetles alone that replaced Fanny, it was also John Henslow. Falling in with Henslow was perhaps the watershed development of Darwin's time at Cambridge, if not his entire life. Henslow, a young, energetic, and popular naturalist and don at Cambridge, gathered around him a circle of students, tutors, and other professors keen to use the emerging tools of science to probe the mysteries of the natural world. They attended his lectures, took tea and dined at his house, and made gala field excursions through the countryside. On these excursions the men, accompanied by wives and sisters, traveled by stagecoach, enjoying not only the stimulation of nature, but also a convivial social event. They stayed in country inns, the evenings cheered by a band brought along for the purpose. The days were for science. Henslow, a field man's field man, paused on these trips, as Darwin later wrote, to "lecture on some plant or other object; and something he could tell us on every insect, shell, or fossil collected, for he had attended to every branch of natural history." Darwin felt at home in this circle and rose in its ranks to become a favorite. His afternoon walks with Henslow along the paths in and about Cambridge discussing science led to his becoming known as "the man who walks with Henslow."

Henslow was also an ordained Anglican clergyman and it seems likely that Darwin began to feel that his father's plan for him to become a country parson was not such a bad idea. In Henslow he likely saw a role model for combining the life of a churchman with the study of natural history. About this time, inspired by Alexander von Humboldt's description of his travels, Darwin developed a plan for a private natural history expedition to the Canary Islands. His father had supported a year in Europe for his big brother Erasmus after graduation, so Darwin expected that he could convince his father to support this trip before he got down to the brass tacks of preparing for ordination. He set about recruiting his college pals for the trip. Henslow felt that his acolyte needed some training in geology, and arranged for Darwin to accompany his good friend, fellow professor, and Anglican, Adam Sedgwick, on a geologic field trip to North Wales after graduation in August of 1831. Shortly before departing on the trip, Darwin wrote a friend, "I am at present mad about Geology." After his return, he wrote Henslow, "My trip with Sedgwick answered most perfectly."

◆

While Darwin was studying at Cambridge, wooing Fanny, and developing his enthusiasm for beetles, FitzRoy was earning a postgraduate degree in hard knocks. One course in particular stands out. At three o'clock in the morning on February 5, 1830, the *Beagle*, engaged in its work of hydrographic surveys, lay anchored near the southern tip of South America, near the western entrance of what had been named the Beagle Channel on the previous expedition under Captain Stokes, and west of the mysterious land called Tierra del Fuego. FitzRoy was awakened with the news that the crew of a whaleboat he had sent to survey Cape Desolation (now Cabo Desolación) had met with serious misfortune. The crew had been sent out a week earlier, and was now overdue. FitzRoy had supposed that poor weather and the diligence of the crew to complete their survey had led to their tardiness. But now he learned that the crew, overnighting on what they had believed was an uninhabited island, had fallen victim to a thieving group of the native people, then called Fuegians, who had made off with their whaleboat and supplies. The sailors bringing the news had arrived in a jury-rigged contraption more resembling an oversize basket than a proper boat. FitzRoy immediately set out in a second whaleboat to rescue his men.

After FitzRoy found his men, he began what turned out to be a multi-week wild goose chase trying to recover the stolen whaleboat.

Initially he trusted the Fuegians, but quickly came to regard them with deep suspicion. He captured a few individuals with the motive of using them as guides, but as they turned out not to be reliable, he decided to hold them as hostages, to trade in exchange for his boat. The other Fuegians were not tempted by the offer. Ultimately he took four hostages back with him to England with the idea that they could be Christianized, learn English and other useful and civilizing skills, and then return to Tierra del Fuego to improve the lot of their people, and most importantly, to improve relations with the British.

Once back in England the situation became more complex. Initially the Admiralty supported his plan. After a year of schooling the Fuegians at a mission school, he sought to return them to their native land. Unfortunately in the meantime the Admiralty's curiosity about the region had diminished and for a time it looked as if he would have to return the Fuegians at his own expense. Finally, with the help of influential relatives and Captain Francis Beaufort, then the new director of hydrography at the Admiralty, he was able to convince the Admiralty that another expedition to the region would be desirable and useful—and enabling him to fulfill his intent to return his captives. After some fits and starts the *Beagle* was again selected for this purpose and FitzRoy oversaw her refitting.

On the previous *Beagle* expedition, only a few days before the theft of the whaleboat, FitzRoy had come upon some rocks that piqued his curiosity, and he resolved that should he ever make another voyage of exploration he would bring a geologist along. The idea appealed to Beaufort, as he was extremely supportive of increasing the scientific brief of the expedition. The Admiralty blanched, however, at the idea of actually paying for a geologist. They needed to find someone to go for free.

But that wasn't all FitzRoy wanted. Remembering the unhappy fate of Pringle Stokes, and after living alone as commander of the *Beagle* for two years in the same cabin where his predecessor had shot himself in the head, FitzRoy wanted a companion, a gentleman with whom he could lighten the tedium, sharing dinner and some intellectual conversation. FitzRoy's superior, Beaufort, was plugged into an informal network of scientists stretching from London to Cambridge, men with a common interest in bringing modernity to both science and government. George Peacock, a Cambridge mathematician, served as one of the network's key nodes and as a principal correspondent with Beaufort. The network immediately began to buzz, recognizing FitzRoy's idea as a golden opportunity. What they needed was a wealthy, gentlemanly geologist willing and able to spend the next two or more years on a tiny ship exploring the coasts of South America, and then onward around the globe.

The first offer went to Darwin's mentor, John Henslow, but his wife had just had their first child and he was too involved at Cambridge. He said no, but recommended another acolyte a few years older than Darwin, one of Darwin's competitors in the world of beetle collecting. However, Henslow's first recommendation had just accepted the post of clergyman in a rural parish, and thus he too, said no. Thoughts then turned to the young Darwin, who at that very moment, in August 1831, was in Wales, learning the skills of geologic fieldwork from Adam Sedgwick, prowling the mountains of Snowdonia. Returning home to Shrewsbury from the field trip, Darwin found a letter waiting for him, a letter opening a possibility beyond his dreams, inviting him on the voyage around the world on the *Beagle*. Darwin's hopes soared.

But three obstacles stood between Darwin and his joining the *Beagle*: his father, his heritage, and his nose. And he wouldn't even realize that his nose had been an issue until much later.

Darwin's father, after all, had finally succeeded in getting his problem son on an acceptable life course: toward a parsonage. The voyage would delay, if not derail this plan. Plus, he would have to foot a significant bill, since not only would Darwin not be paid, but he would also have to pay his own expenses. His father thought it was a terrible idea. He said no, but left an out. If Darwin's uncle, Josiah Wedgewood, could come up with some good arguments, maybe he would reconsider. Darwin headed off to see Uncle Jos, and the two of them cooked up a letter in response to Dr. Darwin's points of concern. The letter lists what Charles felt were his father's eight objections to his participation in the voyage, and Uncle Jos's point-by-point analysis and rebuttal. The thrust of their argument was that the *Beagle* voyage wouldn't make Charles any less acceptable as a clergyman. In an amazingly prescient response to objection number eight, "That it would be a useless undertaking," Uncle Jos argued that, "The undertaking would be useless as regards his profession, but looking at him as a man of enlarged curiosity, it affords him such an opportunity of seeing men and things as happens to few." Finally their arguments carried the day, and Dr. Darwin gave his assent.

Charles Darwin spent an anxious Sunday night and the early hours of Monday, September 5, 1831, clattering in a horse-drawn coach from Cambridge to London. The journey of about fifty miles had been shortened to about six hours, and the road smoothed, by the application of a new surface called "macadam," developed only a decade before. On any day but Sunday he would have had his choice of coaches to London: the *Times*, the *Rocket*, the *Star*, the *Telegraph*, and the *Fly*, among others, but none ran on Sunday, so to reach London by Monday morning he likely had to take the *Rapid*, a coach leaving the George Inn in Cambridge at a quarter past midnight, and arriving at Charing Cross around six in the morning. Maybe he was able to tamp down his flow of adrenaline enough to catch some sleep.

Most of the day on Sunday he had spent with his mentor, Henslow, first making enthusiastic plans, and then, dealing with catastrophe. All the information from Peacock and the Cambridge network about the opening for the geologist aboard the *Beagle* had been so positive. They began to think about how Darwin should prepare for the voyage and what he should take along. Then they received devastating news. An acquaintance of the pair, Alexander Charles Wood, a student of Peacock at Cambridge and FitzRoy's cousin, had written to FitzRoy recommending Darwin. The reply was a disaster. FitzRoy had offered the spot on the *Beagle* to someone else! But despite the bad news, and afraid that he was wasting his time (and likely dreading his father's reaction to this aborted endeavor), Darwin took the coach that night to London anyway.

When Darwin awoke Monday morning, London was abuzz with preparations for the coronation of William IV on Thursday. The preparations provided a welcome relief from the political crisis that gripped the nation. The Whigs had won reelection the previous spring and were struggling to pass a reform bill against Tory opposition. Rioters demanding the right to vote would rise up in only a few weeks. But Darwin was likely oblivious to these political questions. His mind was focused on FitzRoy and the *Beagle*. He probably had a couple of hours at his lodgings in Spring Gardens, only a few steps from the Admiralty, before hurrying to his late morning meeting with FitzRoy on, at best, a few hours' sleep. It had been a wet summer in London, and rain was in the offing. He would have been well advised to take an umbrella.

Darwin was likely anxious and apprehensive, walking in to meet the dashing and energetic captain. FitzRoy was certainly charming, but likely showed no cards at the outset. The prospect was, after all, that FitzRoy would be spending the next three years with this person in a space not much larger than a closet. Wouldn't he want to get to know him and make his own, personal assessment?

Amid the praise for Darwin, FitzRoy's cousin had confided in his letter that Darwin was, of all things, a Whig, with all that it stood for: broader voting rights, supremacy of parliament over the monarchy, opposition to slavery, religious dissent (those oddball Independents, Presbyterians, and Unitarians who spurned the Church of England), and new money. For a staunch Tory and Anglican like FitzRoy, a royal descendent himself, this was a potential deal breaker. On top of that, when Beaufort wrote to FitzRoy that he had found a candidate, "full of zeal and enterprize," he cited Darwin's grandfather, Erasmus, "the well known philosopher and poet" to prove Darwin's respectability. Darwin's grandfather—a polymath, physician, inventor, personal friend of Benjamin Franklin, antagonist of Samuel Johnson, and associate of several pioneering industrialists (including Darwin's other grandfather, Uncle Jos's father, also Josiah Wedgewood)—penned several long poems with scientific as well as sexual themes. Wordsworth admired his poetry, but at least some of it "nauseated" Coleridge. Whether Grandfather Erasmus was well known—or notorious—for his avant-garde, if not risqué, verse and free-flowing ideas (including among others, opposition to the slave trade, education for women, and the transmutation of species) was a matter of opinion, and FitzRoy likely inclined to the latter.

As Darwin entered their meeting, FitzRoy, a student of phrenology and physiognomy—the now-discredited science of judging a person's character from his facial characteristics—saw his nose. FitzRoy thought to himself, he later confessed, "How can anyone with a wide flat nose like that have the fortitude to succeed on an expedition like mine?" Unaware of this shortcoming, Darwin's concern was the bad news about FitzRoy's choice for a companion aboard the *Beagle*.

But whatever his initial doubts, wide flat nose or not, FitzRoy quickly warmed to Darwin. Darwin inquired whether the position was still available, and FitzRoy replied that it was, reporting that not five minutes before the meeting he had received a note from "Mr. Chester" saying that he would not be able to join him. Historians are unsure whether FitzRoy had truly offered the place to someone else, the mysterious "Mr. Chester," or had simply invented the story to provide some cover in case he didn't like Darwin.

FitzRoy explained that the dinners they would share in the mess in his cabin aboard the *Beagle* would be simple and without wine, something that Darwin had grown to enjoy at Cambridge, and that space would be at a premium. But as he warmed to Darwin, FitzRoy offered to share his "many books, all instrument(s), guns" with him. FitzRoy advised that Darwin bring only a minimum of simple clothes, "the fewer and the cheaper the better." And aware of his own needs, FitzRoy asked, "Shall you bear being told that I want the cabin to myself? When I want to be alone—if we treat each other this way, I hope we shall suit, if not probably we should wish each other at the Devil." They talked for almost two hours.

The voyage would last for three years, FitzRoy explained, not two as Darwin earlier had been led to believe. They would stop at Tenerife in the Canary Islands—the object of Darwin's wanderlust—Rio de Janeiro, and Montevideo, explore the coasts of Patagonia and Tierra del Fuego, and then on to Chile. While the ship lay at anchor or was engaged in surveys of the coast, Darwin would be able to explore the countryside. Returning around the world was not guaranteed, but Darwin would have the possibility of leaving the ship to find return passage to England if he desired. While FitzRoy offered to share everything in his tiny cabin, Darwin reported that FitzRoy "thought it his duty to state every thing in the worst point of view." He urged Darwin to come to Plymouth to see the *Beagle* for himself, to take some time to think things over, and to make up his mind. They broke up in early afternoon and agreed to have dinner together later to continue their discussion.

Darwin left the meeting ecstatic, immediately writing to his sister, Susan, about the positive turn of events. "I scarcely thought of going to Town, but here I am & now for more details & much more promising ones.—Cap Fitzroy is [in] town & I have seen him; it is no use attempting to praise him as much as I feel inclined to do, for you would not believe me." Certainly Darwin was charmed. "One thing I am certain of nothing could be more open & kind than he was to me . . . There is something most extremely attractive in his manners, & way of coming straight to the point . . . There is indeed a tide in the affairs of men & I have experienced it, & I had entirely given it up till 1 today."

Later in the day, after his dinner with FitzRoy, his excitement was so great that he wrote to Henslow "Gloria in excelsis is the most moderate beginning I can think of—Things are more prosperous than I should have thought possible."

Darwin's mind was all but made up.

His enthusiasm wasn't diminished by a visit to the *Beagle* where he inspected the tiny cabin where he and two shipmates would hang their hammocks. He floated through what stretched into weeks of frenetic preparations.

During his preparations Darwin received a note of congratulations from an envious beetle-collecting pal from his Cambridge days, a man who would one day be Archdeacon of York. Frederick Watkins wrote to Darwin, congratulating him for being now headed for a place among great naturalists, "Whilst I, luckless wretch, am rusticating in a country Parsonage & shewing people a road I don't know—to Heaven," a fate that Darwin could now defer, if not avoid. He egged Darwin on, cheering "Woe onto ye beetles of South America."

CHAPTER TWO

Field Trip with a Master: The State of Geology

The tour was of decided use in teaching me a little how to make out the geology of a country.

—Charles Darwin

O n the morning of August 5, 1831, surrounded by the sweet summer smells of his father's garden, Charles Darwin and a middle-aged man sat side by side in a light, open, two-wheeled carriage in the lane of the Darwin family home, The Mount. With a snap of the reins the gig clattered out onto the Holyhead Road and turned west toward North Wales, a turn that would resonate in Darwin's life long after.

And this was three weeks before Darwin would receive the unexpected and pivotal letter opening the possibility of his joining the *Beagle*.

The man on the seat beside Darwin was a crusty, but gregarious and eminent Cambridge don. The night before, at dinner, he—a bachelor—had smitten Darwin's older sister, Susan, with his erudition and wit. But his visit to The Mount was not merely social. This trip—arranged by Darwin's Cambridge mentor, John Henslow, and for which Darwin had carefully prepared—had a serious motive. Darwin hoped to learn from this man about the practice of geology in the field. He could not have picked a better teacher. The man beside Darwin was Adam Sedgwick, one of the leading geologists in Britain, the Woodwardian Professor of Geology at Cambridge, and who, only a few months before, had stepped down from his term as president of the Geological Society.

Behind the pair, at the bottom of Frankwell hill, and just across the Welsh Bridge and River Severn lay the town of Shrewsbury. Before them stretched the Holyhead Road, which Darwin's father, a wily investor, had participated in financing. It had been built over the decades since the Acts of Union in 1800 to speed travel between Dublin and London in the newly united kingdoms of Great Britain and Ireland, and largely under the direction of the leading English civil engineer of the day, Thomas Telford. To their left, just as they entered the road, was a tollgate that was likely something of an annoyance to the Darwin family living behind the wall guarding The Mount. Shouts of "Gate!" echoed through the night as travelers roused the gatekeeper and paid their tolls. Tollhouses dotted the road every few miles. Sedgwick and Darwin encountered their first less than two miles down the road at Shelton where Telford had built a new tollhouse only two years before. Here they also passed a splendid but battered oak tree, alleged to have been the lookout for Owain Glyndŵr, leader of the last major Welsh rebellion, as he overlooked the Battle of Shrewsbury in 1403.

But the objective of the pair was not the tumultuous history of the English and the Welsh. It was the history of the earth. In North Wales, Sedgwick intended to continue his sorting out of the relative ages of some particularly difficult rock strata from eras that Sedgwick would, in the coming decades, name the *Cambrian* and *Paleozoic*—terms still in use today. Cambrian he derived from Cambria, according to legend the Roman name for Wales, and Paleozoic, for "ancient life."

Darwin had returned to Shrewsbury from Cambridge two months before, his BA degree in hand and thoughts of an expedition to Tenerife in his head. He spent the summer boning up on geology, and apparently less successfully trying to learn Spanish. In July he had procured one of the basic tools of a field geologist, a simple instrument that combines the functions of a magnetic compass with that of a clinometer, a device that measures the angle of a slope. Despite his protestations of boredom, Darwin had learned the basics of strike and dip, as well as a good deal about minerals, from the lectures of Robert Jameson in Edinburgh. These initially abstruse concepts are really not all that difficult. Think of a flat-lying bed of limestone, for example, then imagine the flat bed tilted. The degree of tilt is the dip. Then imagine a perfectly horizontal line across the surface of the tilted bed, the waterline to which an imaginary lake might rise. The compass direction, or azimuth, of the waterline is the strike. That summer in preparation for his trip with Sedgwick Darwin experimented with his new toy for days, gleefully writing Henslow that he had "put all the tables in my bedroom, at every conceivable angle & direction. I will venture to say I have measured them as accurately as any Geologist going could do." He also tried out his tool in measuring the strikes and dips of some real rock strata near Shrewsbury, his notes showing a persistent tendency to occasionally mix up directions. He also made a rough geologic map of the surroundings of Shrewsbury, but found that it wasn't "so easy as I expected." Now, under the tutelage of Sedgwick, he would become well practiced not only at the art of taking strikes and dips, but of indentifying rocks, describing them in his notebook, and map making.

The road first followed the broad, gentle valley of the River Severn. Then, after crossing the river on Telford's Montford Bridge with its three elliptical arches of red sandstone, they turned more northerly. After a few hours they passed the massive medieval battlements of Chirk Castle, finished in 1310 during the reign of Edward I at the time of battles with the Welsh. Then they crossed Offa's Dyke, a fortification built in the sixth century or before, and the traditional boundary between England and Wales.

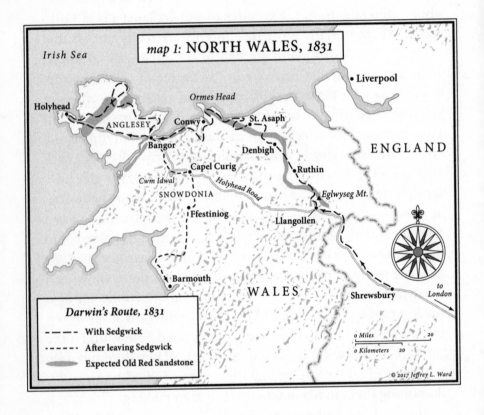

map 1: NORTH WALES, 1831

Irish Sea

Liverpool

Holyhead

Ormes Head

ANGLESEY

Conwy

St. Asaph

Bangor

Denbigh

ENGLAND

Capel Curig

Ruthin

Cwm Idwal

Holyhead Road

SNOWDONIA

Eglwyseg Mt.

Ffestiniog

Llangollen

Barmouth

WALES

Shrewsbury

to London

Darwin's Route, 1831
- - - With Sedgwick
- - - After leaving Sedgwick
Expected Old Red Sandstone

0 Miles 20
0 Kilometers 20

© 2017 Jeffrey L. Ward

The gently rolling English countryside began to give way to the hills and mountains of Wales as they followed the meandering course of the River Dee and entered the Vale of Llangollen. To their right rose the dominating limestone cliffs of Eglwyseg Mountain. They ascended a steep, conical hill set off from the mountain and topped with the ruins of a medieval masonry castle, Castell Dinas Brân, located within the earthworks of a much more ancient hill fort. From here a spectacular view spread before them. To the south they saw the gentle fields of the Vale of Llangollen. To the north, they saw the rugged escarpment along the side of Eglwyseg Moun-tain and the section of outcropping limestone exposed along it. This was their first geological objective. The limestone beds along the escarpment here dip gently to the northeast, and here Darwin likely had his first chance to use his clinometer under Sedgwick's supervision.

The scientific problem before Adam Sedgwick—and his colleague Roderick Murchison who was working in South Wales—was to fit the rocks of Wales into the time sequence that had emerged from the work of William Smith, George Bellas Greenough, and others. Describing and understanding the sequence of strata in the British Isles was one of the central challenges before geologists of the day.

Earlier, thanks to William "Strata" Smith, the surveyor and canal builder, fossils had proven to be the key to tracing layers of rock across the English countryside and to establishing their relative ages. Now Adam Sedgwick and his fellow leaders of the Geological Society were focused on completing this descriptive work and were poised to sort out the sequence of rocks into a real history of the earth and its creatures. Already a pattern was emerging: the oldest rocks contained few, if any, fossils, while the fossils in the younger and younger rocks represented increasingly complex forms of life.

But the rocks in North Wales where Sedgwick was headed belonged to a part of the sequence that was particularly problematic at the time, then called the Transition rocks. To understand how and why these rocks presented a problem requires a bit of history.

◆

In 1669 Nicolas Steno, a Dane who was at the time residing in Italy and eventually became a Catholic bishop, articulated what are perhaps the most basic principles of geology, although today they almost seem like common sense. The first principle was that in a sequence of layered rocks, the layer above is younger than the layer below. Next, he posited that all rock layers were originally horizontal and of unlimited extent. And finally, if a layer or discontinuity in a rock exposure cuts across another layer, than the cross-cutting layer or discontinuity must be younger than the layer that is cut. Versions of these simple ideas were in the minds of Sedgwick and Darwin, just as they are in the mind of every geologist looking at cliff or road cut today.

Then, in the 1700s an Italian geologist, Giovanni Arduino, working on the southern slope of the Alps, gave names to the levels of Steno's layer cake. At the bottom of the pile were massive, crystalline rocks, without fossils and typically without obvious bedding. These were, by Steno's reasoning, the oldest rocks. Arduino called them, in Italian, *monti primari*. In English, these rocks at the bottom of the pile came to be called Primary rocks. Above the Primary rocks were bedded rocks, like limestones, sandstones, and shales, frequently bearing fossils of mollusks and other creatures. These he called *monti secondari*, or Secondary rocks.

In the 1800s, working in the region surrounding Paris, Georges Cuvier and Alexandre Brongniart extended and refined Arduino's terminology. Above the Secondary rocks, containing fossils of relatively simple but extinct life forms, including fish, corals, mollusks, and plants, were younger rocks—consolidated rocks, not unconsolidated gravels as near the surface—but containing fossils of rather complex forms of life. These they called Tertiary rocks, a slight redefinition from Arduino's use of the term. The youngest deposits, the gravels in river valleys that contained most of the fossils of extinct mammals, they called Superficial.

Resting between what were clearly Primary rocks—but below the clearly Secondary rocks—were rocks that sometimes contained fossils of simple life forms, and sometimes apparently none at all. These were the Transition rocks that Sedgwick had come to Wales to study, rocks then at the frontier of understanding. His task was to put these rocks into the sequence of strata, and to make some sense of them.

Murchison, who had previously and productively worked with Sedgwick in Scotland and the eastern Alps, had proposed that he and Sedgwick spend the summer of 1831 together in Wales attacking the Transition rocks. But Sedgwick had been detained by university and national reform politics, and manuscripts demanding completion, including one describing their joint work in the Alps. He was not able to escape from Cambridge until the end of July. Now as he and Darwin began their investigations in North Wales, Murchison was well along in South Wales. The failure of these two to study the same rocks at the same time sowed seeds that would lead to competition and ultimately estrangement, and one of the most famous controversies in British geology, but that was still years in the future.

Now as Sedgwick and Darwin looked out from the ruins of the castle at Dinas Brân at the rocks exposed in the escarpment at Eglwyseg, Sedgwick's eyes were no doubt searching the cliffs of gray limestone, hoping to spy a known marker that he could use to anchor the time sequence of the strata, much like a reader scans a book to find his place by recognizing the last familiar paragraph. An important candidate for the marker was the Old Red Sandstone, a series of rocks found to the northeast in England and Scotland that were then thought to be near the bottom of the Secondary rocks, below the beds of coal with their abundant fossils of plants, that a decade before, had been given the name *Carboniferous*. If he could find the Old Red, he would have a place to start decoding the relationships of the other rocks in the region. Sedgwick and Darwin descended the hill for a closer look at the base of the cliff, but they would not find the Old Red at Eglwyseg.

After looking up Robert Dawson—the topographer and splendid artist then employed in making topographic maps of the area for the Ordnance Survey—for a chat, they spent the night in Llangollen.

The next day, the pair left the River Dee and the Holyhead Road, turning north up the Eglwyseg River. Passing the ruins of the Valle Crucis Abbey, built by Cistercian monks beginning in 1201, Darwin jotted in his notes: "The bank facing the abbey consists of clay slate . . . striking NW and dipping 25° to the NE." Here was Darwin's new clinometer put to work. Along the road beyond the abbey the two were rewarded by a panorama of the limestone escarpment of Eglwyseg Mountain to the east, which Darwin noted "The contrast between this & the more regular slope of the clay slate gives great grandeur to the views." Here Steno's layer cake was tipped, as Darwin's measurements with his clinometer showed, tilted down to the northeast. As they traveled generally to the northwest, along strike as a geologist would say, the limestone cliffs to their right were younger, higher in the geologic section, a part of the known Secondary period. In contrast the clay slates beneath the road, and to their left, were lower in the section, and older, a part of the puzzling Transition rocks that Sedgwick had come to study. If only they could find a definitive contact, a marker, that separated these two sets of rock.

As the horse pulling Sedgwick's gig climbed up the road to Horseshoe Pass, dark clouds filled the sky. Crossing the pass, the pair were "nearly drowned in a thunder storm," but from the crest of the pass, as Sedgwick wrote later "the greywacke hills continued in cloud." The rocks making up those hills were at even lower level in Steno's layer cake, and thus older yet.

"Greywacke" (or graywacke) is a strange word, almost as strange as the rock it describes. For Sedgwick and Darwin the graywacke hills were the hills of mystery, composed of a dark gray, coarse-grained sandstone, with few if any fossils. Could they make any sense of them? Were they truly rocks of the Primary, or could they find some clues that would enable them to decipher a time sequence within them?

Although Darwin didn't write it in his notebook, it is tempting to wonder whether he asked himself—as he and Sedgwick suffered through the torrential downpour and crossed Horseshoe Pass and gazed out at these older rocks—how it was that these layers of rock, with limestone on top, then clay slate, then graywacke, Steno's layer cake, had come to be bent upward, the beds of rock dipping off to the northeast. The partial answer was that the mountains of Snowdonia, the region of Wales to west and north, had been uplifted or elevated. He couldn't have known it then, but this is the same phenomenon that he would soon see in St. Jago, in Patagonia, in the Andes, and later infer in Scotland. It was a phenomenon that would come to fascinate him.

Over the next two days, they traveled mostly northwest, following this contact between the younger, better known Secondary rocks to east, and the mysterious, unknown, older rocks to the west, hoping to find the Old Red. A map made a few years before by George Bellas Greenough—one of the founders of the Geological Society—identified a band of the Old Red separating these rocks.

They descended from Horseshoe Pass to the town of Ruthin, where they spent the night, then on to Denbigh, St. Asaph, then west along the north coast to Conwy and Bangor. They had seen some reddish beds underneath the limestone of the escarpment in one or two places near Llangollen, and again in a spot near Ruthin. At first they thought that these were the Old Red beneath the limestone. But slowly they became convinced that the reddish rocks were not the Old Red. Darwin wrote in his notebook, "From several observations I am sure that the Sandstone does not crop out anywhere near Abe[r]gele. The very colour of the soil under the escarpment, I attribute entirely to the *very* ferruginous clay seams in the rock itself, & not to the supposed sandstone beneath it." Sedgwick too didn't believe these rocks were the Old Red, although he later wrote that these beds "may pass for Old Red for want of better."

Darwin later wrote "after a day or two he sent me across the country in a line parallel to his course, telling me to collect specimens of the rocks, and to note the stratification," a traverse of some twenty miles between St. Asaph and Conwy. "In the evening," Darwin continued, "he discussed what I had seen; and this of course made me exceedingly proud; but I now suspect that it was done merely for the sake of teaching me, and not for the anything of value which I could have told him." Sedgwick, a great teacher at work. And Darwin would turn out to be a star pupil.

Greenough's map showed Old Red Sandstone near Ormes Head on the northern coast of Wales bordering the Irish Sea, but Sedgwick and Darwin found no evidence for its existence there. Sedgwick later wrote to Murchison, "The Old Red all round Orm Head &c. &c. is a pure fiction. At least I can't see a trace of it. There is not a particle of it between Denbigh [about twenty miles southeast of Ormes Head] and the Isle of Anglesea [across the Straits of Menai from the mainland of Wales and about fifteen miles to the west]."

Thomas Telford's spectacular suspension bridge across the Straits of Menai, completed only five years before, connected the Isle of Anglesey with the mainland, and carried the Holyhead Road to its terminus at the port of Holyhead, where a steam packet could be taken across the Irish Sea to Dublin. Darwin's Cambridge mentor, John Henslow, had made a geologic map of the island showing a band of Old Red Sandstone. Here Sedgwick confirmed Henslow's identification of the Old Red, later writing to Murchison, "I spent some days in the Isle of Anglesea in the hopes of learning my lesson for Snowdonia. Henslow's paper is excellent, but the lesson is worth next to nothing; for Anglesea is almost as distinct in structure from Snowdonia, as if they were separated by the Atlantic rather than the straits of Menai." For Darwin, however, the unusual rocks that he likely saw on Anglesey, such as serpentine, a rock found only one other place in England and Wales, would prove to be valuable preparation for what was to come.

After more than two weeks together the pair separated probably at the Menai Bridge near Bangor. Darwin set off on a long traverse to the south across the mountains of Snowdonia to Barmouth, a traverse that took Darwin, on foot, the better part of four days.

The mountains of Snowdonia are not very high compared to the world's great ranges, only some three and a half thousand feet, but they rise directly from sea level. The peaks are rugged and the weather famously foul. Trees are rare or absent at the higher elevations. Darwin's traverse across the mountains of Snowdonia took him through an iconic location, which seemed to hold a special fascination for him, and to which he would return, Cwm Idwal. A *cwm* is to the Welsh, what a *corrie* is to the Scots, and what a *cirque* is to the French people of the Alps. These words all refer to valleys high in the mountains that appear as if they were scooped out of the rocky terrain that surrounds them. Cwm Idwal is a classic example. Steep, rugged cliffs surround a small, crystal clear lake on three sides. It is one of the favorite scenic places in all of the UK, and Darwin fell for its charms as well. But the geology here is challenging, and by himself, without the guiding hand of Sedgwick, Darwin found it difficult. It was the final exam for his crash course in field geology. The rocks "generally consist of an altered slate," he noted. And south of the lake, "There is a very large mass of Basalt protruded out of the Slate. In shape it must be an inverted cone." In short, a complicated mess of what seemed to be both sedimentary and igneous rocks. Darwin did not get 100% on the final, but he acquitted himself well. In the slate Darwin also thought that he found some fossil corals, but Sedgwick couldn't find them when he visited the same site later to see for himself. Sedgwick also concluded that the igneous rock differed "from basalt" in containing too much silica, most commonly seen as the mineral quartz.

Darwin spent two nights in the nearby Capel Curig Inn, a large stone inn that was built along the Holyhead Road to serve the growing flow of visitors to the mountains of Snowdonia, but also was a place where workers in the slate quarries in the surroundings could have a pint. Queen Victoria would stay at the inn in 1870, and it would be renamed the Royal Inn. From Capel Curig Darwin walked and geologized the remaining thirty-six miles to Barmouth in two days, spending one night in Ffestiniog. In Barmouth, he spent a few days with old friends before returning to Shrewsbury to find the fateful letter.

Overall Sedgwick seemed pleased with Darwin's progress and the level of skill that he attained, and took a keen interest in Darwin's findings later on the *Beagle* voyage. Four years later, while Darwin was still on the voyage, Sedgwick would write to an acquaintance, "[Darwin] is doing admirable work in South America and has already sent home a collection beyond all price. It was the best thing in the world for him that he went out on the voyage of discovery. There was some risk of his turning out an idle man, but his character will now be fixed, and if God spares his life he will have a great name among the naturalists of Europe."

◆

We don't know precisely what Sedgwick and Darwin talked about beyond the work of the day as they bounced in the gig along the Welsh roads, or perhaps enjoyed a pint or a bit of wine in the inns where they stayed, but Sedgwick certainly would have had a lot to say if he had wanted, or if Darwin had asked him. Five months before, Sedgwick delivered his Presidential Address at the Annual General Meeting of the Geological Society. His address summarized the state of geology and marked the passing of at least two important waypoints in the history of the science: moving it away from the province of the elite toward a more open meritocracy, and loosening the constraints of Scripture in geologic interpretation.

Up until this time geology had progressed as a science, first through description and classification, and then into the development of a chronology of the history of the earth, the phase in which Sedgwick was an early leader. Even as these components of the science would continue, another component, relatively nascent until then, was about to roar on to the scene, the explanation of how and why this history had come to pass as it did. Sedgwick would go to his grave believing that most if not all of the significant events in geologic history were the result of catastrophes of one sort or another, but Darwin was destined to become a major player in an emerging realm—the theory of how the earth was shaped and changed gradually, by observable processes.

◆

The Geological Society was then the most lively and animated of London's learned societies. In contrast to the Royal Society and others, the Geological Society permitted discussion and spontaneous debate of the papers presented. Its meetings were a hot ticket, attended by members of Parliament, academics, writers, and other opinion leaders who we would not now associate with discussions of the history of the earth.

Six years before, the Society—it had begun as a supper club—received a Royal Charter from King George IV. The Geological Society, and the science it represented, had clearly found its place within the scientific establishment of the day. But the annual meeting in 1831 caught the society in transition, and we owe our glimpse of the important changes that were taking place to Adam Sedgwick.

Sedgwick was a direct, intense, intellectual force, a Yorkshireman, and son of a pastor in the small town of Dent. He did not suffer fools. Writing to Darwin after their fieldwork in North Wales, he railed against the "stupid red-nosed waiter" who had dallied in passing him a letter from Darwin, summarizing Darwin's observations. But he also possessed a sense of fairness and a cerebral honesty that formed the bedrock of his personality.

One day on their trip in North Wales, Sedgwick sent his gig ahead so that he and Darwin could walk through the valley examining the rocks together. Darwin later recounted: "We left Conway early in the morning, and for the first two or three miles of our walk he was gloomy, and hardly spoke a word. He then suddenly burst forth, 'I know the d-d fellow never gave her the sixpence. I'll go back at once,' and turned round to return to Conway." Sedgwick feared that the waiter had not given the chambermaid a tip.

The first remarkable event at this Annual General Meeting was the inaugural presentation of the Society's new award, the Wollaston Medal, to William Smith. Smith was not someone in the Society's inner circle. Indeed, for many years he had been a bête noire to leading elements of the Society. Smith, the surveyor and canal builder, had established the method, at least in England, of identifying strata of rock and their relative position and age, by means of fossils embedded within them. George Bellas Greenough, the first president of the Society, was dismissive of Smith's work, and published his own map, even as many believed that Greenough's map owed much to Smith. Bad blood ran between them. Smith was not a man of means, but neither was Sedgwick, unlike all the previous presidents of the Society. The Society's award to Smith, who in his presentation Sedgwick termed the "Father of English Geology," was a critical step in moving the science of geology from the fraternity of a gentlemen's club to something based on merit.

After the presentation to Smith, the Fellows and guests gathered for dinner and libations at the Crown and Anchor, a popular pub a short walk from the Geological Society's new apartments in Somerset House, London. As Smith later described the festivities, "90 merry Philosophical faces glowed with a most sumptuous dinner." After dinner, they all returned to the massive neoclassic Somerset House, also home to the Royal Society, for Sedgwick's presidential address.

After warming up, congratulating the Fellows on the increase in their number over the previous year, and the firmness of their financial position, Sedgwick launched into a review of the papers presented before the Society in the last twelve months. It was only after this warm-up that the real fireworks began.

Only a few months before, Charles Lyell, a lawyer turned geologist, and soon to be named Professor of Geology at King's College, had published the first volume of his *Principles of Geology*. While decorously praising "nineteen twentieths" of Lyell's book, for the "instruction I received in every chapter . . . and . . . the delight with which I rose from a perusal of the whole," he aimed withering fire on Lyell's focus on theory. "I cannot but regret, that from the very title-page of his work, Mr. Lyell seems to stand forward as a defender of a theory." The wordy, if not awkward, subtitle of Lyell's book irked Sedgwick from the beginning. But the sentiment of the subtitle would later suffuse Darwin's theories both about geology and natural selection, and still resonates powerfully in the twenty-first century. Lyell titled his book, *Principles of Geology: Being an attempt to explain the former changes of the earth's surface, by reference to causes now in operation.* All geologists were familiar with the erosion of rivers and streams, the lavas and ash that flowed from volcanoes, the "actual causes" or vera causa in the language of Newton. The processes at work to which they could be witness themselves. But Lyell went much further. He argued that these processes were indeed adequate, through the eons, to account for all that the geologists saw. No catastrophes necessary, only gradual change over the vast expanse of time.

For the finale, Sedgwick displayed his profound intellectual honesty, addressing the vexing question of the superficial gravels that covered the British Isles and much of Europe. Sedgwick—who was also an Anglican churchman—and many others had long considered these "diluvial gravels" as remnants of the Mosaic flood, "Bearing on this difficult question, there is, I think, one great negative conclusion now incontestably established—that the masses of diluvial gravel, scattered over the surface of the earth, do not belong to one violent and transitory period.

"Having been myself a believer," he continued, "and to the best of my power, a propagator of what I now regard as a philosophical heresy . . . I think it is right, as one of my last acts before I quit this Chair, thus publicly to read my recantation." This statement by a man of Sedgwick's clearly expressed and acknowledged faith, constituted a key step away from the use of a literal interpretation of the Bible as a template for the thinking of serious geologists, and toward the separation of faith from logic in understanding the history of the earth.

◆

Thanks to Sedgwick, young Darwin developed the skills of field geology, learning to recognize and map different varieties of sedimentary, igneous, and metamorphic rocks. Although Sedgwick's attempt to find and confirm the existence of the Old Red Sandstone in North Wales resulted in a negative conclusion—it wasn't there—the work was far from fruitless. During Darwin's time on the *Beagle*, Sedgwick continued his work in North Wales, as did Murchison in South Wales. The rocks in the two areas had similarities and differences. In North Wales, Sedgwick's rocks were complicated by systematic fracturing and he found few fossils, while in South Wales, Murchison found abundant fossils and the Old Red Sandstone. Sadly the relationship between the two men deteriorated severely and even turned a bit nasty. Sedgwick defined the period he called the Cambrian. Murchison defined the Silurian. The relationship really turned sour when Murchison proposed to subsume Sedgwick's earlier *Cambrian* period into his own *Silurian*. Much later, the dispute was finally resolved when an intermediate period, the *Ordovician*, was defined, and these three periods were accepted to form the early part of the Paleozoic era, another term coined by Sedgwick.

◆

Between the wings of the Sedgwick Museum of Earth Sciences at Cambridge stands a bronze statue of the great man himself. In Sedgwick's right hand, dropped to his side, he holds a rock hammer. But in his left hand he holds up, as if offering to visitors for their inspection, a piece of shale. And within the piece of shale is a fossil, a trilobite.

There may be no more iconic fossil than a trilobite. Trilobites come in many varieties, but the most familiar suggest a greatly enlarged form of their modern fellow arthropods, the roly-poly or pill bug. The overall shape of their multi-segmented body is an oval, frequently slightly wider at the head, with a prominent central lobe and delicate spines radiating outward from the segments. We don't know whether Darwin and Sedgwick encountered a fossil trilobite on their trip to North Wales, indeed it seems a bit unlikely, but trilobites would figure in the futures of both men, for Sedgwick in sorting out the chronology of the Paleozoic, and for Darwin as a perplexing example of a creature that evolved to a quite sophisticated state at what seemed to be a very early point in geologic time. The untangling of the Transition rocks, and the recognition of the forms of life that defined the Cambrian, Ordovician, Silurian, and Devonian Periods, would illuminate that part of geologic time when life on earth really took off, and firmly established itself upon the planet.

◆

Notwithstanding Sedgwick's criticism, Charles Lyell would clear the way to apply the techniques and disciplines of geology to sort out the ups and downs, the changes of climate, and other events of earth history in even these young deposits. It seems doubtful that Sedgwick and Darwin talked of these matters as they trekked the mountains and valleys of Snowdonia in August 1831, focusing on the older rocks, but it would not be long before Darwin began to examine the islands, plains, and mountains that he visited on the *Beagle* expedition with skills and concepts learned from Sedgwick, Henslow, and even Jameson, along with a Lyellian eye to unraveling their origin.

CHAPTER THREE

Setting Out:
The Adventures Begin

> *It then first dawned on me that I might perhaps write a book on the geology of the various countries visited, and this made me thrill with delight.*
>
> —Charles Darwin

amuel Johnson said that "being in a ship is being in a jail, with the chance of being drowned." Darwin might have substituted the certainty of seasickness for the chance of being drowned. After interminable delays caused by foul weather and adverse winds, the *Beagle* set sail on December 27, 1831. Darwin immediately became seasick. Not just a passing nausea. Three weeks of debilitating seasickness.

His discomfort was leavened by the anticipation of visiting Tenerife in the Canary Islands. He'd dreamed of visiting Tenerife and testing his naturalist's mettle, dating back to his days at Cambridge and his reading of Alexander von Humboldt's gripping tales of exploration on the island. Humboldt's *Personal Narrative of a Journey to the Equinoctial Regions of the New Continent* describes his explorations from 1799 to 1804 in South America. Its pages burst with history, romance, insightful observations, and theorizing about natural history at the nexus of biology, meteorology, geology, and geography. Humboldt sought to explain the geographic distribution of plants, the inner workings of volcanoes, and the rising of the Andes. Darwin was smitten.

Humboldt's narrative leads off with a stirring account of his ascent of the volcano Teide on Tenerife. Soaring more than 12,000 feet from the waters of the Atlantic, Teide presents perhaps the most striking evidence of the volcanic history to which the island owes its origin. Darwin had long desired to follow in Humboldt's footsteps. In vain, he'd tried to organize some Cambridge friends, including his mentor, John Henslow, to undertake a journey to the Canary Islands, but he couldn't get it to come together. Among the many attractions of his chance to join the *Beagle*, was the planned stop at Tenerife.

As his seasickness waxed and waned and Darwin gained "dear bought experience" of this cursed plague those first weeks, Humboldt's narrative served as an unorthodox treatment. He spent afternoons lying on a sofa chatting with FitzRoy or reading Humboldt's vivid accounts of adventure and tropical scenery. Darwin's diary records periods when this novel therapy cheered "the heart of a sea-sick man." But, alas, the relief was only temporary. Other entries record days of "great & unceasing suffering."

To Darwin's utter dismay, the health officials in the port of Santa Cruz, the main port on Tenerife—alarmed that the *Beagle* crew might be bringing cholera as an unlisted passenger—denied permission to land, pending a twelve day quarantine. The impatient FitzRoy would brook no such delay, weighed anchor, and set sail for the Cape Verde Islands to avoid losing any time. Darwin was crushed, wailing, "We have left perhaps one of the most interesting places in the world, just at the moment when we were near enough for every object to create, without satisfying, our utmost curiosity."

FitzRoy noted Darwin's chagrin, "This was a great disappointment to Mr. Darwin, who had cherished a hope of visiting the Peak.

To see it—to anchor and be on the point of landing, yet be obliged to turn away without the slightest prospect of beholding Tenerife again—was indeed to him a real calamity."

So it was with even greater resolve and enthusiasm that Darwin addressed his attention to his next chance to land, on the island the British then called St. Jago (now Santiago) in the Cape Verde Islands. Before leaving England, Darwin asked the advice of his field geology mentor, Adam Sedgwick, about geology books that he should take on the voyage. Sedgwick recommended several books to Darwin, including Humboldt's personal narrative that Darwin selected as his treatment for seasickness. He also recommended a work by the French geologist, Jean-François d'Aubuisson, ("one of the best tho' full of Wernerian nonsense"*), and a book by the English geologist Charles Daubeny, *A Description of Active and Extinct Volcanos.*

Initially Darwin worried that St. Jago would be a poor second to Tenerife, lacking the beauty that Humboldt and others described, but there was a definite upside to visiting St. Jago. If Darwin—as he lay in his hammock fighting seasickness—read Daubeny's discussion of the Cape Verde Islands, it likely thrilled him. "The structure of the Cape de Verde Islands is too imperfectly known to detain us long; they are said to consist principally of volcanic matter . . ." Darwin was headed for virgin territory.

At St. Jago the *Beagle* anchored in the harbor of Praya (now Praia) near the small, barren, flat-topped Quail Island. Here FitzRoy's crew set up camp to read their chronometers, to determine the longitude, and to make other astronomical and magnetic measurements. And as FitzRoy's team set about their work, Darwin began his "geologizing" with gusto. FitzRoy seemed bemused by his new

---

* Abraham Werner was a leading German geologist who believed that almost all rocks, excepting those of obviously volcanic origin and the very young soils and gravels, owed their formation to deposition at sea. This school of thought came to be known as "Neptunism." Sedgwick was allied with a school of thought traceable to James Hutton, which argued that some rocks, especially granites and basalts, had once been molten, a school of thought known as "plutonism."

friend's youthful, geologic enthusiasm, "The vicinity of Port Praya offers little that is agreeable to the eye of an ordinary visitor, though interesting enough to a geologist."

It would be hard to imagine a place more different from his verdant native Shropshire, or indeed anywhere that Darwin had been before, than the barren volcanic island that rose before him. But when he landed and walked through a grove of coconut trees, he was enraptured. He "first saw the glory of tropical vegetation" and returned to the ship in high spirits. "It has been for me a glorious day, like giving a blind man eyes."

Just as the new flora and fauna thrilled his senses, so too did the rocks and the challenge of extracting their story. Unlike the ancient rocks of Britain and Wales, these volcanic rocks appeared little altered from the day that they erupted. And in contrast to the mellow English countryside, here the landscape was young, and the processes responsible for shaping it—volcanic activity, vertical movements, and rapid erosion—still displayed their virility. The task before him was to understand it all.

In his enthusiastic rush about Quail Island, he quickly brought into play the skills and ideas he had learned from Henslow, Sedgwick, and his reading. He saw in the cliffs surrounding the island a kind of geologic sandwich, two layers of black volcanic lava, separated by a twenty-foot-thick layer of bright white limey sand and shells. The white layer particularly intrigued him, jotting in his notebook, "it has the exact appearance, as if a piece of the present *beach* was lifted up & cut off." The similarity of the fossil shells within the layer—limpets, oysters, periwinkles, and sea urchins—to the creatures that he found living in the tide pools at the bottom of the cliffs, spoke to Darwin of the youth of his fossil beach. Shells found in the sedimentary rocks of England were, for the most part, remnants of animals long extinct. Here, however, the shells in the rocks that he found were of species, that as he noted, were "as far as my knowledge goes . . . the same as those of the present day." Darwin immediately recognized that the rocks he hammered on Quail Island were far younger than those he'd explored with Henslow and Sedgwick in England and Wales. A narrative of how this sandwich formed began to take shape within his head.

As he extended his surveys to the main island of St. Jago, he again recognized his familiar sandwich in the sea cliffs along the coast. He found that the nearby Red Hill was a volcanic cone from which had flowed additional lavas onto the top of the upper layer of black lava, and were therefore even younger still. The valleys in which he found the lush vegetation that so thrilled him were cut into these volcanic deposits. When he traced the flows of lava down the sides of Signal Post Hill, yet another volcanic vent, he discovered that these flows filled parts of a nearby valley, proving that they had formed later than the valley.

Walking along the beach, Darwin followed his now familiar stack of black lava with its bright white stripe of limestone. Then, the limestone began to plunge downward. Beneath Signal Post Hill, his fossil beach actually dropped below sea level, then returned to its former elevation farther along his traverse.

SIGNAL POST HILL.

Darwin's cross-section of Signal Post Hill

Nearby Darwin found a cave in the sea cliff, a spot with a picturesque view of the tide pools below and the ocean beyond. He sat at this spot, sifting through his observations, fitting together the geologic story of the island. As he parsed the possibilities, his explanation came together. "The geology of St. Jago is very strikingly simple: a stream of lava formerly flowed over the bed of the sea, formed of . . . recent shells and corals, which it has baked into a hard white rock." Then—Darwin noting that the whole island was uplifted relative to sea level—he reflected on the drooping stratum, "But the line of white rock revealed to me a new and important fact, namely that there had been afterwards subsidence around the craters, which had since been in action, and had poured forth lava." Here the pivotal thought struck him. "It then first dawned on me that I might perhaps write a book on the geology of the various countries visited, and this made me thrill with delight."

◆

As Darwin wandered the beach, FitzRoy's crew labored with their instruments. One of the first tasks assigned to FitzRoy was to straighten out the longitudes of South America. In his memorandum laying out the instructions for the expedition, Beaufort wrote: "A considerable difference still exists in the longitude of Rio de Janeiro, as determined by Captains King, Beechey, and Foster, on the one hand, and Captain W. F. Owen, Baron Roussin, and the Portuguese astronomers, on the other." The longitude of Rio was critical because it served as the reference for all the other longitudes in South America. Beaufort needed to know which of the preceding surveys was correct. The procedure to find the error, or what surveyors call a "bust," was to repeat the observations of longitude at the intermediate points, trying to find the first place at which they disagreed with the previous surveys. If all went well as the expedition proceeded, Beaufort and FitzRoy hoped that the *Beagle* would continue a chain of longitude measurements around the globe.

It seems like it should be so simple. You just go outside during the middle of a nice day, check your watch when the sun reaches its highest point in the sky, calculate the difference with the time in Greenwich, and voilà, that's your longitude. But aspects of this problem have challenged thinkers ranging from Ptolemy to Einstein. I confess that earlier I swept a few pesky, but critical, details under the rug.

The first practical issue is to figure out the exact moment of noon. FitzRoy used a tried-and-true technique called "equal altitudes" to solve this part of the problem, a technique that provides, as an added benefit, a reliable means for determining the direction of true north to which he could compare his measurements of magnetic north. It is actually quite difficult to tell when the sun is at its highest point, particularly in the tropics, because this is also the time of day when the sun's elevation changes the most slowly. The *Beagle*'s officers would note the time when the sun first rose above a particular angle of elevation or altitude, then the time when the sun set below the same angle; the average, halfway between the two, is noon. Or noon, as astronomers would say, "apparent solar time," the time one would read on a sundial.

But there is a second, more subtle, problem. If the earth's orbit around the sun were a perfect circle, and the earth's axis were perpendicular to orbit, that would be all that there is to it. Time read from a sundial and from a steadily ticking clock would always agree. But the earth's axis is not perpendicular to the orbit, it's tilted at 23½° (after all, many of us earthlings do very much enjoy our seasons), and the orbit about the sun is an ellipse not a circle. The effects of these two facts conspire to change the exact time of the sun's highest point during the day slightly through the course of the year. Ptolemy knew about this complication in the second century A.D., even as he still tried to explain it in terms of an earth-centered, rather than sun-centered solar system. Globes and sundials often bear the mark of a strange-looking, asymmetrical figure eight, called an "analemma," the curve that would be traced by plotting the position of the sun in the sky at the same time every day for a whole year. To get around this variation astronomers define "mean solar time," the time measured by a steadily ticking clock. Using this steady time, a navigator can use a chronometer to measure the same longitude for an island in January as in July. Fortunately for FitzRoy and the officers of the *Beagle*, all the calculations relating these two ways of keeping time, apparent solar time and mean solar time, had already been done for them. All they had to do was look up the correction in the *Nautical Almanac and Astronomical Ephemeris for the Year 1832*. For January 20, they would have added 11 minutes, 11.2 seconds to the sundial time to get the clock time.

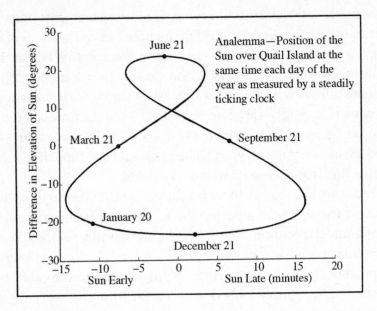

Analemma Chart

FitzRoy's measurement of longitude at Quail Island, 1 hour, 34 minutes, 00.3 seconds (23°30′4.5″) west of Greenwich, was close enough to the previous determinations to show that the bust in longitudes at Rio de Janeiro would have to be sought further down the line, beyond the Cape Verde Islands. The measurement was also within less than half a mile of the position determined today.

◆

I was fortunate to have a window seat on the overnight LAN Chile flight to Santiago, finally embarked on my own voyage. So many things ran through my thoughts. It might have made better sense to follow Darwin's footsteps in a more chronological fashion, beginning at the beginning, and to follow along step by step. But the chance to go to Chile with Brian Atwater was too good to pass up. I could get my feet wet with someone I knew, working on one of the phenomena that had intrigued Darwin, but also something that I knew a little bit about, earthquakes, their traces, and their origins. I would be starting in the middle, but I had to start some-where. Best of all, I was actually underway. My idea transformed from daydream to reality. After all the arranging and planning, departure was a cathartic relief.

I reflected on the other differences between my trip and Darwin's. Darwin set out for what he thought would be three years, but turned out to be almost five. On this initial foray I'd be gone a little more than three weeks. I remembered that as Darwin hung his hammock before going to sleep, he had to remove a drawer to make space for his feet. I wouldn't complain about the legroom on the plane. I'd be in Santiago in about twelve hours. It took Darwin and FitzRoy about three years to get that far, but of course they and the *Beagle* had lots to do along the way . . .

I awoke and slid up the window shade to reveal a brilliant red-orange sunrise radiating from behind a bulwark of dark, jagged peaks, peaks towering almost to the altitude of the cruising jetliner. The Andes. I had arrived.

◆

"*Chile es un país muy seismico,*" the priest said, "Chile is a very seismic country."

The priest spoke as he explained the damage that his nineteenth-century, corridor-style church buildings had suffered from an earthquake in 1985. Brian and I stood in the fresh early morning air in the church courtyard opposite the main plaza in Licantén, ten miles inland from the Pacific beach. As we waited for our two colleagues to finish gathering their things after breakfast, we had wandered across the street from our *pensión* to have a look at the buildings, and the subject of the damage had come up in a chat with the caretaker. The 1985 earthquake occurred well to the north, near Valparaíso. We were surprised that buildings this far away, some 130 miles, had been damaged, but the church buildings were built of adobe, one of the worst forms of construction for "a very seismic country."

A very seismic country indeed. In 1960 southern Chile was rocked by the largest earthquake recorded since the invention of the seismograph, with a magnitude of 9.5, larger even than the great Alaskan earthquake of 1964, larger than the terrible earthquake off Sumatra of 2004, and larger than the Japanese earthquake of 2011. Chile is one of the most seismic countries in the world. And most of the deaths in the 1960 earthquake were caused by a devastating tsunami.

Brian had first come to Chile in 1988 to learn what fingerprints the 1960 earthquake and tsunami might have left, fingerprints that a geologist might use to recognize a prehistoric tsunami. In the 1980s Brian had found suspicious sand layers in estuaries along the coast of Washington and Oregon. He believed that these layers were deposited by tsunamis that preceded European occupation of the region, but he wanted to make his arguments more convincing. He wanted to compare what he saw in the Pacific Northwest with what had happened in Chile in 1960.

Brian persuaded his employer, the United States Geological Survey (USGS), that this effort deserved a research grant. The grant was not large and Brian, whose frugality is legendary among his colleagues, wanted to stretch it as far as he could. To save money, he chose not to rent a car or truck. Instead, each day from his base in the tiny farming and fishing community of Maullín in south-central Chile, he dragged his shovels and augers onto a local bus and jounced along the dirt roads out to one of the areas that he wished to investigate.

This decision had two marvelous consequences. First, with the money Brian saved by not renting a car, he was able to hire Chilean assistants to help in the field, digging pits, and interviewing survivors. Now, two decades later, one of those field assistants, Marco Cisternas, was our host in absentia for this trip. Marco, now a professor at the Pontifical Catholic University of Valparaíso, had organized our trip, but then had received an offer too good to refuse, a chance to visit some remote islands off the southern coast of Chilean Patagonia that had been affected by the 1960 earthquake. He had made the arrangements for our trip, put us in the hands of his young associate, Marcelo Lagos, then departed for the south before our arrival.

Secondly, riding in the bouncing buses, Brian met and talked with the local people. One contact he met on a bus directed him to a man whose fields had been covered with sand by the 1960 tsunami. The site turned into a goldmine for tsunami geologists. Marco, who now leads the work, has shown that the site holds evidence for as many as five major tsunamis and two additional large earthquakes within the last 2,000 years, in addition to the one from 1960. Indeed, our objective for this trip was to scout for locations that might yield similar information about the history of tsunamis along the coast of much more densely populated central Chile.

◆

Our two colleagues appeared with their gear, and Brian and I excused ourselves from the priest, and joined them to finish loading our silvery crew cab pickup or *camioneta*. Marcelo was a young physical geographer and tsunami researcher. The fourth member of our team was Yuki Sawai, one of the top young tsunami geologists in Japan, especially noted for his discovery of evidence for an ancient super tsunami on the island of Hokkaido.

I was the hanger-on, a seismologist trying to learn how modern geologists develop this kind of evidence, and hoping to compare this modern thinking with Charles Darwin's "geologizing," as he called it, on the expedition of the *Beagle*. In 1835, some hundred and forty miles to the south of where we loaded the truck, Darwin and FitzRoy witnessed the effects of a massive earthquake and tsunami in the cities of Concepción and Talcahuano. Although Darwin and FitzRoy made note of the extensive flotsam and trash dumped on the land by the tsunami, they said nothing about deposits of natural materials. Tsunamis, however, also leave characteristic traces of these materials. Beach sand, especially, can be carried by a tsunami into places where—according to the normal, everyday processes of deposition—it shouldn't be.

From Licantén we drove down the valley of the Río Mataquito to the Pacific coast and began to search for sites where evidence of a great tsunami might be preserved. We continued north along a flat terrace behind the beach. The terrace, wedged between the beach and the sharply rising hills of the Chilean coast range, was covered with fields and vacation *cabañas*. We passed through crowds of Chilean vacationers, some in cars, some in buses, some on motorcycles, all in sunglasses, many in bathing suits, here to enjoy a few summer days at the beach. Stalls at the tiny fish market in the village of Iloca overflowed with fish and shellfish, and teemed with vacationers, eyeing their beloved *mariscos* (or seafood).

Anyone looking through the windows of the truck would have immediately suspected that we were not headed for a vacation at the beach. We all sported floppy sunhats and orange or red vests like those worn by highway construction workers, our pockets loaded with pencils and rulers. As we drove, Brian marveled at the landforms, asking rhetorically, "How would you make that terrace but by the sea?" Fossil sea stacks, now high and dry, were special favorites. Aside from being surrounded by fields and pastures rather than crashing waves, they looked for all the world like those along the coast of Washington and Oregon.

Brian's easygoing manner belied the intensity of his commitment to reducing the risk from tsunamis to people living in coastal areas. Following his discovery of the potential for enormous tsunamis along the coasts of the Pacific Northwest, Brian became an agitator for tsunami preparedness and education.

Brian and Marcelo joked constantly as we drove, Brian speaking extremely enthusiastic, if not always fluent, *castellano*, as Chileans prefer to describe their language. He carried a small notebook in which he recorded interesting and memorable phrases and expressions. Marcelo's rapid-fire one-liners were priceless—when I could follow them from the back seat. Marcelo worked his way

map 2: CHILE—Valdivia to North of Concepción

74° 73° 72° 71° 70°

Pacific Ocean

34°

Rancagua ● Termas de Cauquenes

Yaquil ● ● San Fernando

Lipimávida ●
Iloca ● ● Licantén

35°

Constitución ●

FitzRoy and crew charted this
coast including the harbors of
Valdivia, Isla Mocha, Isla Santa
María, Bahía de Concepción,
and Constitución, 1835.

CHILE

36°

ANDES MOUNTAINS

Quiriquina Island
Talcahuano ● Bahía de
Concepción
● Chillán
Concepción ● Penco

Isla
Santa María

37°

Golfo de
Arauco

Encampment of
Challenger crew
Río Biobío

ARGENTINA

✕—Wreck of HMS Challenger

38°

Isla Mocha
Tirúa ●

Except for Islas Santa María and Mocha,
Chiloé and nearby islands, and an enclave
around Valdivia, the land in Chile south
of the Río Biobío was largely under the
control of the Mapuche until the latter
half of the 19th century.

39°

0   Miles   50   100
0   Kilometers   100

● Valdivia

40°

© 2017 Jeffrey L. Ward

through college as the host of a surfing program on TV, in which he interviewed surfers on the beach just in from their rides, and pretty girls strolling in bikinis. He promises us that on this field trip we will see lots of beautiful Argentinean girls wearing bikinis reminiscent of dental floss, but so far we've missed them. Yuki is mostly silent, staring out the window, tuned out of the hilarity.

But all was not youthful frivolity. As we drove, Brian's eyes scanned the countryside, his mind constantly building a narrative of how the landscape was formed, searching for a set of circumstances that might be propitious for finding the fingerprints of a tsunami. Marco had identified some possible candidate sites from satellite and air photos, especially some estuaries—marshes at the mouths of rivers and creeks. There the sediments collecting in the quiet waters are typically a very fine-grained mud. In such circumstances, a layer of beach sand dumped by a tsunami can stand out as if it were raising a red flag. Could we spot other situations where the watery intruders had left their prints?

◆

Dust swirled as our truck bounced to a halt. Nearby, beside an old Toyota pickup, stood a neatly dressed, silver-haired man in a blue V-neck sweater. Marcelo, long black hair streaming from beneath his floppy sun hat, leapt from our truck, and with the enthusiasm and optimism of a polished salesman, strode toward him.

The man stood in a lane running to a low shed. Patches of vegetables protected by rough wire fences surrounded him. Squash plants bloomed bright yellow. A lone black and white cow grazed in the field nearby. Dry hills covered with eucalyptus rose behind. A hand-painted sign read PAPAS VENDO, potatoes for sale.

A gorgeous black sand beach spread behind us toward the Pacific Ocean, the murmur of breaking waves faintly audible.

"¡Buenos días, caballero! I am Marcelo Lagos from the Pontifical Catholic University in Santiago. My colleagues and I are investigating ancient earthquakes and tsunamis along the coast of Chile. I wonder if I might speak with you for a moment? What is your name?"

"Señor Ruíz."

"Señor Ruíz, as I said, my colleagues and I . . ." continued Marcelo, launching into his rapid-fire explanation. His pitch concluded with the request for what he and his strange-looking colleagues actually wanted to do. Dig a hole in Señor Ruíz's field.

To my amazement Señor Ruíz agreed. We lugged our shovels, soil auger, packs, and notebooks over to the field and began to dig.

◆

Brian's interest was drawn to Señor Ruíz's field because of a creek and the subtle fan that it created as it emerged from the hills behind the beach. His thinking went something like this. In the summer, the creek as we saw it was a mere trickle, but in the rainy season, the creek drained a large area of the hills above the beach. Heavy rain would wash soil from the hills. Usually the flow of the creek would be confined to the streambed, but sometimes the flow would exceed the creek's capacity, overflow the banks, and drop its load of sediment on the potato field, building this small fan. The most powerful of these floods would be made up not only of water, but of a thick soup of mud that geologists call a debris flow, depositing a sheet of mire across the field.

The field was at an elevation of only ten or twenty feet above the beach—and only about two hundred yards from it. Any tsunami worth its salt should have dumped lots of beach sand here. With luck, this beach sand would have been covered relatively quickly, geologically speaking, by more soil and debris flows washed down from the hills, before it could be eroded away. This was Brian's theory. Our job was to dig a hole and test it out.

Scraping a few cow pies aside, and under the watchful eye of Señor Ruíz, we began to dig. The digging was not as random as you might expect. Neatness counts. We preserved the sod to cover the hole after we refilled it. We wanted to end up with two walls of the rectangular pit in the sunlight to optimize the viewing and photography of the deposits. We used the "slicer shovel" to get these two surfaces smooth and flat, laid back just a little from vertical, and ordinary shovels to remove the bulk of the material. After digging down almost three feet, we stopped to log the pit.

◆

Many professions have their special tools. The doctor has a stethoscope, the carpenter a hammer. The tools that no tsunami geologist can be without are, first, an old army folding trenching shovel—preferably with its handle painted with alternating red and white stripes so that it can double as a scale in photographs— and, second, a wicked-looking Japanese gardening tool called a *nejiri gama*. The nejiri gama is the size of a hand trowel, but with a sharp, curved blade. It is intended for chopping the weeds out of a garden.

Prior to the trip, I sought the advice of a geologist acquaintance about how to acquire these two essential tools. He counseled that the folding shovel should be military surplus, and that it be as old as possible. "The new ones won't even last a day!" he said. Then he took pity on me, opened a drawer in his office and presented me with a brand new nejiri gama, explaining that someone had brought him several from Japan, and that there was no chance that I could find one on my own in the U.S.

Walking into a surplus store in search of an "old" trenching shovel, I was bowled over by the familiar musty, but mysterious, smell that seems to contain elements of mildewed canvas, petroleum solvents, and preservatives. On my search for the shovel aisle, I encountered an item that I did not recall from by boyhood days of prowling austere surplus stores, a display of ladies' panties in brown and green camo pattern. In bold capital letters, the words BOOTY CAMP were printed across the front. The thought flashed through my head that I might buy a pair for my wife, but I quickly thought better of it. Finally, after rejecting many lightweight, freshly painted, but cheesy, shovels, I found the real thing. It was heavy, a little beat up, and its blade scratched and worn. This was what I was looking for!

Brian later explained that the best shovels were from World War II and were stamped with "1944" or "1945" for the year of manufacture. Supposedly, many GIs in the Pacific, when in hand-to-hand combat, had favored the trenching shovel (with its blade locked at a 90° angle to the handle, like a hoe) over a bayonet, citing its great balance and the ease of moving on to the next blow without the delay required in extracting a bayonet. Our use was not to be so dramatic.

◆

At our pit in Señor Ruíz's field, after cutting the two sunlit faces of the pit back with the trenching shovel, Brian uses his nejiri gama to carefully smooth and clean the surfaces. The trick is to use the sharp blade of the nejiri gama to bring out the texture of the dirt, but not to add any suggestion of bedding that is not really there.

Then we all take out our notebooks and describe what we see. We note the GPS coordinates of the site, the size and texture of the sediments in the pit, and draw sketch maps of the surroundings. Marcelo asks Señor Ruíz whether the place has a specific name. "Yes," he replies, "Lipimávida." All very professional.

This is "quiet time." We each work for the most part silently, trying not to disturb one another. I am quietly embarrassed, knowing that my own notes are feeble. But for the others, their notes are detailed and vivid. In five years when they look at their notebooks, they will want the descriptions of the dirt to bring the expedition back to life, and to distinguish this particular hole from every other. They diligently describe every detail, from the composition and consistency of the soil, to the grain size of the sediment at each level in the pit. Marcelo carefully measures the depth to different subtle features. Brian draws lots of cartoon-style pictures and maps, and records variations of his ideas about how the deposits might have formed.

Yuki describes every detail using three different systems of soil description. He teases the soil with his nejiri gama. He rubs the dirt between his fingers to estimate the size and texture of the particles; silt is gritty, clay is slippery. He draws and annotates a geologic section of the hole that is so neat and complete it looks as if it could be scanned and used directly as an illustration to a scientific paper.

The rest of us wait as Yuki finishes.

But it looks like a dud.

All we can see in the hole, from just below the roots of the grass to the bottom of the hole is brownish dirt. Brian calls it a *diamict*, a mix of sand and silt with lots of flecks of mica. It seems a very good bet that this poorly sorted stuff was, as Brian suspected, deposited here by debris flows of soil derived from the weathering and decomposition of the bedrock in the hills rising above us, a metamorphic rock called schist, with lots of shiny mica.

But no sign of anything interesting.

After logging the hole in our notebooks, we pull the soil auger out of its case. The auger itself is a kind of cylindrical frame about a foot long with two blades at the bottom end to bore through the soil. To a screw end at the top end of the auger, we attach a threaded pipe and handle. Placing the auger in the bottom of the hole, we rotate it with the handle, collecting a sample of the first few inches of the material below the bottom of our pit. We extract the auger and knock out the sample from the frame, placing it along a meter stick to keep track of the position of each sample from the auger hole. The first few samples are the same boring stuff, the diamict.

Then the sound of the auger rotating in the hole changes slightly, to that of a sandy scraping. The sample that emerges is not the same old, monotonous, brown diamict, but black beach sand! Our little group tenses with quiet excitement. And even more so when, after about four inches of beach sand, the samples again return to our old friend the diamict.

"A geologist's dream, black beach sand derived from the Andes, alternating with debris flows derived from local schist!" enthuses Brian.

This is just the sort of thing we are looking for. We have a candidate tsunami deposit.

"Beautiful stuff, huh?" says Brian, scrutinizing a sample of the sand.

Buoyed by this discovery, we deepen our pit to fully display the layer of beach sand and the sharp contacts above and below it. Could our candidate be from the tsunami associated with the earthquake Darwin witnessed in 1835? Maybe . . . but the reports of that earthquake and tsunami suggest that most of the action was farther south. The list of earthquakes and tsunamis that have devastated Chilean coast is a long one. There is no shortage of events that could have been responsible.

Our layer of black beach sand—the grains of which began life as volcanic rocks in the Andes before embarking on their own voyage down to the Pacific, swept along first by the Río Mataquito, then northward along the coast by the coastal currents, now sandwiched between deposits of diamict derived from the schist of the local hill slopes—is still only a candidate for a tsunami deposit. We try to imagine all the possibilities. The sand contained chunks of the diamict embedded within it, chunks that I soon knowingly refer to as "rip-up clasts," pieces of the diamict that could have been plucked or picked up by the swiftly flowing waters of a tsunami. But sometimes during the strong shaking of an earthquake, sand saturated with water can liquefy. Perhaps our layer was squirted into place by this process, or maybe it is the remnant of a deposit from a severe winter storm, or something left by an earlier course of the creek. These explanations seemed a little unlikely given the flat-lying deposit with sharp contacts above and below that we see in our pit, but more investigation is required, more pits need to be dug, or even trenches with a backhoe. We are in reconnaissance mode. We leave this more detailed study of the potato field for Marco.

But I can't help wondering. Perhaps, just perhaps, this field holds a trace of Darwin's tsunami.

CHAPTER FOUR

The First Years of the Voyage

To the southward of the Rio de la Plata, the real
work of the survey will begin.

—Francis Beaufort

A fter stopping at St. Jago in the Cape Verde Islands, and St. Paul Rock and Fernando de Noronha, islands in the Atlantic along the way, the *Beagle* reached South America in late February of 1832. FitzRoy's first order of business—as directed by the memorandum from Captain Beaufort—was to sort out the longitude of Rio de Janeiro. The existing determinations were a bit of a mess, but Beaufort had a plan for sorting them out.

In 1781 the Portuguese astronomer, Bento Sanches Dorta, first determined the longitude by comparing his measurements of the satellites of Jupiter from Rio with those of a colleague in Lisbon. Beginning in the 1820s, the French captain and statesman Baron Roussin and four expeditions from the British Admiralty measured the longitude with chronometers and sightings of the moon, with results varying over about ten minutes of longitude (about eleven miles), enough to drive Beaufort to distraction.

FitzRoy's British predecessors were an exceptional group. Captain William Fitzwilliam Owen with the HMS *Leven* and HMS

Barracouta surveyed the coasts of Atlantic and Indian Oceans from 1820 to 1826, focusing on Africa, Arabia, and Madagascar, visiting Rio along the way. One of his officers, Richard Owen, wrote—as part of the expedition report—"An Essay on the Management and Use of Chronometers."

Captain F. W. Beechey in the HMS *Blossom* stopped in Rio while taking the long way to the Arctic in search of the Northwest Passage in 1825–28. He sailed south from England through the Atlantic, around South America, and up the Pacific coast of the Americas through "Beering's Strait" to the Arctic Ocean. Ultimately his expedition reached and named Point Barrow, Alaska, after Sir John Barrow, the second secretary of the Admiralty who long promoted the exploration of the Canadian Arctic.

Captain King, leader of the previous expedition of the *Beagle* and *Adventure* from 1826–30 also determined the longitude with chronometers. FitzRoy's predecessor on the *Beagle*, the ill-fated Captain Pringle Stokes, who was known to be a skilled observer, also determined the longitude from the rather intricate procedure known as lunars—determining the "absolute" time from tracking the position of the Moon relative to the stars and comparing it with the local sun time.

Perhaps the most scientific of the four was Captain Henry Foster in the HMS *Chanticleer*. Foster was sent to the southern Atlantic to survey the coasts and to measure gravity with a pendulum balance, visiting Rio along the way. In February 1831, just as FitzRoy was overseeing the refitting of the *Beagle*, Foster was wrapping up his expedition to map the Isthmus of Panama. He employed a novel technique using rockets to make simultaneous observations to determine the meridian distance, or difference in longitude, across the narrow but mountainous isthmus. Returning from the successful completion of his measurements, tragedy struck. According to the report of the expedition, he was in a native canoe floating down the Chagres River (now site of the Panama Canal) back toward the Atlantic, "reclining beneath the awning of the canoe in conversation [when] Captain Foster crept out [on] the after part of it. Being outside of it with his feet resting on the gunwale, he incautiously seated himself on the awning, which had no sooner received his weight than it gave way, and he was precipitated into the river." Two of his men dove in after him, but were unable to save him. Foster, already a member of the Royal Society, drowned at the age of thirty-four.

FitzRoy hoped to make more accurate measurements by careful attention to detail. Chronometers tended to drift, that is, to lose or gain time. If the drift were steady, it would be a simple matter to correct for it. Proper procedure required staying in one place for several days, and to make observations of the sun to measure the drift, or "rate" the chronometers. The real problem was irregular drift. As FitzRoy studied this problem, he became convinced that variations of temperature and mechanical vibration were the principal enemies of a constant rate. His solution to both of these problems was sawdust.

FitzRoy placed his chronometers in a closed area on the *Beagle* between the decks that was only to be opened for accessing them. Each chronometer was "suspended in gimbals . . . within a wooden box, each was placed in sawdust, divided and retained by partitions, upon one of two wide shelves. The sawdust was about three inches thick below, as well as at the sides of each box." The sawdust provided terrific thermal insulation, as well as cushioning the chronometers from vibration, both of which could cause drift. Although he placed the chronometers as close to the center of motion of the ship as he could, FitzRoy believed that the rolling and bobbing of the ship was not the primary problem, rather it was the shorter and sharper vibrations. "Suspending chronometers, as on board the *Chanticleer*, not only alters their rate, but makes them go less regularly; and when fixed to a solid substance, as on board the *Adventure*, they feel the vibrations caused by people running on the decks, by shocks, or by a chain cable running out." The mechanical isolation provided by the sawdust was the solution. "Cushions, hair, wool, or any such substance, is preferable to a solid bed; but, perhaps, there is nothing better than coarse dry saw-dust." He even devised a novel approach for testing the efficacy of this arrangement. "Placed in this manner, neither the running of men upon deck, nor firing guns, nor the running out of chain-cables, caused the slightest vibration in the chronometers, as I often proved by scattering powder upon their glasses and watching it with a magnifying glass, while the vessel herself was vibrating to some jar or shock."

The stops at Praya on St. Jago and at Fernando de Noronha were both part of Beaufort's strategy to find the bust in the chain of measurements leading to the longitude of Rio. Praya formed a key link in Owen's chain of measurements, as did Fernando de Noronha in those of Foster. FitzRoy's determinations, however, were tolerably close to those of his predecessors at both of these locations. The problem lay further down the chain.

The *Beagle's* first stop on the actual continent of South America was at Bahia (now Salvador), Brazil. Then the *Beagle* headed to Rio. "At Rio de Janeiro," Beaufort directed FitzRoy, "the time necessary for watering, &c. will, no doubt, be employed by the commander in every species of observation that can assist in deciding the longitude of Villegagnon Island."

FitzRoy was quick to realize that the measurement of difference in longitude between Bahia and Rio differed by an amount equivalent to more than four miles from that of the Frenchman, Roussin. Abruptly, he turned the *Beagle* around to head back for Bahia, to check his measurement, then turned around again and sailed back to Rio.

This back and forth yielded three separate and independent measurements of the time difference in longitude between Fort San Pedro in Bahia and the well at the fort on Villegagnon Island in Rio. The measurements on his chronometers agreed to within one second of time.

We measure time in units of hours, minutes, and seconds. We commonly measure angles—the way we almost always think about longitude today—in units of degrees, as well as minutes and seconds. The two different uses of these same words (minutes and seconds) with different meanings can be a little confusing. But there is a simple relationship between the two ways of thinking about longitude. It takes one day of twenty-four hours for the earth to complete one rotation, that is, to rotate through an angle of 360 degrees. Thus in one second of time the earth rotates through an angle equal to fifteen seconds of arc. At the equator this corresponds to a distance a bit less than three-tenths of a mile, and the distance decreases toward the poles. At the latitude of Rio, about 22°54′S, FitzRoy's three measurements of the difference in longitude between Bahia and Rio—all within one second of time or fifteen seconds of arc—agreed to within only a hair more than a quarter of a mile. This agreement gave him significant confidence, first, that his techniques were sound, and second, that his calculation for the longitude of Rio was accurate. Indeed, his calculation of 43°8′45″W is within a mile of the modern value.

But it was poor old Pringle Stokes and his lunars who had been right on the money. With a calculation of 43°9′W; Stokes's determination was the best of FitzRoy's predecessors.

◆

As FitzRoy stewed about longitude, Darwin quickly established his own method of working. While the *Beagle* was in port, or conducting a lengthy survey in one place, he would roam the countryside. Darwin's explorations at Bahia were curtailed by a sore knee, but he saw enough to conclude that, "The whole neighboring country around Bahia, when viewed from a distance appears like a very level plain of about 300 ft elevation. At a few points near the sea beach, a level terrace of about 20 ft height may [be] traced."

He also noticed that, "The whole country moreover is intersected by very numerous, winding flat-bottomed valleys, which although not absolutely wide, are so when compared to the small size of the rivulet which they conduct." A few months later, after his first view of Patagonia, he would reflect back, "The structure of the country [around Bahia] precisely resembles that of the plains of Patagonia, or any other soft formation which during gradual elevation has been modelled by the sea." Even Darwin's first views of South America began to evoke the notion of uplift.

◆

In Rio, Darwin received his first bundle of letters from home. The news was not all good.

Although Darwin's interest in his sometime girlfriend Fanny Owen seems to have been overshadowed by natural history and then by his opportunity to join the *Beagle*, she was definitely somewhere in his thoughts. Their relationship had cooled. She had other suitors, and before departing England his eye may have wandered to other ladies. Nonetheless Darwin received a letter from Fanny while he awaited the *Beagle*'s departure at Plymouth that was at once flirty, sassy, and a bit nostalgic. It was if she already knew, but wasn't quite ready to say, that she wouldn't be waiting for him. She described helping to arrange her sister's wedding. "I was the **Under-taker** and managed the whole affair from *cutting up* of a *Ton* of cake to making *gallons* of *Rum Punch*." She mocked the groom's nervous stutter during the vows. "The word *ch ch ch ch ch-erish* did stick some time but that was the only one." Then at the dinner afterward, "I President of a side table, and *didn't* I pass the Champagne I never allowed a glass to be empty a moment . . . all my *gentleman* became so much *more elevated* than those at the other table that I began to be in fright lest they should expose themselves." But no one became "too much elevated except Mr. B. O. who rose to propose the *Ealth of our Ost & Ostess* and caused much fun." She concluded the letter on a plaintive note, "How I do wish that you had not this horrible **Beettle** taste as you might have staid *"asy"* [easy?] with us here I cannot bear to part with you for so long."

In Rio, Darwin received the news from his sisters that Fanny would be marrying Robert Biddulph, a wealthy politician and the master of Chirk Castle, the castle that he and Sedgwick passed on their way from Shrewsbury to Llangollen. As Darwin set off in the *Beagle*, Fanny had made up her mind. Reading all this news in Rio made Darwin's head spin. He ended a section in a letter to his sister Caroline, "I find that my thought & feelings & sentences are in such a maze, that between crying & laughing I wish you all a good night." He later received a note from Fanny, concluding, "Remember you will always find me the same sincere friend I have been to you ever since we were *Housemaid & Postillion* together." Whatever Darwin's lingering thoughts about Fanny might have been, she was no longer a possibility. This bridge was burned. Now he would allow himself to be completely consumed by exploration for the next four and a half years.

◆

After settling the question of the longitude of Rio, FitzRoy and the *Beagle* set about what Beaufort had called "the real work of the expedition." Using Montevideo at the mouth of the Río de la Plata as a base, from mid-1832 until late 1833, the *Beagle* shuttled between Montevideo and Cape Horn, including two stops at the Falkland Islands. FitzRoy used the months of the austral summer, December to March, for his surveys in the far south. He leased two small ships and purchased a third, for which he reprised the name *Adventure*. He put these other ships under the command of his senior officers and divided up the responsibility for surveying different parts of the coast among the four ships. The region from Bahía Blanca to Río Negro in northern Patagonia received special attention, as did the previously uncharted portions of the waters near Cape Horn.

During his excursions from the *Beagle*, Darwin ranged far afield. While FitzRoy fussed with the longitude of Rio, Darwin met slaveholders and slaves in rural Brazil. Later he explored the land of Uruguay and the nearby area in Argentina across the Río de la Plata around Buenos Aires, more than thousand miles to the southwest of Rio de Janeiro. In the Pampas of Argentina he ranged from Río Negro, a further five hundred miles to the south of Buenos Aires, and inland and northward more than two hundred miles. He traveled mostly on horseback, sleeping in the open with gauchos, or staying with local ranchers in their haciendas. He visited General Juan Manuel de Rosas (the future ruler of Argentina) and his forces as they waged war against the indigenous people of northern Patagonia. Everywhere he ventured, he collected and described the flora, fauna, and geology that he encountered along the way. He described the salt flats around Bahía Blanca, dined on armadillo, and after a restless night on the ground grumbled about the *tuco-tuco*, a small rodent that made "its odd little grunt beneath my head, during half the night." He marveled at the flightless, ostrich-like rheas—but even ate one of a smaller species of rhea for dinner that he had heard about—and long sought in vain—before discovering his mistake. "The bird was cooked and eaten before my memory returned. Fortunately the head, neck, legs, wings, many of the larger feathers, and a large part of the skin, had been preserved."[*]

Whatever rock, fossil, landscape, rodent, bird, or beetle that he found, he wanted to tell its story. What it was made of, how it was structured, where it was found. Beyond that, what was its behavior, how did it get to be the way it was? How was the rock or landscape formed? How did the creature's physical features and behavior benefit it? If the fossil was of an extinct species, why might it have gone extinct?

Reaching "those immense plains of Buenos Ayres, dominated by the Pampas," he developed a narrative for the "great thickness of a red argillaceous [that is, clayey] earth" covering the plains. "For its origin we must look to a period when the estuary of the Plata . . . covered all the surrounding areas with its brackish waters." And uplift played a key role in his story. "Signs of the gradual elevation of the land can in many places be discovered on the shores of the river; and it is probable that the red earthy mass is, geologically speaking, of no very ancient date."

The elevation of the continent was an undercurrent in his geologic thinking during the first years of the *Beagle* voyage, but his discoveries of new and exotic fossils—fossils of large mammals

---

[*] Based on the salvaged remains, the species is now referred to as Darwin's rhea.

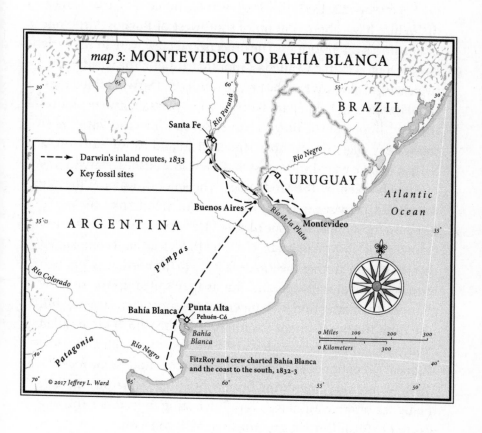

map 3: MONTEVIDEO TO BAHÍA BLANCA

- - -► Darwin's inland routes, 1833
◇ Key fossil sites

BRAZIL

URUGUAY

ARGENTINA

Atlantic
Ocean

Santa Fe

Río Paraná

Río Negro

Buenos Aires

Río de la Plata

Montevideo

Pampas

Río Colorado

Bahía Blanca Punta Alta
                Pehuén-Có
             Bahía
             Blanca

Patagonia

Río Negro

FitzRoy and crew charted Bahía Blanca
and the coast to the south, 1832-3

0 Miles   100   200   300
0 Kilometers   300

© 2017 Jeffrey L. Ward

unknown in Europe, Asia, or Africa—amazed the fellows of the Geological Society and did much to establish Darwin's reputation as a serious geologist.

◆

On September 22, 1832, the *Beagle* was anchored off Bahía Blanca, or "White Bay," about 350 miles southwest of Buenos Aires, one of Beaufort's top priorities. The officers and crew were in the early days of a project that would last months. At noon the weather was overcast and gloomy, but the breeze was light. Darwin joined Captain FitzRoy and Lieutenant Sulivan for a "pleasant cruize around the Bay" in one of the *Beagle*'s boats. Beyond the sand dunes near the water's edge, the Pampas stretched inland to the horizon, a flattish plain covered with grass and a few low bushes. Darwin was not much taken by the landscape. "The smooth water & the sky were indistinctly separated by the ribbon of mud-banks: the whole formed a most unpicturesque picture." He wrote later, "I never knew before, what a horrid ugly object a sand hillock is," and complained that the terrain around the Río de la Plata to the north was "enough to make any naturalist groan." But as they sailed up the shallow channel, their attention was drawn to one spot where a gentle hill rose from the beach about ten miles from the ship, a place called Punta Alta. There they landed and explored the beach and a low cliff near the entrance to a creek. Here Darwin found "some rocks.— These . . . are very interesting from containing numerous shells & the bones of large animals." Very interesting indeed, for this site was to establish Darwin's stature as a fossil collector.

Over the next three weeks Darwin would find the fossil remains of several different creatures. He really got into it. "There is nothing like geology; the pleasure of the first day's partridge-shooting or first day's hunting cannot be compared to finding a fine group of fossil bones, which tell their stories of former times with almost a living tongue." One find particularly intrigued him. "To my great joy I found the head of some large animal, embedded in soft rock.—It took me nearly 3 hours to get it out: As far as I am able to judge, it is allied to the Rhinoceros." FitzRoy teased Darwin, who the crew had nicknamed "Philos," short for philosopher:

> My friend's attention was soon attracted to some low cliffs near Punta Alta, where he found some of those huge fossil bones . . . and notwithstanding our smiles at the cargoes of apparent rubbish which he frequently brought on board, he and his servant used their pick-axes in earnest, and brought away what have since proved to be most interesting and valuable remains of extinct animals.

In August 1833 Darwin returned to Punta Alta, where he collected even more fossils. He shipped his haul back to England, where they came to be studied by Richard Owen, the foremost British paleontologist of his day, and the man who gave us the word *dinosaur*. In all, Darwin found at Punta Alta the remains of nine different types of "great quadruped," including four kinds of giant ground sloth.

◆

Giant ground sloths just don't get the respect they deserve. Some species were as large as elephants. They were limited to the Americas and are all now extinct, but—in their day—some were as ubiquitous as bears are in the wilder parts of North America today. Ground sloths played a fascinating role in the beginnings of paleontology, especially in the Americas. They likely coexisted briefly with the first humans to enter the New World. And the story of the discovery of the ground sloths involves not only Charles Darwin, but the French naturalist Georges Cuvier, and a sitting president and vice president of the United States, Thomas Jefferson.

The story began in 1789 when bones of some gigantic animal were discovered near Luján, west of Buenos Aires. The bones were sent to the Royal Museum in Madrid. But somehow unpublished plates with detailed drawings of the assembled skeleton found their way to Cuvier in Paris. Whether this was unacknowledged cooperation with the Spaniard who assembled the skeleton or scientific thievery remains murky. In any case Cuvier published the first paper describing the fossil skeleton in 1796. He named the animal *Megatherium* or "huge beast." It was, after all, twelve feet long by six feet tall. Cuvier wrote, "This animal differs, in the ensemble of its characters, from all known animals." But he also recognized that certain of its features, its "indicative characters," particularly its teeth (no incisors, all molars), and its claws, were similar to sloths, armadillos, and anteaters. "The great thickness of the sides of the lower jaw, which even surpasses that of an elephant, seems to indicate that [*Megatherium*] doubtless did not content itself with leaves, but—like the elephant and the rhinoceros—broke and crushed the branches themselves." He was also persuaded by the similarities, despite the great difference in size, between the skull of his *Megatherium* and those of modern tree sloths.

As Cuvier was preparing to publish his results, some laborers, digging for saltpeter in the floor of a cave in what is now West Virginia, "at a depth of two or three feet, came to some bones, the size and form of which bespoke an animal unknown to them . . . The importance of the discovery was not known to those who made it, yet it excited conversation in the neighborhood, and led persons of vague curiosity to seek and take away the bones." The bones were recognized as something special, and were sent to the smartest guy around, Thomas Jefferson, then serving as vice president. The pace of government being somewhat different in the 1790s than now, Jefferson had the spare time he needed to pursue his interest in the budding field of paleontology. At first Jefferson thought the bones were from a lion-like animal. He called it *Megalonyx*, or "giant claw." But in 1797 as his paper describing the discovery for the *Transactions of the American Philosophical Society*—running to fifteen pages—was in press, he learned of Cuvier's paper on the *Megatherium*. Jefferson had only the leg and foot bones of his *Megalonyx*, but he realized that these two creatures were most probably related. He added a note to his article pointing out this similarity. But he also suggested that until the question was resolved of whether the two animals were indeed the same, that the name *Megalonyx* be preserved.

Later as president, still curious about the giant ground sloths— and philosophically disinclined toward the concept of extinction, Jefferson encouraged Lewis and Clark to keep an eye out for living specimens as they explored the Louisiana Purchase. Lewis and Clark came back empty-handed, but subsequently fossil remains of the bear-size *Megalonyx jeffersonii*, or Jefferson's ground sloth as it came to be known, have been found from Florida to Alaska, in the La Brea tar pits in California, and most recently during the excavations for a reservoir near a ski resort in the mountains of Colorado.

◆

Darwin correctly assessed that a portion of his bones from Punta Alta were those of Cuvier's *Megatherium*, but when he sent the collection to Richard Owen, Owen concluded that a part represented something similar to Jefferson's *Megalonyx*, plus two new species of giant sloth, one he subsequently named for Darwin, *Mylodon darwinii*, and another, *Scelidotherium*, both more than ten feet long when still kicking.

Some of the sloths could rear up on their hind legs to browse on higher leaves and branches. Others were grazers, finding their sustenance at ground level. In addition Darwin found the remains of two kinds of a related animal, the *Glyptodont*. These strange animals looked like giant armadillos, carried around with them a boney carapace, and were almost as large as a small compact car in some species. Some have a tail culminating in what looks like a spiked mace.

But the animal that seemed to intrigue Darwin most of all, "perhaps one of the strangest animals ever discovered," was the *Toxodon*, not a relative of the sloths, but another extinct animal unique to South America, a "gnawer" that Owen and Darwin suspected to be an aquatic creature, perhaps related to a manatee. *Toxodon* is about the size of, and likely shared the habits of, the modern—and unrelated—rhinos and hippos. It did not have a horn like a rhino, or the peg-like incisors of a hippo. Rather it had its own distinctive array of four arched upper incisors arranged in a horizontal row, which led Owen to bestow its name, meaning "arched tooth." Notwithstanding Owen and Darwin's suspicions, the *Toxodon* likely lived in open, savannah-like country, and, despite its ferocious-looking teeth, dined on a mixture of grass and leaves. It was likely the most common large-hoofed mammal in South America of its day, which ended not more than about 12,000 years ago.[*]

In the same layers of gravel and mud where Darwin found his strange quadrupeds, he also collected numerous fossil shells of more common sea creatures: mussels, clams, snails, and other mollusks. These he sent to the Frenchman and disciple of Cuvier, Alcide d'Orbigny, who had returned to Paris from South America with more than ten thousand natural history specimens just as Darwin was beginning his explorations. D'Orbigny informed Darwin that all of the twenty-three species of shells he'd sent were recent, leading Darwin to conclude that his great quadrupeds all "lived whilst the sea was peopled with most of its present inhabitants." In other words, the ground sloths, the *Glyptodonts*, and the *Toxodon*, had lived in South America quite recently—from a geologic point of view—and thus their extinction, too, was quite recent.

More than a century would pass before a number could be put to the time when Darwin's creatures vanished, and before their origin was related to the causes of the elevation and motions of their continent that so intrigued him.

◆

---

[*] Ages given in this book younger than 50,000 years are estimated calendar years before 1950 A.D. Carbon-14 ages are corrected to calibrated calendar years using the chart given by Roberts, *Holocene*, p. 352, based on the Intercal calibration curves. For consistency, I have also adopted dates for other events given by Roberts. Most dates are rounded to the nearest thousand.

Dr. Teresa Manera is always in motion. A short, sturdy woman, with short-cropped, silvery-gray hair and a quick smile, she radiates energy. When we first met she wore a powder blue polo shirt from a kids' science program bearing the insignia "Mini-Darwin Argentina," jeans, and black tennis shoes. Her glasses hung askew from a chain around her neck, and she was none too happy about the obligatory first stop on our excursion to Darwin's Punta Alta, a call on the commander of the Belgrano Naval Base.

"The cost of our visit," she explained. A paleontologist, she has studied Darwin's notes and the maps made by the officers of the *Beagle* to fix the location of Darwin's fossil finds: right in the middle of Argentina's largest naval base. She needed permission for us to visit the site, and with the news that an American geologist would be coming along, the commander of the base apparently decided that he wanted to size us up.

Captain Temperoni, an urbane and gracious naval officer, his English polished by a tour as a liaison officer with the U.S. Navy in Virginia Beach, greeted us in his office. One of his staff officers and the woman in charge of the base museum joined for good measure. "Okay, right," he interjected as Teresa explained what Darwin found on Captain Temperoni's base 180 years before, and as I described my quest to visit the sites of his geologic discoveries. We had a lot to see and were anxious to get going. The interview, though pleasant, ate up valuable time. Nonetheless Teresa subdued her impatience and, switching to Spanish, took advantage of the opportunity to brief the commander on the progress of her proposal to make an area adjacent to the base a World Heritage Site in honor of Darwin and his fossils.

Today Darwin would not recognize the site of his discoveries. It is totally destroyed, Teresa explained, as we strolled across the grass. The sea cliff is graded and covered with lawn, the beach is covered with fill, eucalyptus trees rise from what were once dunes, large cranes mounted on docks arch skyward, and atop the gentle hill that first attracted the attention of the *Beagle* crew now stands a brick building that has variously been used as a signal tower and the tank house for a water distribution system. No trace of the natural landscape remains.

But a few miles away the coast remains nearly untouched. There the rocky beach, the low cliff, the sand dunes, the small bushes, and even wet spots in the pampas grass all remain just as when Darwin visited and collected his bones.

Teresa has been visiting these beaches since she was a little girl, her introduction to paleontology being the search for scutes, or fossilized chunks of the bony *Glyptodont* shell, about the size of a large walnut. Her father built a beach house in the tiny resort community of Pehuén-Có near Punta Alta during the 1950s. She and her family visited there almost every weekend. After completing her doctorate in mining geology at the university in Bahía Blanca, she took a leave from professional geology in the late 1970s to raise her three daughters. Nonetheless, she and her husband, Roque Bianco, a physician in the city of Punta Alta adjacent to the navy base, became serious fossil collectors, and amassed a remarkable collection. As her interest in fossils, her first love in geology, resurged, Teresa began to work with Silvia Aramayo, a paleontologist at the University del Sur in nearby Bahía Blanca.

Then came the footprints.

"My husband found them," Teresa admitted with a shy smile. One weekend morning in October, 1986, Teresa's husband Roque rose early to take a drive along the beach in his jeep near Pehuén-Có. The wind was howling a gale out of the southwest, scouring the sand from the beach, exposing the rocky platform beneath. As the jeep bounced over the rocky surface, he was utterly amazed to see an abundance of footprints in the newly exposed rock, including one line of footprints, each track the size of a flattened basketball. He rushed back to share the news. That afternoon Teresa called her friend, Silvia. The largest footprints, some three feet or more across, turned out to be those of the giant ground sloth, *Megatherium*. These tracks, and others from an array of its contemporaries, date from roughly 14,000 years ago. The pair published their first paper on the footprints a few months later.

It might seem strange that something as seemingly ephemeral as a footprint might become a fossil. It is really not all that unusual. Footprints or other traces of an animal or plant in soft sand or mud can be quickly covered by other sediment carried by water or blown by the wind. Then if the sediment dries and hardens before these same agents of erosion have a chance to erase the traces, the sediments can be covered again and again, and eventually become cemented into rock.

In the late 1990s the fossils collected by Teresa and her husband became the core of the Museo Municipal de Ciencias Naturales "Carlos Darwin," located at 123 Urquiza in Punta Alta. Its exterior walls are painted with colorful life-size and larger murals of the animals whose fossils are displayed within: the sloths, mammoths, mastodons, and saber-toothed tiger. The museum is a monument not only to Teresa's commitment to explain natural history and Darwin to the public, but to her bulldog tenacity. "The current mayor doesn't love the museum, like the mayor who came before him," she sighed as we toured the neat exhibit cases and interpretive displays, alluding to the continuing struggle to fund the museum. The building housing the museum was formerly a social club, and on the stage where bands used to play, a model of *Megatherium* rises on its hind legs.

Naming the museum for "Carlos" rather than "Charles" Darwin, struck me as odd. I had been corrected a few times when in my bumbling Spanish I had referred to Darwin as "Carlos." We use the English name, Charles, I had been told by Spanish-speaking friends.

"When we began the museum, it was not long after the Falklands War. So many local people [from the Argentine Navy] had been killed, it didn't seem right to use an English name," she explained. The issue of the Falklands was still very much alive during my visit, as signs prominently displayed on the navy base in Spanish read "The Malvinas [Falklands] are Argentine!"

In 2004 Teresa received an international Rolex Award for her activism in securing protection for the site where the footprints were threatened by beach traffic and overenthusiastic visitors. Today the site is a provincial park with rangers on duty to protect and explain the significance of these exceptional traces of the past. But not yet satisfied, Teresa persists on in her quest to have this beach named a World Heritage Site.

◆

Farther north during his travels in Uruguay Darwin was also on the lookout for fossils. His interest was amply rewarded, as it was here he would find his *Toxodon*. To help me trace his discoveries I contacted Daniel Perea, a paleontologist at the Universidad de la República in Montevideo. Daniel and his colleague Valeria Mesa picked me up at the hotel in Montevideo in a pickup truck from the university. Daniel is a short man with thinning, slicked-back hair, a neat goatee, and a shy smile. He appeared dressed in a maroon T-shirt, shorts, sandals, and sporting a fanny pack. Daniel wrote the book on the fossils of Uruguay, titled appropriately enough *Fósiles de Uruguay*. He graciously agreed to show me the most important fossil sites that Darwin visited in his country. Valeria, after already finishing a PhD in geology, was working on one in paleontology with Daniel. A slender young woman in a white tank top, with long dark hair, she was studying dinosaur footprints in central Uruguay. For my benefit they agreed to fast-forward from the time of the dinosaurs to the time of Darwin's ground sloths, another of Daniel's interests.

Uruguay is not a big country, but it is proud and well-to-do. The people are soccer fanatics. Montevideo hosted the very first soccer World Cup. In a way this is fitting, because until its independence, Uruguay, wedged between Argentina and Brazil, was fought over like a soccer ball, first by Spain and Portugal, and then by Argentina and Brazil. Spanish is now the official language, but many people also speak Portuguese.

Today Uruguay is a stable democracy. At the time of my visit, its colorful but humble president was José Mujica. With a silvery mane and drooping mustache, he presented a slightly disheveled appearance in public, wearing an open-collared shirt and disdaining a necktie even on formal occasions. His principal means of transport was an aging VW Beetle. An urban guerilla with the Tupamaros movement during the 1960s, he was subsequently imprisoned, including spending two years at the bottom of a well. But as a candidate of a left-wing coalition called the Broad Front he won the election for the presidency in 2009. Uruguayans are serious about their politics, but with their election of Mujica they seemed to show a sense of humor as well.

With Daniel at the wheel we drove west from Montevideo on the main road to Colonia, following the route that Darwin followed in 1832. From Colonia Darwin traveled north to visit an *estancia* (ranch) along the Río Negro where he found some bones of what he thought were *Megatherium* in the bank. He set out to return along a direct line for Montevideo, but "having heard of some giant's bones at a neighbouring farm-house on the Sarandis, a small stream entering the Río Negro, I rode there . . . and purchased for the value of eighteen pence, the head of an animal equalling in size that of the hippopotamus. Mr. Owen . . . has called this very extraordinary animal, Toxodon." Driving north, Daniel explained that *sarandí* is the name of a common bush in Uruguay that grows close to rivers and streams, so there are many arroyos that might have had this name. Nonetheless, we hoped to get as close as we could to where Darwin found his *Toxodon*.

As we drove Valeria replenished Daniel's maté gourd from a thermos of hot water. Uruguay leads the world in the consumption of tea-like maté, surpassing even Argentina. I asked Daniel and Valeria what they thought had happened to the ground sloths. They took a cautious view, allowing that the changing environment and other factors may have had a role, but declined to back a specific cause. I probed them about theories that the arrival of humans might have played a role, but they remained dubious.

We finally reached our objective, the Estancia La Porteña. Crossing a wooden cattle guard, we drove down a long, tree-lined drive to a spreading, one-story house with a tennis court in front.

We circled to the side in the hope of finding a foreman who could give us permission to look around. We found a man working behind some enormous farm machinery. He told us that the *dueño* (the owner) was at home and that we should check with him. As our truck bounded across the shaded grass to the large house, a large man with rumpled white shorts, wavy white hair, and glasses hanging from a cord around his neck, emerged from the house. He waved, an iPhone pressed to his ear.

When the man completed his conversation, Daniel explained who we were, what we wanted, and where we hoped to go. To my surprise, the man not only gave us permission, but he wanted to join us. He quickly reentered the house, then reappeared from around back with a white sun hat, astride a bouncing four-wheeler, motioning for us to follow.

We caromed off in the truck trying to keep pace with our new friend, Javier, as he raced ahead down the dirt track, his black-and-white border collie struggling to keep up. His hat flew off. Daniel stopped the truck and I retrieved the hat from the grass, as Javier sped ahead.

Javier was a retired engineer and international businessman from Buenos Aires. He had purchased the estancia six years before. He enjoys the calm of the estancia. He also enjoys talking. He used to sell cable for power lines and undersea communications. I asked about his crops. On his three thousand acres he raises mostly soybeans in the fields spreading over gently rolling hills. (Uruguay had recently enacted a law mandating that diesel fuel be mixed with 5% biodiesel, increasing the local demand for soy beans.) He also raised a little corn. Javier spoke excellent English, but he couldn't remember the English word for another of his crops, *cebada*. "It's used in beer," he said. Barley, we discovered referring to the dictionary on an iPhone. "And some cattle, of course."

To our amazingly great good fortune, Javier was not only a Darwin buff, but also a fossil collector. He knew a great deal about what Darwin and seen and done in the area, and what he had found. He promised to show us his collection of fossils after we looked around the estancia. He led us to a hill overlooking the Río Negro, thought to be the one Darwin described as "Sierra del Pedro Flaco." Darwin reflected on the scene, "The view of the Río Negro from the Sierra was the most picturesque which I any where saw. The river, broad, deep, and rapid, wound at the foot of a rocky precipitous cliff: a belt of wood followed its course, and the horizon was terminated by the distant undulations of the plain of turf." A monument, constructed "by a German," Javier told us, stood to mark Darwin's visit.

After taking in the view, we proceeded on an even rougher track down the hill. Javier stopped ahead of us, and held up four fingers, signaling Daniel to shift the truck into four-wheel drive, before he guided the lurching four-wheeler through a gully. Daniel followed carefully, but successfully.

We stopped by the side of a small stream. It was here that Javier himself had collected most of his fossils. The stream had cut down through a layer of reddish soil, just as Darwin described, "reddish earthy mass containing a few small calcareous concretions . . . even where it is of little thickness, and where it might readily be mistaken for detritus produced from the underlying granites, remains of large quadrupeds have several times been discovered." I asked Daniel if this was a good place to find fossils from the time of the ground sloths.

"A very good place!" he nodded. Had Darwin searched right here, I wondered.

In the half an hour we stayed by the creek I scoured the creek banks, hoping to find something exciting myself, but regrettably I found nothing interesting. The best time to search would be just after a storm, or other period of fresh erosion of the creek banks. Daniel did find a thin layer of volcanic ash, blown over from an eruption in the Andes. He took a sample that might be helpful in establishing the age of these sediments.

Finally we headed back to Javier's house where he promised us a cold drink to go with our sandwiches. The summer sun was hot, and a cold drink sounded great. Javier's house, in the style of a classical estancia home, was built in a U-shape around an inner courtyard. As we entered the cool dining room, our eyes were immediately drawn to a shelf extending beneath a large bay window. It was covered with pieces of fossil bone.

We gathered around the window. Javier picked up the bones and tested his identifications on Daniel. The thighbone and scutes of a glyptodont. Some bones from an extinct species of *Equus*, or horse. A piece of ancient guanaco. Daniel knew them all well, gently correcting Javier, as needed. But then, there it was. Daniel picked it up, holding one large bone, orienting the bone—as big as his lower arm—as it would have been in the living animal. It was the lower jaw of a *Toxodon*. Daniel pointed to the spot where it would have attached to the skull, a skull just like the one Darwin purchased at an estancia somewhere nearby for eighteen pence, nearly two centuries years before.

CHAPTER FIVE

Patagonia:
The Great Workshop of Nature

Such is the history ... by which the present condition
of Patagonia has, I believe, been determined ... a
steady but very gradual elevation, extending over
a wide area, and interrupted at long intervals by
periods of repose.

—Charles Darwin

During the later half of 1832 and 1833, FitzRoy and the *Beagle* plied the eastern coast of southern South America, filling in gaps in the Admiralty's surveys from Bahía Blanca to Cape Horn. Since it was first visited by the round-the-world expedition of Ferdinand Magellan, this region of South America, Patagonia, has fascinated Europeans, drawn by its harsh, windswept plains, its jagged, glacier-covered peaks, its peculiar wildlife, and wild tales of its indigenous inhabitants.

As Charles Darwin caught his first glimpses of this land from the deck of the *Beagle*, he was intrigued by the flat-topped hills and terraces that he could see rising from the coast like giant stair steps. He was struck by the apparent consistency of their elevations over long stretches of the shore. Then, as 1833 came to an end, the *Beagle* began surveys of the ports of southern Patagonia.

This was Darwin's chance to go ashore and to explore this land for himself. He set about examining these step-like terraces and the deposits upon them. He measured and compared their elevations. And he began to piece together a story of how these terraces had formed.

◆

As David Catling and I bounced in our little Ford EcoSport along the dusty road on a terrace above the coast of eastern Patagonia, our eyes searched the flat-topped hills above, trying to distinguish the different levels of terraces that Darwin had described. A strip of flat, sandy plain, dotted with scrubby bushes, stretched before us. To our left, a cliff fell sharply ninety feet to a gravelly beach and the foaming breakers of the South Atlantic. Above stretched a leaden sky. To our right rose a flat-topped hill. The wind howled.

David is a professor at the University of Washington in Seattle. A tall, bespectacled Englishman with a high forehead and sandy hair, he was educated as a physicist at Oxford, but his interests range widely, from the origin of life on earth to the emerging field of astrobiology, or life beyond the earth. With a student he published a paper inferring the density of the earth's atmosphere 3.6 billion years ago from the pock marks left by ancient raindrops, recorded in rocks of the Australian outback. But most importantly, David shared my interest in Darwin and his geology. We teamed up to follow Darwin's path through Patagonia.

Driving back along the top of the sea cliff, I looked out over the rough, rolling waves, the wind-tossed white caps, and the scudding clouds. I commented to David that I couldn't imagine how much courage it took to sail on this ocean in the *Beagle*, a mere ninety feet long, and twenty-four feet wide. In January 1833, when trying to round Cape Horn, the *Beagle* was caught in a "whole" gale and was struck by three great waves. The third hit the ship on the side, rolling her until about a third of the upper deck was underwater. "The sea filled our decks so deep, that if another [wave] had followed it is not difficult to guess the result," Darwin wrote. Only a whaleboat was lost, but some of Darwin's collections in the upper cabins on deck were "much injured."

David's reply—dripping with the unemotional worldliness at which exacting Brits excel—came quickly, "It's the south Atlantic. You wouldn't expect a mill pond."

We had started the morning in Puerto San Julián—a small town on the eastern coast of Patagonia—whose current humble status belies its storied past. Ferdinand Magellan and his three ships wintered over here in 1520 before finding the straits that now bear his name and the passage to the Pacific Ocean. It was also here that Magellan outwitted the mutiny that almost led to his undoing.

Questioning Magellan's judgment about continuing to the cold and stormy south in search of a passage to the East Indies and fearful of the unknown, the captains of three of Magellan's five ships conspired to mutiny against him. Magellan skillfully maneuvered to isolate and overcome the mutineers. What followed was a brutal period of execution, torture, and intimidation, until Magellan had purged his ranks of the most disloyal, and cowed the remainder to follow his command.

We passed the starkly named Gallows Point, where Magellan hung the remains of the leaders of the rebellion after they had been drawn and quartered. Later, when he sailed off from San Julian, he left another two of the conspirators on the beach, never to be seen again.

Fifty-nine years later Sir Francis Drake also visited Puerto San Julián on what would become the second circumnavigation. His men found the bleached bones of Magellan's victims and the remains of the gallows. Drake, too, overcame a mutiny here, executing his former friend, Thomas Doherty, near the same spot. Puerto San Julián must have felt like the last stop before doom to those who sailed from a comfortable home, carried away by romantic dreams of riches in faraway lands, but who lacked the resolve to persist in the face of danger, deprivation, and fear of the unknown.

Puerto San Julián became an almost compulsory stop for the English, Dutch, and other explorers and state-sponsored pirates who pilfered the Spanish spoils from the New World. One of the British commanders, decrying Patagonia's lack of trees, was famously quoted as never seeing "a stick of wood in this country, large enough to make a handle for a hatchet."

Earlier that morning, David and I found—next to the bay—a grave surrounded by a neat but rusting iron fence. On the tombstone was inscribed the name of Lieutenant Robert Sholl, an officer from the first *Beagle* expedition who had died here after a brief illness. His grave is treated with considerably more respect than those of the various mutineers.

◆

Both at Puerto San Julián and Puerto Deseado, Darwin traced and measured the elevation of the terraces, or plains as he called them. Somewhere near Puerto San Julián, on the plain at an elevation of ninety feet—no one knows exactly where—in an "earthy mass . . . of a pale reddish colour," Darwin found the bones of a creature that he thought was a mammoth, "several of the vertebrae in a chain, and nearly all the bones of one of the limbs, even to the smallest bones of the foot." From the positions of the bones, he concluded that "the skeleton was certainly united by its flesh or ligaments, when enveloped in the mud." Later, back in London, Robert Owen determined that the large camel-like creature was a previously unknown and extinct species that he named *Macrauchenia patachonica*. The *Macrauchenia* was a big animal, reaching ten feet tall and weighing close to a ton. Today, the most common large wild mammal in Patagonia is the guanaco, from which the llama was domesticated. The guanaco is also camel-like, but smaller, typically about six feet tall and reaching about two hundred fifty pounds, with long legs and a long neck, light brown body, and white underbody and neck. They

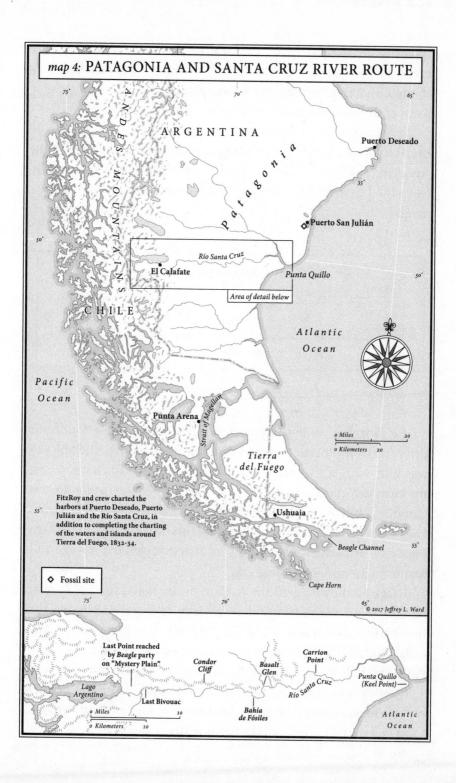

map 4: PATAGONIA AND SANTA CRUZ RIVER ROUTE

ARGENTINA

Patagonia

Puerto Deseado

ANDES MOUNTAINS

Patagonia

Puerto San Julián

Río Santa Cruz

El Calafate

Punta Quillo

CHILE

Atlantic
Ocean

Pacific
Ocean

Area of detail below

Punta Arena

Strait of Magellan

Tierra
del Fuego

FitzRoy and crew charted the
harbors at Puerto Deseado, Puerto
Julián and the Río Santa Cruz, in
addition to completing the charting
of the waters and islands around
Tierra del Fuego, 1832–34.

Ushuaia

Beagle Channel

◇ Fossil site

Cape Horn

© 2017 Jeffrey L. Ward

0 Miles   20
0 Kilometers   20

Last Point reached
by Beagle party
on "Mystery Plain"

Condor
Cliff

Basalt
Glen

Carrion
Point

Lago
Argentino

Last Bivouac

Bahía
de Fósiles

Río Santa Cruz

Punta Quillo
(Keel Point)

Atlantic
Ocean

0 Miles   10
0 Kilometers   10

typically live in small herds. Ever watchful, when alarmed one will make a low cry, and the entire herd will lope off to safety. Owen thought that the *Macrauchenia* was likely related to the guanacos, a piece of information that Darwin would store away.

David was determined to find the site of Darwin's discovery of the *Macrauchenia*. Every time we passed the exposure of anything "reddish," we pulled over to search for some bones that might resemble Darwin's. Despite repeated attempts, our efforts came to naught.

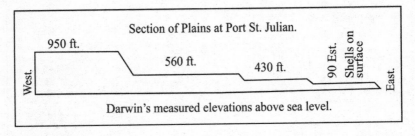

Section of Plains at Port St. Julian.

950 ft.

560 ft.

430 ft.

90 Est.

Shells on surface

West.

East.

Darwin's measured elevations above sea level.

Darwin's Sketch of the Plains at Puerto San Julián

The morning we left Puerto San Julián to begin the four-hour drive north to Puerto Deseado, the sun was shining, and the sky was blue with scattered clouds. Along a gravel road that led past the curiously named Cabo Curioso, we stopped the car for a view, climbing a hill overlooking the harbor, a small remnant of Darwin's lowest surface. A sign informed us that this beach was named Playa Pigafetta, after the Italian who chronicled Magellan's voyage and first brought notice to Patagonia.

Before FitzRoy entered the harbor, he anchored the *Beagle* near here, to scout the shifting shoals prior to taking the ship into the inner bay. Darwin seized the opportunity to wander the nearby hills. Perhaps this was when he found the fossils of *Macrauchenia*. From the hilltop we scanned the surroundings with binoculars.

Then we were amazed to find a shiny fossilized bone poking out of the sand beneath some bushes. Digging, we found more. Most were long bones, one something like a shoulder bone, then a few vertebrae. It seemed too good to be true. David enthused, "I am happy. We didn't find the red mud but we did find something. I have these fine vertebrae and something. . . . Oh, I wish I'd done some proper anatomy!"

That evening I sent an email with some pictures of our fossil bones to an Argentine paleontologist, Sergio Vizcaíno, explaining that there was a remote chance that we might have found something interesting, together with the coordinates from the GPS. The response was deflating.

> *Dear Rob,*
>
> *It seems that the bones you found belong to guanacos. However, you do not hesitate in sending pictures of whatever you suspect might be a fossil vertebrate.*
>
> *Good luck!*
> *Sergio*

So much for pushing back the frontiers of paleontology.

◆

Arriving in Puerto Deseado, we found our hotel perched above the river and small harbor. Looking across the river, I recognized the spot almost as if I had been there before. I had indeed seen it before. FitzRoy's artist, Conrad Martens, painted this exact scene. This was where the *Beagle* crew spent Christmas Day 1833, playing a raucous game called "slinging the monkey."

Martens's watercolor shows one sailor, the "monkey," hanging upside down, suspended by a rope tied to his ankles and affixed above to a tripod made of long oars. The rope was adjusted so that the monkey's head did not touch the ground, but so that he could use his arms to help propel and direct his swings. The other players stood outside a circle drawn on the ground near the limit of the monkey's reach. In one hand the monkey held a piece of chalk, trying to mark the players outside the ring. If he succeeded, he was released and replaced by the chalk-marked sailor. The others held knotted handkerchiefs, attempting to hit, or "baste," the monkey, and score points. The rowdy game required strength and skill, not to mention an ability to maintain one's wits while swinging wildly by the ankles. It seems a safe bet that Darwin, the gentleman, did not participate.

But in the center of Martens's sketch was the giveaway that allowed me to recognize the spot, an odd pair of volcanic needles that the party called "Tower Rock." I asked the hotel clerk about the rocks. She told me that it was now called *"Piedra Toba."* My dictionary showed *toba* as the Spanish word for *tuff* or *tufa*.

"The volcanic rock is tuff. Tufa is a sedimentary rock," huffed David, who has little patience for imprecision. For English-speaking geologists, it is a sophomoric mistake to mix up these two very different kinds of rock. Sadly, in Spanish there is only one word for both.

In 1520 a party of Magellan's men discovered the mouth of a large river sixty miles to the south of Puerto San Julián, which they named the Santa Cruz. But it was not until April of 1834 that FitzRoy put the Río Santa Cruz on the map.

The *Beagle* had struck a rock at Puerto Deseado, and after a rainy and difficult side trip to and from the Falkland Islands, FitzRoy wanted to determine the extent of the damage and make the necessary repairs before he headed though the Straits of Magellan and on to Chile. A beach at the mouth of the Río Santa Cruz seemed to be an ideal location for the repair.

The ship's crew turned the *Beagle* on its side on the beach as the tide fell. Fortunately, FitzRoy's adept carpenter, Mr. May, was able to replace a part of the *Beagle*'s false keel, and "a few sheets of copper [that] were a good deal rubbed" before the tide rose again. FitzRoy named the place "Keel Point."

The gravel crunched noisily beneath our feet as David and I wandered the steep beach at Punta Quilla, the phonetic transcription of FitzRoy's name now adopted for the point. "A peaceful place to replace your keel, if you had to," David remarked. Once again I felt as if I were walking into the picture that Conrad Martens painted of the *Beagle* lying on her side at this very place, where a lone fisherman now tended his lines. Darwin wrote "Nothing could be more favorable than both the weather & place for this rather ticklish operation." As Mr. May worked on the *Beagle*, Darwin examined the rocks and fossils in the cliff above the beach, and roamed the level plain above, finding it "dry & sterile in the extreme."

FitzRoy also had a second objective. On the previous *Beagle* expedition, when Captain Pringle Stokes visited this harbor, he explored about thirty miles up the Río Santa Cruz. FitzRoy was determined to go farther, "During the former voyage of the *Beagle*, Captain Stokes had ascended the rapid current as far as a heavy boat could be taken; but his account served only to stimulate our curiosity, and decided my following his example." FitzRoy wanted to find the source of the river, somewhere in the mountains, beyond the plains.

On the morning of April 18, 1833, FitzRoy, Darwin, Martens, and twenty-two of the *Beagle*'s other officers and men, set off in three whaleboats up the estuary of the river on a flood tide. Their goal was to reach the Andes.

The first day was easy. They sailed, then rowed, their whaleboats to the west. By nightfall they had almost reached the limit of the tides. Then they encountered the river's real challenge. "Perhaps its most remarkable feature is the constant rapidity of its current . . . which runs at the rate of four to six knots an hour [about 4½ to 7 miles per hour]," Darwin wrote.

The next day, the current was too strong to either row or sail, and the party began to "track," or pull, the boats. At FitzRoy's direction, all but two men scrambled ashore from each boat. The three boats were strung together. Half of the men ashore donned harnesses made of canvas and affixed to a rope, the other end of which was tied to the line of boats. As these men dragged the boats, the men still aboard steered them along the shore, avoiding obstacles. In this manner, "one-half of our party relieving the other about once an hour, every one willingly taking his turn at the track rope, we made steady progress against the stream of the river."

Only John Lort Stokes, the officer to whom FitzRoy had assigned the task of making a map of the river, was excused. They proceeded in this way for the next sixteen days. "Many were the thorny bushes through which one half of the party on the rope dragged their companions. Once in motion no mercy was shewn: if the leading man could pass, all the rest were bound to follow. Many were the duckings, and not trifling the wear and tear of clothes, shoes, and skin."

Although the valley heads more or less directly west, the party's course up the meandering river almost doubled the distance, a source of considerable frustration. The river valley varies from about five to ten miles in width and is cut into the nearly horizontal Patagonia plain. The sides of the valley, Darwin noted, rose from the river to the plain above in steps, much as he had seen along the coast.

FitzRoy wrote of the plain above, "brownish yellow is the prevailing colour. . . . Here and there, in hollow places and ravines, a few shrubby bushes are seen. But over the wide desolation of the stony barren waste not a tree . . . can be discerned." Fitzroy would not have been any more successful than his predecessors in finding a handle for a hatchet. But the sparse yet distinctive wildlife caught his eye. "Scattered herds of ever-wary guanacoes, startled at man's approach, neighing, stamping, and tossing their elegant heads; a few ostriches striding along in the distant horizon, and here and there a solitary condor soaring in the sky." On the geologic character of the plain, he asked "Is it not remarkable that water-worn shingle stones, and diluvial accumulations, compose the greater portion of these plains? On how vast a scale, and of what duration must have been the action of those waters which smoothed the shingle stones now buried in the deserts of Patagonia?" These were questions that FitzRoy would return to in the years ahead.

They found a boat hook lost by the expedition led by Pringle Stokes. Then they entered terra incognita, and saw smoke from the fires of Indians. At night, there was a severe frost and "some of the party felt the cold."

About fifty miles west of Keel Point, Darwin found something that surprised him, something he had not seen before on the eastern plains of South America, a boulder, "7 feet in circumference." Not just any boulder, an erratic boulder.

◆

For a geologist an erratic boulder, or simply an erratic, is a boulder distinctly out of place, transported a considerable distance from the bedrock from which it was derived. Often, owing to distinctive minerals or other characteristics, it is possible to trace an erratic to its source, sometimes hundreds of miles away. Darwin recognized at once that this boulder was a long way from home, likely from the Andes.

As a boy, Darwin learned of a famous erratic boulder in Shrewsbury. "An old Mr. Cotton in Shropshire," he wrote, "who knew a good deal about rocks, had pointed out to me . . . a well-known large erratic boulder . . . called the 'bell-stone'; he told me that there was no rock of the same kind nearer than Cumberland or Scotland, and he solemnly assured me that the world would come to an end before any one would be able to explain how this stone came where it now lay."

Nonetheless, as a student at Edinburgh, Darwin learned of the then-current theory that boulders could be transported on icebergs. "I felt the keenest delight when I first read of the action of icebergs in transporting boulders, and I gloried in the progress of Geology." The idea was that chunks of ice broken off from coastal glaciers as icebergs could carry blocks of rock until the ice melted. Then the block would be dropped, only to be found when the land rose about he level of sea. Unfortunately for Darwin, the transportation of erratic boulders by icebergs became something of a fixation. He collected reports of blocks of rocks seen floating on ice, and submitted a note about the sighting of a block of rock on an iceberg to the Royal Geographical Society. He wrote an entire paper on the erratic boulders of South America and their transportation by icebergs for the Geological Society. Later he even suspected that seeds might be transported by icebergs. He mentions icebergs in nearly two hundred of his letters. As one British geologist put it to me, "He was besotted with icebergs."

◆

The party saw tracks of horses and the lines of spears dragged on the ground, then lots of tracks, including those of men, children, and dogs at a place where the Indians apparently crossed the river. They found a dead guanaco lying in the water, which after due consideration and a "few doubtful looks it was considered by the greater number better than salt meat, & soon cut up & in the evening ate."

Darwin took his turns pulling the boats. On April 22, he reported succinctly in his notebook, "Two spells." He also served as a scout, walking on one of the higher terraces, getting a look ahead, watching for Indians, examining and pondering the landscape. He traced his plains or terraces, rising from the river in as many as five steps. It did not seem to him that the excavation of the valley had been the work of the river alone. "My great puzzle how a river could form so perfect a plain as 2^d [the second level up from the river] & cemented even its higher parts—draining of sea ???" He began to think that the valley of the Río Santa Cruz had once been a channel of the sea.

Both Darwin and FitzRoy had mountain barometers with them to measure elevations. A third was left with the *Beagle* at sea level to be read at certain times each day. These barometers, although called "portable" at the time, were about three feet long and weighed in excess of ten pounds, and were slung in a case with a strap over the shoulder. When making a reading the barometer was hung from a tripod or held vertically. A cistern at the base, filled with mercury, was exposed to atmospheric pressure, forcing the mercury to rise in a tube extending upward, in proportion to the atmospheric pressure. The level of the mercury in the tube could be read from a scale engraved along the side.

The preferred method of measuring the difference in elevation between two points was to make simultaneous measurements at the two points, but that was not always possible. Each measurement required reading the temperatures of both the air and the mercury itself. Later, using a procedure and tables developed by the businessman and astronomer, Francis Baily, Darwin would perform—by hand—a tedious series of calculations, requiring logarithms, that led to the difference in elevation. Darwin also recorded elevations determined from measurements with a theodolite, likely made by his cabin mate, John Lort Stokes.

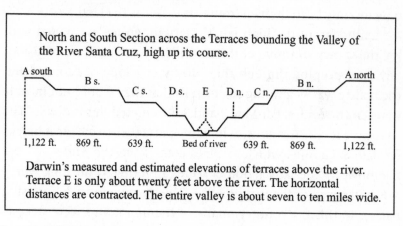

North and South Section across the Terraces bounding the Valley of the River Santa Cruz, high up its course.

Darwin's measured and estimated elevations of terraces above the river. Terrace E is only about twenty feet above the river. The horizontal distances are contracted. The entire valley is about seven to ten miles wide.

Darwin's Sketch of the Terraces in Valley of Río Santa Cruz

The landscape changed only very slowly as they plodded up the river, but finally his patience was rewarded. "This day I found, for the first time, some interesting work; the plains are here capped by a field of Lava." Large blocks of lava, fallen from the cliffs above, littered the sides of the valley and even the river. A collision with one block knocked holes in the side of one of the boats, necessitating an unplanned stop for repair.

Bit by bit Darwin found support for his idea that the river valley had once been a seaway, noting for example, "This plain . . . is *remarkably* level & gravel *much* white-washed so . . . that I think it must have formed beneath sea." Beneath the uppermost plain, capped in places by lava, he found beds of rock containing oyster shells, just as he had found in the cliff at Keel Point.

On April 30 the valley seemed to widen and open to the west. "Like St[rait] of Magellan?" he wondered. Here Darwin found some really large erratic boulders, "generally of an angular form . . . The size of some was immense . . . [one] five yards on each side and projected five feet above the ground; a second . . . sixty feet in circumference, and stood six feet above the ground." Among them were "great blocks of Slate & Granite, which in former periods of commotion have come from the Andes." At the base of a cliff he found three seashells, two "yet partially retaining their color," arguing to him that the sea had only recently been drained from this valley as the land rose.

Finally, "from the high land," Darwin wrote, "we hailed with joy the snowy summits of the Cordilleras, as they were occasionally seen peeping through their dusky envelope of Clouds." But the reality was beginning to creep in that they would not be able to reach the Andes. For May 2 and 3, he entered in his diary, "The river here is very tortuous, & in many parts there were great blocks of Slate & Granite, which . . . had come from the Andes . . . We had however the satisfaction of seeing in full view the long North & South range of the Cordilleras.—They form a lofty & imposing barrier to this flat country; many of the mountains were steep & pointed cones, & and these were clothed in snow.—We looked at them with great regret, for it was evident we had not time to reach them; We were obliged to imagine their nature & grandeur, instead of standing as we had hoped, on one of their pinnacles & looking down at the plains below."

For eighteen days the party had trudged westward, dragging their three whaleboats up the river. Some nights the temperatures fell below freezing. The going was rough. The rocky banks of the river wore out their shoes, and the thorny bushes tore at their clothes.

The men were weary. Their food was almost exhausted. FitzRoy had cut their rations of biscuits in half. Even Darwin complained, it "was very unpleasant after our hard work. . . . It was quite ridiculous how invariably the conversation in the evening turned upon all sorts, qualities & kinds of food."

On May 4, "The Captain decided to take the boats no further." Leaving the boats at a campsite on the river, FitzRoy and party set out on foot. But unlike the flat plain topping the hills above them, the rolling ground here denied them a full view. They would ascend to the crest of one hillock, only to be faced with the sight of another. "We crossed a desert plain which forms the head of the valley of the S. Cruz, but could not see the base of the mountains," Darwin wrote. He suspected that the Río Santa Cruz turned to the north, draining the eastern slopes of the Andes, but his curiosity was unrequited. "We took a farewell look at the Cordilleras which probably in this part had never been viewed by other European eyes, & then returned to our tents.—At the furthest point we were about 140 miles from the Atlantic, & about 60 from the nearest inlet of the Pacific." On their map, FitzRoy and Stokes labeled this last stretch "Mystery Plain" and those beyond "Disappointment Plains."

Their return down the river to the waiting *Beagle* sped past in a blur. It took only three days.

◆

Milthon was a gentle bear of a man, neatly dressed with quick green eyes, well-trimmed beard, and broad muscular shoulders. Those big shoulders are good, I thought, since my upper body could be charitably described as puny. David and I met Milthon in the lobby of our hotel in El Calafate at the eastern base of the Patagonian Andes. Most of the visitors to this popular tourist site are here to see the spectacular glaciers and their handiwork, the sculpted granite peaks, including the most famous, Cerro FitzRoy, named for Robert FitzRoy.

But David and I were not here to see the glaciers or mountains. And for us, Milthon and his skill and experience as a kayak guide were the keys to reaching our goal. His sturdy stature and sun-browned face and hands, even his high-tech outdoor watch, reassured me that I was dealing with someone who knew what he was doing. Our goal was to visit the Río Santa Cruz. But unlike the party from the *Beagle*, we planned to skip the grueling ascent of the river. Our plan was, with Milthon's help, to kayak downstream. We would explore the uppermost part of the river near Lago Argentino with an ATV, then put in the kayaks at a place the *Beagle* party named Condor Cliff.

"Okay, my friends," Milthon said, as he opened a discussion of our plans. Neither David nor I had more experience than a few days paddling kayaks, but Milthon exuded confidence and assured us that we could do this. My doubts lingered.

Two days later Milthon and his friend, Flaco (or Skinny), arrived at the hotel to pick us up in Flaco's well-used Chevy SUV. Two sea kayaks were lashed on top. Behind was a trailer with an ATV. This eclectic mix of transportation options was our answer to FitzRoy's three whaleboats.

We stopped first at a spot overlooking the turquoise blue waters of Lago Argentino, near its outlet into the Río Santa Cruz. We stood not more than about six miles as the crow flies beyond where FitzRoy, Darwin, and the party halted to make their last measurements of latitude and longitude, before turning back, without reaching—or even seeing—the lake.

Darwin noted that the water of the river was a "fine blue colour, but with a slight milky tinge, and not so transparent as at first sight would have been expected." He didn't realize that this milky translucence was the consequence of the river's origin among the glaciers of the Andes and the fine particles of glacial flour that the river carried within it.

Returning to Flaco's truck, we noticed oily streaks radiating from the hubcap of the left rear wheel, suggesting that its bearing just blew a seal. Undaunted, we continued on.

For about fifty miles we followed a dusty gravel road paralleling the river, across the wide, rolling Mystery Plain, using the ATV to bounce across the rolling plain, and take a couple of peeks down into river.

◆

Much has happened along the Río Santa Cruz since the time of *Beagle* expedition, but the future seems to hold even more drastic change. The first big change came at the end of the nineteenth century, when large-scale sheep ranching came to southern Patagonia and the region of the Río Santa Cruz. Virtually all the land passed into private hands. Large estancias were established, many run by English, Scottish, Welsh, or German owners. A typical estancia would control a vast area of land where sheep were grazed, overseen by gauchos who camped with them. The owners would build a house and a few buildings where the sheep would be gathered and sheared, and the wool collected to be sent on its way to the woolen mills of Europe and the United States. The industry reached its peak just before World War II, when the surrounding province had more than 1,500 ranches and 7.5 million sheep.

Then demand fell, and the numbers dropped. Many estancias, especially the most remote like those along the Río Santa Cruz were abandoned. By 2000, the number of ranches in the entire Province of Santa Cruz was down to less than six hundred, with less than two million sheep. The intense grazing of sheep significantly altered the fragile ecology, damaging the soil of this dry, desert environment. Scrubby native bushes called *mata* have flourished, while the native grasses are now greatly diminished.

The area around El Calafate and Lago Argentino, the source of the river, is now popular with tourists from around the world who come to see the peaks and glaciers of the Patagonian Andes. El Chaltén, the jumping-off point for climbers aiming for the famous peaks of Cerro Torre and Cerro FitzRoy—and hikers seeking only a look—is seventy miles to the north as the crow flies. In the early

1900s rainbow trout were introduced into the Río Santa Cruz, and at present anglers come from far and wide to fish for enormous trout ranging up to twenty pounds.

And now the government of Argentina has contracted with Chinese firms to build two giant hydroelectric dams on the river.

◆

We arrived at an abandoned estancia at the base of Condor Cliff in the late afternoon. David and I set out to reach to its crest, just as Darwin described, capped with lava. After a long climb, we found a gap in the cliff and scrambled to the remarkably flat plain above. As we looked out, anxious to absorb the view that Darwin enjoyed, we realized that the weather was taking a decided turn for the worse. It was beginning to rain. We hurried down the slope, flagging down a truck from a seismic crew—that had spent the day exploring for oil and gas—to hitch a ride, just as the rain began in earnest. The truck was the first vehicle that we had seen in hours, and its occupants would be the last people outside our little group that we would see for three days.

Aside from a kitchen table and four chairs, the house at the abandoned estancia contained no furniture. While musty and not exactly clean, Milthon rendered it habitable, and with Flaco's help cooked a delicious lamb dinner on a fire outside in the yard. The next morning, after a dry and comfortable night in my toasty sleeping bag on the kitchen floor—having pushed from my mind any thoughts about small creatures whose domain I might be sharing—I awoke to the sound of rain. Not the best weather for beginning a kayak trip.

A few hours later on the bank of the Río Santa Cruz, David and I pulled on the yellow and gray Gore-Tex dry suits that would be our garb for the next few days, a far cry from the wool, cotton, or even canvas outwear likely worn by members of the Beagle party. If any of them had seen us, they would surely have thought that we were from outer space. We wished Flaco good luck for his trip back to El Calafate with the problematic wheel bearing, and David and I climbed into our two-person kayak. Under Milthon's watchful eye, we paddled tentatively into the current. David, being larger and professing somewhat more experience, took the stern seat, and I the bow. Thus began several hours of learning how to coordinate our paddling so that the kayak headed more or less in the direction that we wanted to go.

The Río Santa Cruz is one to two hundred yards wide, and carries only about one-twentieth of the water of the Mississippi. It is not white water. But to me—at kayak level—with the river's flow near its annual peak, with its mildly turbulent whirlpools, eddies, and cauliflower-textured upwelling, the river looked as if it were gently boiling. While I feel at home in the mountains, floating in fast-flowing water scares the dickens out of me. Our sea kayaks seemed mostly stable, but every time I climbed in, I looked forward to the moment when—I hoped—I could put my feet back on shore.

The first day we paddled almost twenty-five miles down the river, rewarded with occasional sightings of guanacos. The rain let up, but the cold and wind did not. We warmed ourselves with some hot soup for lunch. After some stops to search for erratic boulders, we pitched our tents on the riverbank for the night.

Our first stop the next day—with better weather—was at a place Darwin missed. He must have been scouting on the plains above the north side of the river, or perhaps he was in the team pulling the boats on the opposite bank. Otherwise, he would have certainly seen this site, now called Bahía de los Fósiles, or the Bay of Fossils. Here, in a small area of badlands on the south bank of the river, we found bones literally sticking out of the soft sediments. Milthon watched in amusement as David and I poked and dug, trying to recover a few of the fragile fossils. This site was discovered more than fifty years after Darwin's visit by the Argentine paleontologist Carlos Ameghino in 1886. Here and nearby, fossils of ground sloths, *Glyptodonts*, and even a kind of *Toxodon* have been found. However the species of these animals found here were much, much earlier (now we know as much as 15–20 million years earlier), and also much smaller than the species of similar creatures that Darwin found at Punta Alta and in Uruguay. The ground sloth found here, *Hapalops*, was only the size of a sheep, and the *Toxodon* the size of a tapir, whereas Darwin's giant ground sloth, the *Mylodon*, was the size of a large bear, and his *Toxodon* the size of a hippo. And while most of Darwin's species went extinct relatively recently, roughly 12,000 years ago, these species had vanished eons before.

If these fossils had been discovered while Darwin was still alive— some not more than two hundred yards from where he certainly passed floating down the river—he would have been bitterly disappointed at not finding them himself, but ultimately delighted at how well they fit into his framework of evolution. The fossils found here were the evolutionary predecessors to those he found on the *Beagle* expedition.

For lunch, we went ashore at the deserted but not abandoned Estancia Barrancosa, peering through the eerie windows of the main house at the neatly covered furniture, book shelves, and player piano inside. The walls inside a shed were covered with newspapers from 1931. Tombstones on graves behind the house told, in German, of the family of Arturo Behr who had built the estancia and lived here. Arturo died in 1941, but his determined widow soldiered on for another thirty years.

Only two and a half miles farther down the river, at Basalt Glen, we once again found ourselves in an illustration drawn by Conrad Martens. Here a tributary joins the Río Santa Cruz from the north, and now though little more than a trickle, it has cut down through the lava-capped plain above to form an oasis of green beneath dark gray cliffs. FitzRoy included a spectacular drawing by Martens of this aptly named glen—where the *Beagle* party camped—in his narrative of the expedition. While Milthon and David explored around the bed of the stream, I scrambled up to the plain three hundred feet above, spooking some wary guanacos as I came over the lip. From here Martens's view spread before me, minus only the puma that he added for effect.

Then I saw them.

Amid the chunks of basalt that littered the top of the steep slope, were rounded boulders of granite and other plutonic rocks up to two feet or so in diameter. These boulders were definitely a long way from home. The closest bedrock of this type was perhaps a hundred miles away, yet here were these boulders sitting on the surface about three hundred feet above the bed of the river. How the heck did they get there? Darwin might have argued for transportation on an ice raft, but these boulders were well-rounded, suggesting a much rougher ride, as if tumbled by flowing water.

From my vantage point of the highest plain, I could also see the intermediate terraces that Darwin had measured much more clearly than we were able to see them at river level. I took some photos of the terraces, and of the erratic boulders, with my rock hammer for scale, of course, then descended the slope to the inlet, pondering the origin of the boulders.

Later paddling in the late afternoon light as it turned a reddish gold, we spied something splashing in the water. Was it a fish or a bird? We paddled over. "Armadillo!" exclaimed Milthon, the first to reach the thrashing creature. Had it fallen into the river? It was hard to imagine that it had begun its swim on purpose, though it was definitely holding its own. Then again, Daniel Perea, the Uruguayan paleontologist, had told me that the ground sloths, the armadillo's distant relatives, were actually very good swimmers.

After another night of camping, and another long day winding through the meanders of the river, absorbing the views of Darwin's terraces, we reached the estancia where Flaco was to pick us up. It was attended by a lone arthritic gaucho, Don Arteaga, who kindly allowed us to take refuge in one of the estancia's empty houses. He also shared with us some *torta frita*, a kind of fried bread and a traditional favorite of the gauchos. We built a fire. "I see a fire, and I want to eat meat," said Milthon, a true Argentine. After dinner, I chose to pitch my tent in the courtyard, and as the night progressed and I discovered what the torta frita did to my digestive system, this choice proved quite fortunate.

I was relieved to finally be out of the kayak and my dry suit. We had seen the river, the terraces and plain above, and the erratic boulders, much as Darwin and FitzRoy had seen them, albeit without the herculean effort of pulling whaleboats up the river. I could only wonder what might have happened differently, had the *Beagle* party been able to push on to discover Lago Argentino. If Darwin had seen the source of the Río Santa Cruz among the glaciers above, it might well have changed the history of geology.

◆

"Every one excepting myself had cause to be dissatisfied; but to me the ascent afforded a most interesting section of the great tertiary formation of Patagonia," Darwin reflected as he returned to the *Beagle*. As the little ship turned south to round the southern tip of South America, Darwin carefully assembled his facts and put his ideas about the uplift of Patagonia down on paper.

The step-like terraces along the coast extended, at nearly the same height, for hundreds of miles. Terraces also extended up the Río Santa Cruz. Along the coast, this slope was about the same as the slope of the sea bottom just offshore. And there each step was topped by a thick layer of gravel derived from the rocks of the Andes. Modern, or nearly modern, shells were scattered on the upper surfaces of the lower steps along the coast, and he even found shells on Mystery Plain.

These were Darwin's facts. His first instinct to explain them was with some kind of catastrophic uplift, thinking in the mold of Sedgwick and Greenough. But for Darwin, it didn't add up. He saw no evidence of faults or other deformation that he expected if the terraces had been created in a series of catastrophic jerks of tens or hundreds of feet.

Instead, he turned to Lyell. Guided by *Principles of Geology,* Darwin wrote, "I came to another, and I hope more satisfactory conclusion." It was clear "that the whole coast has been elevated to a considerable height within the recent period." The coast, he suggested, was uplifted gradually, rather than in a series of giant upward jerks. Earthquakes might play a role, he supposed, but included within a "series of lesser and scarcely sensible movements." He saw no requirement whatever for catastrophic events with uplifts reaching hundreds of feet.

Darwin imagined the gently sloping bed of the ocean, elevated at a slow but steady rate. The result, he supposed, would be a surface sloping upward at an angle controlled by the rate of uplift and without any cliffs on interruptions.

If, however, this gradual uplift were interrupted for some period, the waves and currents would wear away the land, and "there would be formed a line of cliff." And the longer the pause in uplift, thought Darwin, the higher the cliff. In this brilliant turnabout, in Darwin's view, the height of the cliffs did not measure the magnitude of a spasmodic, catastrophic jerk, but instead, measured the length of the pause in the uplift. "Let the elevations recommence, and another sloping bank . . . must be formed, which again will be broken by as many lines of cliff, as there shall be periods of rest in the action of the subterranean forces."

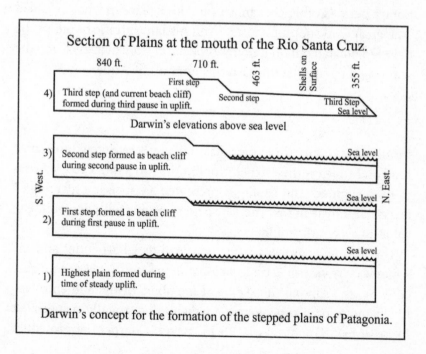

Darwin's concept for the formation of the stepped plains of Patagonia.

The valley of the Río Santa Cruz and especially Mystery Plain fit neatly into the same picture. Darwin jotted in his notebook, "There cannot be any doubt, but that the sea has excavated this great valley & that by the successive elevation the step-like plains were deposited in its bosom & the escarpments being its former coasts."

The setting of Mystery Plain, the valley of the Río Santa Cruz, all together reminded Darwin of Obstruction Sound on the west coast of Tierra del Fuego. He hypothesized that if "the land was depressed from 1000 to 2000 feet an inlet of the sea would cross the Andes." Thus the valley of the Río Santa Cruz, he concluded, had formerly, and not so long ago, been a channel of the sea connecting the Atlantic with the Pacific, not so unlike the Straits of Magellan or the Beagle Channel farther to the south.

Darwin channeled his new mentor, Charles Lyell, to come up with an explanation for the step-like terraces on the coast of Patagonia, and the valley of the Río Santa Cruz without the need for any catastrophe. The explanation was gradual uplift—perhaps including, but not requiring, earthquakes—interrupted by temporary periods of repose, an action taking place in what he called "the great workshop of nature." And his large erratic boulders, he believed, floated out on icebergs broken off from glaciers when the sea lapped up against the eastern slopes of the Andes.

◆

In July, 1834, only three months after the *Beagle* crew ran out of food and time on their expedition up the Río Santa Cruz, a Swiss mining engineer and geologist presented a paper at a meeting of Swiss naturalists in Lucerne. The paper was met with derision. The subject was erratic boulders.

The mining engineer's name was Jean de Charpentier and his paper was *"Notice sur la cause probable du transport des blocs erratiques de la Suisse."* Among the skeptical members of the audience was likely another Swiss geologist, a paleontologist really, a specialist in fossil fish, Louis Agassiz. In his paper, Charpentier presented the heretical idea that the large erratic blocks and boulders that littered the fields and pastures of Switzerland were transported neither by water, nor icebergs at a higher level of sea level, but by glaciers themselves.

Charpentier argued that the frequent grooves and scratches on the exposed bedrock of the Rhône Valley showed that glaciers had once extended far beyond their current limits and, indeed, had covered much of Switzerland at some time in the past. These angular erratic blocks had been carried by the glaciers, just as the rubble of modern moraines was carried by the active glaciers of the Alps.

Whether or not Agassiz heard the original presentation of the paper, we know that he was not a convert to Charpentier's ideas until, to satisfy his wife, in 1836 he accepted an invitation to spend a summer vacation at Charpentier's home in the sunny lower reaches of the Rhône valley. Charpentier enjoyed a reputation as a charming host and purveyor of fine wines. Also Charpentier's wife, like Agassiz's, was German, and Agassiz hoped that his wife, burdened by a small child, might enjoy the company of a kindred spirit. The family had a wonderful time.

But more importantly, Agassiz not only heard the arguments for the glacial theories directly from Charpentier, but he saw the evidence in the field himself. Agassiz became not only an ardent convert to Charpentier's hypothesis, but an apostle and evangelist.

A collision with Darwin was in the making.

CHAPTER SIX

Maria Graham and the Debate on the Causes of Elevation

Among the subjects which have for some years past engaged the thoughts of geologists, none perhaps has excited so general and intense an interest as the Theory of Elevation.

—George Bellas Greenough

On February 21, 1834, back in London, six weeks before Darwin and FitzRoy began their epic journey up the Río Santa Cruz, a group of men gathered to hear a lecture. The men, all men, were again attending the President's Address to the Geological Society. The subject of the talk, the theory of elevation, an earthquake in Chile, and the alleged mendacity of a woman.

Just at this time, as FitzRoy, and Darwin, and the men of the *Beagle* were completing their surveys of the east coast of Tierra del Fuego, fighting tides up to thirty feet and surging currents near the entrance to the Straits of Magellan, watching sperm whales leaping out of the water in San Sebastián Bay, and meeting with native Fuegians wearing guanaco skins—or nothing at all—the men of the Geological Society enjoyed another dinner at the Crown and Anchor and a lecture at Somerset House.

The speaker this year was George Bellas Greenough, returning to serve his third stretch as president of the Society, and the same figure who had made life so difficult for William Smith. Greenough, whose topic was the building of continents and mountains, was an accomplished solicitor and rhetorician. He knew how to shape an argument. Once again, the real target of the attack was Charles Lyell, but Lyell went unmentioned. Instead, Greenough chose what he must have thought to be an easier mark, an English widow, Mrs. Maria Graham. How wrong he was. The argument would not be settled that February evening, nor for nearly a century and a half. But very soon Darwin and FitzRoy would have some facts to contribute to the debate that would back up Maria Graham's side of the story.

◆

Geologists had a very big problem. If the fossil creatures found in Steno's layers had actually once lived in the sea, by what mechanism were these rocks, and the fossilized creatures within them, elevated to the heights of mountains? This is the subject of tectonics, how and why the rocks of the earth's crust become so deformed, distorted, broken, and uplifted into the mountains of Snowdonia, the Alps, and elsewhere. Much later, John McPhee, in his book *Basin and Range*, posed this paramount issue of geology, "If by some fiat, I had to restrict all this writing to one sentence; this is the one I would choose: the summit of Mount Everest is marine limestone."

In the 1830s the word *tectonics* and had not yet been coined, and the debate focused on what at the time was called "the causes of elevation." Some argued that the plains, hills, and mountains where these fossil creatures occurred had been produced once and for all at the time of creation; others argued that the mountains had been lifted up, folded, and broken, in a series of global upheavals. But while the likes of Cuvier, Sedgwick, Greenough, and many others argued for a series of cataclysms, Lyell, in his *Principles of Geology*, argued that the landscape was the result of processes that continue up to today.

In adopting this approach, Lyell was following the path that had been blazed by Francis Bacon and Isaac Newton—and in Lyell's time was being examined and articulated by his contemporaries—John Herschel and William Whewell. Subtleties of their arguments and differences aside, the thrust of their guidance enjoined natural philosophers (or *scientists* as Whewell called them) to find the real agents at work behind natural phenomena, the vera causa, or actual causes. Whewell described this change of heart, "the transition from an implicit trust in the internal powers of man's mind to a professed dependence upon external observation." If Lyell believed that continents and mountains were going up, where was his vera causa? Lyell, too, had been trained in the law, and earthquakes became this solicitor-turned-geologist's Exhibit A.

◆

Maria Graham was the daughter of an English naval officer who eventually was promoted to admiral. Sailing to India with her father after his retirement from the navy, she met a handsome young naval officer, Captain Thomas Graham. They married and returned to London. While her husband went off to fight the last of the Napoleonic wars, Maria—not one to sit on her hands—wrote books about her time in India and about a French painter. In 1819 she traveled to Italy with her husband, and subsequently published her journal describing their time in the countryside east of Rome, including a sympathetic portrait of the Italian banditti—a subject revisited a few years later by Washington Irving in *Tales of a Traveler*—comparing them to Robin Hood and his men. Had Greenough researched his target more carefully he might have been warned off by a pledge Maria made in the preface to her *Three Months Passed in the Mountains East of Rome, During the Year 1819*, "The notices of the banditti might have been more full and more romantic, but the writer scrupulously rejected all accounts of them, upon the truth of which she could not rely, thinking it better to give one authentic fact, than twenty doubtful, though more interesting, tales."

Eventually, Maria got the chance to accompany her husband, then the captain of his own vessel, the HMS *Doris*, to South America. Sadly, Captain Graham didn't make it. He died of a fever off Cape Horn in April, 1822. The ship, without her captain, made her way to Valparaíso, where Maria was taken under the wing of the dazzling Admiral Lord Cochrane, who had served together with her husband when both were midshipmen on the HMS *Thetis*.

Lord Cochrane, himself a naval hero of the Napoleonic wars and former member of Parliament, had left England under the cloud of a stock scandal and gone to South America to assist the local patriots in their wars of liberation. Bernardo O'Higgins,* the founder of the Republic of Chile, invited him to organize the navy of the emerging state. This Lord Cochrane successfully achieved with gusto. When he arrived in Chile, the city of Valdivia in the south-central part of the country remained a Spanish stronghold, isolated from the rest of the country by territory held by the Mapuche, the indigenous people

---

* The story of Bernardo O'Higgins, and how the George Washington of Chile came to be of Irish descent, is a very rich tale in itself. Bernardo was the illegitimate son of Ambrose O'Higgins, an Irishman whose family had lost its land. Ambrose found his way first to Spain, then—as an Irishman and a Catholic—he was permitted to immigrate to Spanish South America in 1751. After a brush with the inquisition in Peru, he joined the Spanish army in what is now Argentina, pioneered a postal route across the Andes from Mendoza to Santiago, led battles against the Mapuche in Chile, became Governor of Chile, and ultimately Viceroy of Peru, which at the time oversaw the territories of both Chile and Peru. Along the way he fathered Bernardo in Chillán, a city in the central valley of Chile, 225 miles south of Santiago. At the time of Bernardo's conception, Ambrose was fifty-seven, and Bernardo's mother, Isabel Riquelme, eighteen. The pair did not marry. Servants of the Spanish Crown such as Ambrose were forbidden to have personal contact with the natives of the colonies, so to have married would have been the end of his career. Bernardo was raised first by his mother's family, then by a friend of his father. Ambrosio, as he came to be known, supported Bernardo financially; however, the two never met in person. Bernardo studied in Peru and England and then spent three years in

that the Spanish had been unable to overcome. Valdivia, some twelve miles up the Río Valdivia, was heavily defended from the sea by a well-manned network of stone forts arrayed on the shores of the estuary at the entrance to the river. The forts had been built to repel earlier seaborne threats, Dutch and British pirates, but now stood between the Chilean Republicans and the town. In 1820, in a brilliant display of tactics, Cochrane launched a nighttime amphibious attack approaching the first forts from behind, catching the Spanish royalist defenders in total surprise. In the ensuing confusion, the chain of defenses collapsed and Valdivia surrendered to the Republicans.

Although Cochrane was unable to overcome the Spanish at San Carlos (now Ancud) on Chilóe Island farther to the south, he is still regarded as a national hero in Chile. He is remembered as Admiral Lord Cochrane, the founder of the *Armada de Chile*, the Chilean Navy—one of strongest institutions in the country. In almost every city, town, or village in Chile, from metropolis to hamlet, from Santiago to Tirúa, there is a street named after Cochrane, be it a sweeping boulevard or a muddy lane pocked with potholes.

When Maria Graham arrived in Valparaíso, Lord Cochrane was near the peak of his Chilean glory. To reward his service, Bernardo

---

Spain. With the death of his father, Bernardo inherited a large hacienda and surrounding land in Chile. He returned in 1802 to become a wealthy landowner. Chile acquired its independence in a tortuous process that began with opposition to the French domination of the Spain in 1810, and concluded with the declaration of the republic in 1818. Bernardo became a leader in the anti-royalist forces in 1813, at one point giving a now oft-quoted order, "Lads! Live with honor, or die with glory! He who is brave, follow me!" Then, after a period in which his forces retreated to Argentina, Bernardo joined forces with José de San Martín, and the two led an army that crossed the Andes by a route that Darwin would follow two decades later, and reinvaded Chile. It is said that among the brave who followed Bernardo on his campaigns was his mother Isabel, who cooked his meals. After defeating the royalists, he served as supreme director of Chile from 1817—inching the country toward democracy and promoting national development—until 1823 when he was deposed by a conservative coup, after which he went to Peru accompanied by his ever-faithful mother. Despite significant controversy about aspects of his reign, including his failure to prevent the execution of a key rival, Bernardo O'Higgins is celebrated as the liberator of Chile.

O'Higgins and the government of Chile awarded Cochrane a house and ranch near the town of Quintero, thirty miles north of Valparaíso, and were also in the process of building a spacious house for him on a hill above the harbor at Valparaíso, overlooking not only the bay, but the headquarters of the Armada.

Maria was an accomplished diarist, as well as a talented artist, and would publish an illustrated account of her time in Chile. Although Maria was only recently widowed, and Cochrane had a beautiful younger wife staying in Santiago—a day's journey or two away—it is clear that Maria was much taken with the Admiral. "Though not handsome, Lord Cochrane has an expression of countenance which induces you, when you have once looked, to look again and again," she wrote. "His conversation, when he breaks his habitual silence, is rich and varied; on subjects connected with his profession or his pursuits, clear and animated; and if ever I met with genius, I should say it was pre-eminent in Lord Cochrane."

In November of 1822 Maria was staying at a house belonging to Cochrane about a mile from the coast at Quintero. On the night of November 19 a powerful earthquake struck. Maria recorded every detail of the event.

It was a "very still and clear moonlight night," she began. Then she described the shaking of the earthquake, lasting "three minutes," the sensation of the shaking, "that of the earth being suddenly heaved in a direction from north to south, the falling down again; a transverse also being now and then felt." In the following days she rode on horseback, sometimes in the company of Lord Cochrane, through the surrounding countryside and along the beach, noting with care what she saw. Near Viña del Mar, on the bay just north of Valparaíso, she described a strange phenomenon, "the alluvial valley . . . was covered with cones of earth about four feet high, occasioned by the water and sand which had been forced up through funnel-shaped hollows beneath them; the whole surface being thus reduced to the consistence of quick-sand." This is a vivid and accurate description of what a twenty-first-century geologist would recognize immediately as the classic symptoms of soil liquefaction induced by the strong shaking of an earthquake.

Apparently having got wind of her story, Mr. H. Warburton, Esq., a member of the Geological Society, wrote to Mrs. Graham after she eventually found her way back to London, asking for a more complete account. She provided him with her response in 1824, which he inserted into the Transactions of the Society. The portion of Maria Graham's detailed—and on the face of it, utterly objective—account that drew Greenough's ire, was her description of what she saw along the coast:

> It appeared on the morning of the 20th, that the whole line of coast, from the north to south, to the distance of one hundred miles, had been raised above its former level. I perceived, from a small hill near Quintero, that the old wreck of a ship, which before could not be approached, was now accessible from the land, although its place on the shore had not been shifted. The alteration of the level at Valparaíso was about three feet, and some rocks were thus newly exposed on which the fishermen collected the scollop [sic]-fish, which were not know[n] to exist before the Earthquake. At Quintero, the elevation was about four feet. When I went to examine the coast with Lord Cochrane, although it was high water, I found the ancient bed of the sea laid bare, and dry, with beds of oysters, muscles [sic], and other shells, adhering to the rocks on which they grew, the fish being all dead, and exhaling most offensive effluvia. I found good reason to believe that the coast had been raised by Earthquakes, at former periods, in a similar manner, several ancient lines of beach, consisting of shingle, mixed with shells, extending in a parallel direction to the shore, to the height of fifty feet above the sea. The country has, in former years, been visited by Earthquakes, the last of any consequence having been ninety-three years ago.

Now, a decade later in 1834, in Greenough's address, "although deeply sensible of the honor that the lady [Maria Graham] conferred on the Society by her obliging compliance with the request which elicited her narrative," he begins, "it is only the importance of its contents which could induce me to subject them to the test of rigid examination." Attacking Graham's statement that the line of coast for a distance of a hundred miles had been "raised above its former level," Greenough struck out, "But by what standard was the former level ascertained? Who, on the morrow of so fearful a catastrophe, could command the sufficient leisure and calmness to determine and compute a series of changes, which extended 100 miles in length, and embraced an area of 100,000 square miles." Going on he demanded, "By what means did the surveyors acquaint themselves with what had been the levels and contours before the catastrophe took place, by which, as we are told, all the landmarks were removed, and the soundings at sea completely changed?"

Although Greenough clearly intended his questions to be rhetorical ones, there is an obvious answer that he chose to overlook: the level of the sea. The range of the tides along the coast of central Chile is about six feet, and as was already well known, the various species of mollusks, crustaceans, algae, and other plants and animals that inhabited the coast, as in the British Isles, each enjoyed a particular range of elevations with respect to the tides. Indeed this knowledge had been critical to those who relied on the sea for their sustenance for thousands of years. Even today, Chilean artisanal fishermen, when asked to describe the range of the tides, will point to specific rocks or points on the shore to indicate the levels reached by the tides during different parts of the month, and different times of the year. To those who live near, work on, or eat at the mercy of the sea, its level and its changes are keenly observed.

In the 1830s it was still considered inappropriate for a woman to attend, much less participate in, the vaunted discussions of such an august body as the Geological Society. But undaunted by Greenough's fallacious, if not malicious, assault, Maria wrote a letter of response to the Society in which she gave at least as well as she got.

By then she had remarried—to a prominent landscape artist, Augustus Wall Callcott—and had become Maria Callcott. In an apparent nod to propriety that seems odd today, in her letter responding to Greenough, she referred to herself in the third person, "This attack implies, in the first place, a suspicion of wilful[l] falsification on the part of Mrs. Callcott.—Secondly, it charges her with that high coloring, which 'ignorance, terror, and exaggeration, are apt to indulge.'" "And thirdly, in case Mrs. Callcott should be prepared to rebut the first and second charges," she continued, "the insinuation . . . [in an attempt] to throw discredit on her whole statement," that she had reported incorrectly the times of the earthquakes. Indeed in her letter Mrs. Callcott does a fully creditable job of defending herself, pointing out, for example, though not a geologist, she had no stake in what she was reporting, that she did not claim "such an absurdity" as to have attempted or accomplished "a regular geological survey of an estimated area of 100,000 square miles on the morrow of that fateful catastrophe, or at any other time." From the perspective of nearly two centuries later, perhaps the most offensive aspect of Greenough's attempt to discredit Maria's observations was the undertone that if only a man, "a naval officer or naturalist," as he wrote, had made the observations, or at least corroborated them, they might be taken seriously.

However feckless Greenough's dismissal of Maria's observations, the conclusion of his prosecution was clear. "If I am to pronounce a verdict according to the evidence," he wrote, " I believe that there is not yet one well authenticated instance in any part of the world, of a non-volcanic rock having been seen to rise above its natural level in consequence of an earthquake."

And this was the nub of the argument.

In truth, Maria Graham Callcott's sin in Greenough's eyes may not have been inaccuracy. Her sin was having been being quoted by Charles Lyell in his *Principles of Geology, Volume I*, as evidence for his notion that mountains were uplifted by the subterranean intrusion of masses of molten rock (what we now call magma), and that earthquakes and volcanoes were somehow symptoms of this process.

◆

As we have seen, Charles Lyell, Darwin's mentor in absentia on the *Beagle*, was the leading spokesman for an idea that actually went back to James Hutton. Namely, if you want to understand the present state of the earth, you need to look at the processes that are going on all around you. Lyell argued that these processes were enough, given sufficient time, to produce the landscape. But Lyell's views were at the time still extreme, far from the mainstream of geologic thinking. Others, led by Cuvier until his death in 1832, didn't believe these ideas for a minute. Cuvier and his followers believed in a series of catastrophes.

The debate extended beyond elevation to the formation of the landscape and the history of the creatures that lived upon it. Trickling streams, even if they did occasionally flood, were, Lyell's opponents responded, simply inadequate to carve valleys from solid rock. The explanation of dramatic landscapes required some kind of catastrophe, not just systematic erosion. Cuvier was led to this conclusion in part because of his studies of the fossils of extinct mammals. Why would all these animals have gone extinct without a catastrophe? But when and why might such a catastrophe have happened? At the beginning of the earth? Many thought that the deformation was the relic of some subsequent, but ancient, cataclysmic upheaval, likely an act of God. But, if not, where on the earth today were the forces that could create the magnitude of deformation so visible in the great mountain ranges and lifting fossils of extinct sea creatures to such heights?

In contrast, Lyell believed that the continents came about when land rose from the sea, the ocean basins when the land sank, and that these movements were ongoing. But what was the cause of this process? What was the evidence that such processes were underway? Clearly erupting volcanoes could add land or even create it, but they did not occur everywhere. Something else was needed.

Lyell's bet was on earthquakes, "In the course of the last century . . . a considerable number of instances are recorded of the solid surface, whether covered by water or not, having been permanently sunk or upraised by the power of earthquakes." Lyell studied the reports of the earthquakes of the eighteenth and early nineteenth centuries, especially those in Italy, which by the standards of the day had been particularly well documented. Unfortunately much of the documentation focused on secondary effects: landslides, soil slumps, fissuring—not on the evidence for changes in land level, which Lyell believed to be the primary cause for these accompanying effects.

Lyell, nonetheless, needed more evidence to buttress his arguments. As he admitted, it remained scanty and often problematic, "the difficulty of proving that the general level has undergone any change unless the sea coast happens to have participated in the principal movement." Even then, he continued—still failing to consider the predilection of the various species of tidal creatures to choose their own favored elevation—"the scientific investigator has not sufficient topographical knowledge to discover whether the extent of beach has diminished or increased and he who has the necessary local information scarcely ever feels any interest in ascertaining the amount of the rise or fall of the ground."

Maria Graham's description of the changes in the elevation of the land at the time of the earthquake in 1822 was an exception. It was—in Lyell's eyes—crystal clear. But there was also other evidence that the level of the land was changing. Two other examples gnawed at Lyell and his fellows at the Geological Society, as they pondered the question of how mountains and continents were raised. One was a Greek ruin on the shore of the Bay of Naples. The other was the apparently rising coast of Sweden. Both of these examples seemed to show that the level of the land, or sea, changed during recent history.

◆

In 1832 Lyell used an engraving of a Greek ruin, the Temple of Serapis, as the frontispiece of *Principles of Geology*. Argument about this monument, in the town of Pozzuoli on the northern shore of the Bay of Naples, had bubbled for decades, but the facts were not in dispute. Darwin, even as he sailed on the *Beagle*, reflected on the Temple as he thought through his analysis of the evidence for uplift that he had found at his first stop on the island of St. Jago.

The facts were shown vividly in the frontispiece to Lyell's book. The engraving shows three massive, marble columns rising about forty feet from the flooded ruins of a broad marble terrace. Wispy cirrus clouds streak the sky above the treed hills in the background. In the foreground to the left, on a pile of ruins rising from the flooded terrace, sits a woman in a large hat, gazing at the columns.

But most importantly of all, about one-third of the way up each of the three columns, at the same height on each, is a dark band, with a width of more than six feet, of small holes.

Lyell visited the Temple himself in 1828, and carefully examined the three pillars and the pits excavated in them. He found shells of the culprits: "At the bottom of the cavities, many shells are still found." The holes were as deep as about four inches, and many still contained the shells of the holes' creators, a boring intertidal mollusk and relative of the mussel, with the fancy name of *Lithodomus lithophaga*. As Lyell pieced together the story, "We must, consequently, infer a long-continued immersion of the pillars in seawater, at a time when the lower part was covered up and protected by strata of tuff and the rubbish of buildings; the highest part, at the same time, projecting above the waters, and being consequently weathered, but not materially injured."

The Temple, later determined to have been a market place and now more properly referred to as the Macellum of Pozzuoli, was originally constructed in the late first century or early second century A.D. It was rebuilt several times following sackings of the city, but its ruins were eventually buried by ash from volcanoes erupting nearby, then rediscovered in the mid–eighteenth century, and excavated—motivated at least in part by the desires of King Charles of Naples to decorate his new castle with antiquities.

For geologists the argument about the temple revolved around the evidence that it provided for changing sea level. Two conclusions seemed abundantly clear. First, as John Playfair—the Scottish geologist who played the role of intellectual amanuensis for James Hutton—wrote in 1802, "The pavement of the Temple of Serapis [is] now somewhat lower than the high-water-mark, though it cannot be supposed that this edifice, when built, was exposed to the inconvenience of having its floor frequently under water." Clearly, when the Temple was built the water level was lower. The second aspect of the history of the water level, that to which the traces of the mollusks bore witness, was that at some time since the construction of the columns, the water level was sixteen feet higher. As Lyell wrote, "This celebrated monument of antiquity affords, by itself alone, unequivocal evidence that the relative level of the land and sea has changed twice . . . since the Christian era; and each movement, both of elevation and subsidence, has exceeded twenty feet." The evidence for change was certainly compelling.

How to explain this seemingly bizarre history? Three theories emerged. The first put forward by Scipione Breislak, an Italian geologist and the first to ponder this puzzle in 1798, proposed that the sea level in the Mediterranean Sea had changed over this time. The second possibility proposed by the poet and polymath, Johann Wolfgang von Goethe, and supported by the English geologist, Charles Daubeny, was that somehow a pond had formed around the protruding portion of pillars while they had been buried, thus providing the mollusks a watery home while they did their work. And the third possibility, proposed by Playfair, but championed by Lyell and Charles Babbage—the latter now memorialized as the inventor of the programmable computer, but also an active member of the Geological Society and who had personally investigated the Temple—was that the elevation of the land had changed.

So even as Darwin sailed along the east coast of South America, half a world away, he was thinking about the evidence for the uplift that he saw on the coastline at St. Jago. How did it connect with the Temple of Serapis? Darwin initially took Breislak's point of view, "I at one time felt inclined to think that the sea [at St. Jago] must have sunk instead of the land raised.—but as this supposes the fall of the whole Atlantic it is clearly impossible." Similarly there was no evidence to support Breislak's notion that there had been widespread changes in the level of the Mediterranean since the construction of the Temple of Serapis.

Daubeny supported Goethe's rather ad hoc explanation of the pond, arguing that the elevation and subsidence of the land during earthquakes would have toppled the columns. But Darwin wasn't buying it: "Dr. Daubeny when mentioning the present state of the temple of Serapis, doubts, the possibility of a surface of country being raised without cracking buildings on it.—I feel sure at St. Jago in some places a town might have been raised without injuring a home." And what clinched the argument for Darwin was the variation in the amount of uplift along the stretch of beach that he walked at St. Jago: "the different height of upheaval prove it is not by subsidence of water." Thus Darwin, in thinking about St. Jago, joined the reasoning of Playfair, Lyell, and Babbage regarding the Temple of Serapis. He became convinced that it was the level of the land that had changed. But whether these movements were gradual or sudden was still a matter of conjecture.

◆

In 1491, just as Columbus was trying to round up support for his voyage of exploration to India, the people of Östhammar, a small Swedish town on the coast of the Gulf of Bothnia, faced a serious problem. The sea had receded from their town to the extent that it was no longer reachable by boat. They petitioned the powers that be, requesting permission to move the town, declaring, "During recent years the land has grown outside the town at the sea so that where, some years ago, a small cargo boat could come from the sea to the town of Östhammar not even a fishing boat can go nowadays. And the land is still growing and rising every year." The bereft citizens of Östhammar were in fact allowed to move their town to a new location where they had access to the sea, but that didn't stop the land from rising. Today the level of the land, relative to the sea, is about three feet higher still. Nor was the phenomenon limited to Östhammar. The sea seemed to be retreating all around the Baltic Sea, and especially around the Gulf of Bothnia.

Two of the best Swedish thinkers of the eighteenth century—Anders Celsius, the physicist and astronomer, and Carl Linnaeus, the botanist and zoologist—turned their attention to this problem in the 1700s. Celsius investigated the question of changing sea level during his travels in Sweden. In 1743 he summarized his findings in a paper, "Remark on the Water Decrease in the Baltic Sea as Well as the Western Sea." In his paper, illustrated with drawings, he documented the story of a peasant named Rik-Nils (or Rich Nils) who lived on the Gulf of Bothnia and hunted for seals on a rock. Over his lifetime, the water level dropped to the extent that the seals could no longer reach their customary resting places, so with the help of a fire he excavated on the rock a new and lower spot for them to enjoy the sun. After Rik-Nils's death, the water level continued to fall. Celsius estimated that in 168 years the water level had fallen eight feet. Also, in a generous act to provide data for future investigators, Celsius arranged to have a line showing mean sea level, and the date "1731" chiseled on a rock at Lövgrund, not far away.

Linnaeus also traveled extensively around Sweden, and on the island of Gotland in the Gulf of Bothnia he found a sequence of seventy-seven beach ridges all parallel to the coast. In a paper published in 1745, he concluded that these ridges were old shorelines left high and dry by the retreating sea. Neither Celsius nor Linnaeus had a particularly compelling explanation for their perplexing observations. Celsius considered two possibilities. First, he wondered, could it be that some of the water evaporated from the sea was consumed by plants, and thus not ultimately returned via streams and rivers? Or he suggested that perhaps there were simply some holes somewhere in the bottom of the sea.*

Reports of the falling sea level around the Baltic reached England and perplexed Lyell and his fellows of the Geological Society. But what complicated the puzzle for Lyell was that—unlike the region around the Temple of Serapis where volcanoes and earthquakes abounded—in the region of the Baltic there were no earthquakes, no volcanoes, no unusual or cataclysmic events of any sort. Many of Lyell's antagonists extrapolated from the observations of sinking relative sea level in Scandinavia to indicate that global sea level was declining, some inferring that this had been going on since the beginning of time, or at least since the Deluge. Lyell, of course, desired a geologic explanation. But there was no apparent cause, no vera causa. For a time he simply chose to disbelieve the reports. But even as FitzRoy and Darwin were finishing their excursion up the Río Santa Cruz, Lyell was preparing to sail to Scandinavia to see for himself.

◆

Before arriving in Chile on my first visit to join Brian Atwater in the search for the deposits of ancient tsunamis, I had never heard of Maria Graham or Lord Cochrane, but my innocence quickly came to an end. We read Maria's reports of the 1822 earthquake in hopes of finding clues that might lead our search. Marco Cisternas had previously arranged for us to visit a base of the Chilean Air Force that now spreads across the wetlands near the bay at Quintero, and just below the house where Maria was staying at the time of the earthquake. Maria's lovely drawing of the site before the construction of the Air Force base appears in her *Journal of a Residence.*

---

* We cannot pass the contributions of both Celsius and Linnaeus to the uplift debate without recalling that while Celsius gave us the original centigrade scale for measuring temperature, it was Linnaeus who reversed the Celsius scale, giving us the centigrade scale that we have today. Celsius originally fixed zero degrees at the boiling point of water and one hundred degrees at its freezing point.

Using a hand-powered coring tool in the swampy deposits beside the runways, we found layers of sand that implied ancient earthquakes and tsunamis. Somewhat to my disappointment, however, subsequent work showed that these tsunamis were much earlier, and that none of them was the result of Maria's earthquake in 1822. Nonetheless we remained fascinated by her accounts, their stunning accuracy, and the narrative of her standing up to Greenough and the male-only Geological Society. And, to tell the truth, the titillating ramifications of the dashing naval hero rescuing the damsel in distress was more than we could resist.

On a subsequent visit, on our route to and from the field site through a cluster of houses called Valle Alegre, about five miles inland from Lord Cochrane's house in Quintero, my colleagues and I discovered an old ranch house, that to our surprise had belonged to Lord Cochrane and was being restored as a small museum. The house had in fact been visited by Maria Graham, and charmingly remained much as it had been in the 1820s. It was built in the style of the time with wide porches, white plaster walls, and a classic red tile roof. In the back was a long row of stables. Upon entering, we met Don Claudio Castro, the force behind the museum, together with a colleague, Doña Laura Spencer Ossa. Both were authorities on—and adulating fans of—Cochrane and his time in Chile. They graciously showed us around the house and the Cochrane artifacts and materials that they had collected—a ship's wheel, anchors, and numerous small objects belonging to Cochrane or his time, even the mast believed to be from one of the ships that Cochrane had captured in Peru. They humored almost all our many questions about Cochrane and Maria Graham.

There is no question that Cochrane was a naval hero, and that his contributions in the Napoleonic wars and in the Chilean war of independence were on the side of the angels. Nonetheless some aspects of his judgment and behavior in financial and political dealings are open to question—and not entirely beyond reproach. In particular, Don Claudio was not amused by my question, posed in halting Spanish, about Cochrane's motivations in coming to Chile, telling me that my question was rather foolish, and that Cochrane was an honorable man, a democrat, and a patriot. Likewise, Don Claudio and Doña Laura were quick to deny our perhaps adolescent suspicion that Lord Cochrane and Maria Graham might have become romantically involved. Doña Laura politely, but with an edge of indignation, declared firmly that the two had not been *revolcados en la cama* (wallowing in bed).

◆

Without a doubt Maria Graham Callcott deserves a place of honor in the debate on the causes of elevation, and indeed in the history of geology. But notwithstanding the evidence that she provided for Lyell's point of view, her observations were insufficient to sway the views of Greenough and the conservative members of the Geological Society who opposed Lyell. But within a year of Greenough's address bashing this courageous reporter, Greenough's preference for the reports of "a naval officer or naturalist"—as opposed to those of a mere woman—would be satisfied in spades. Robert FitzRoy would be the naval officer. Charles Darwin would be the naturalist. The substance of the reports, however, would not be any more to his liking.

CHAPTER SEVEN

Darwin's Earthquake

To my mind since leaving England we have scarcely
beheld any one other sight so deeply interesting.
The Earthquake & Volcano are parts of one of the
greatest phenomena to which the world is subject.

—Charles Darwin

On June 10, 1834 the *Beagle* exited the Magdalen Channel of the Straits of Magellan, entering—at long last—the Pacific Ocean. The men aboard would not see the Atlantic again until they would round the southern tip of Africa two years into the future. At first the weather was fair enough for Darwin to see the rugged coast, the sight of which was "enough to make a landsman dream for a week about death, peril, & shipwreck." Then the weather turned against them. FitzRoy and the crew fought "furious gales from the North" and the "great sea" as they struggled to reach their first port of call, San Carlos on Chile's Isla Grande de Chilóe; then onto Valparaíso, for resupply and a respite from surveying to compile and draft their charts.

In Valparaíso Darwin stayed with a school friend from Shrewsbury, Richard Corfield, who had come to Chile to seek his fortune as a shipping agent and merchant. Darwin was surprised and delighted by the literacy of the English expatriate residents of Valparaíso, meeting several "who have read works on geology & other branches of science," including at least one who took Darwin aback: "It was as surprising as pleasant to be asked, what I thought of Lyell's Geology."

The middle of August Darwin set out on a six-week geological excursion to the foothills of the Andes, Santiago, and the central valley of Chile. He arranged to meet his friend Corfield in Santiago, who, he wrote to his sister, Caroline, is "going up to admire the beauties of nature, in the form of Signoritas, whilst I hope to admire them amongst the Andes." Whatever Darwin thought of the "Signoritas" he met with Corfield in Santiago, he did not think much of the geology he saw on the first leg of his trip. He wrote FitzRoy from Santiago, "there is nothing of particular interest—all the rocks have been frizzled melted and bedevilled in every possible fashion."

On the first night of the longer, second leg of the trip to the south of Santiago, he enjoyed a stay at "a very nice hacienda" where he did meet "several very pretty Signoritas" who "turned up their charming eyes in pious horror" at their perception of his irreverence. He attempted to convince them that he too was a Christian, but his defense of the Anglican Church apparently fell far short. "The absurdity of a Bishop having a wife particularly struck them, they scarcely knew whether to be most amused or horrified at such an atrocity."

He visited the Termas [hot springs] de Cauquenes, located in the valley of the Río Cachapoal as it descends through the foothills of the Andes. As in Patagonia, Darwin was much impressed by the "fringes of gravel" or terraces along this and other river valleys draining both the eastern and western slopes of the Andes. These terraces sloped down the river valleys, and also gently toward the center of the valleys. He could not imagine that these terraces had been formed by rivers: "To suppose that as the land now stands, the rivers deposited the shingle along the course of every valley, and all their main branches, appears to me preposterous." On the contrary,

map 5: VALPARAÍSO/ SANTIAGO/MENDOZA

Villavicencio

Jahuel

Aconcagua

Uspallata

Quintero

Cumbre

Mendoza

Valparaíso

Quillota

Puente del Inca

La Campana

ARGENTINA

Casablanca

ANDES MOUNTAINS

Santiago

Pacific
Ocean

Piuquenes

Portillo

Río Maipo

Navidad

CHILE

Rancagua

Termas de Cauquenes

FitzRoy and crew
charted this coast
including the harbors
of Valparaíso and
Quintero, 1834-5.

0 Miles   25   50

© 2017 Jeffrey L. Ward

Yaquil   San Fernando

0 Kilometers   50

Darwin's Routes

1834

1835

Darwin argued, "these same rivers not only are now removing and have removed much of this deposit, but are everywhere tending to cut deep and narrow gorges in the hard underlying rocks." He concluded, "These fringes of gravel . . . I cannot doubt were modeled by the agency of the sea." These valleys again reminded Darwin of Obstruction Sound which cuts through the southern tip of the Andes just north of the Straits of Magellan, and once again he concluded that these terraces had formed as the land rose from the sea.

His plan was to make a loop to the south returning to Valparaíso up the coast, but after another week he became ill, and finally had to call a carriage to take him to Valparaíso where he took to bed for a month.

During Darwin's illness, he missed perhaps the greatest personal drama of the entire expedition. FitzRoy was overworked and exhausted. The stress of eight months of steady surveying and exploring, topped by bringing the *Beagle* through the tumultuous waters off Tierra del Fuego had worn him down. Then he learned that the Admiralty would not reimburse him for the expenses that he had incurred in purchasing, outfitting, and staffing the HMS *Adventure*. He sold the *Adventure* at a loss. All this constituted significant blows both to his personal fortune and to his self-esteem. Perhaps the biggest punch came from Beaufort and the Admiralty appearing to question his judgment. FitzRoy seemed unable to put his treatment by the Admiralty in the context of the political changes brought about by the ascendency of the Whigs and Lord Grey, the passage of the Reform Act of 1832, and the reforms made in the navy during the early 1830s. Or perhaps these changes contributed to his distress. In any case he took it all personally. He went into a state of depression.

Shadows lingering from the suicides of FitzRoy's predecessor, Pringle Stokes, and of his uncle, Lord Castlereagh, fell over him. FitzRoy knew himself well enough that he thought that he was losing his mind. He sought help from Mr. Bynoe, the *Beagle's* doctor. But Bynoe's diagnosis, that he was exhausted and overwrought, did not satisfy him. "I am so surrounded by troubles of every kind that I can only write a short and very stupid letter," he wrote to his sister Fanny in early November. "My brains are more confused than even they used to be in London." Beyond the normally intense pressures of commanding a ship on a mission, the almost daily risk of the lives of nearly one hundred men, FitzRoy had been severely disrespected by his superiors.

He wrote a letter to Beaufort, resigning his command.

The implications for the officers and crew were enormous. First of all, the terms of reference for the expedition stated that, if for any reason FitzRoy was unable to serve as captain, the ship was required to return to England via Cape Horn, turning back from a circumnavigation of the globe. Fortunately his loyal officers, particularly Mr. Wickham—who would have succeeded FitzRoy in command—were able to calm FitzRoy down to the point that he agreed to continue, and to retrieve his letter of resignation before it left Valparaíso. Wickham successfully argued that the *Beagle's* instructions required only surveys of the west coast of South America for which he had time, and that the most useful course was to return via the Pacific, completing the chain of longitudes around the globe. In the end, FitzRoy agreed to this, and to everyone's satisfaction, a reduced plan of surveys was adopted. But as FitzRoy admitted much later, even he was ready to get home: "I confess that my own feelings and health were so much altered in consequence—so deprived of their former elasticity and soundness—that I could myself no longer bear the thoughts of such a prolonged separation from my country, as I had encouraged others to think lightly of." Eventually both Darwin and FitzRoy regained their equilibrium, and Darwin was thrilled to have the date for the completion of the voyage clarified. In November the *Beagle* left Valparaíso, heading south once again to perform surveys of Chilóe, the islands to the south, and the mainland of Chile between Chilóe and Valparaíso.

While FitzRoy's breakdown and near resignation constituted the most significant personal drama of the expedition, the most significant natural event was in the offing. On February 20, 1835, as the *Beagle* was working its way back to the north after its surveys of Chilóe and the Chonos Islands, and while engaged in surveys at Valdivia, the giant earthquake struck. Darwin and his servant, Syms Covington, felt the swaying in the woods outside of town. Within the town itself, Darwin wrote, "Although the houses, from being built of wood, did not fall, yet they were so violently shaken that the boards creaked and rattled. The people rushed out of doors in the greatest alarm." The town survived with only minor damage. But there was damage beyond the physical: "A bad earthquake at once destroys the oldest associations: the world, the very emblem of all that is solid, has moved beneath our feet like a crust over a fluid," he observed. It would be nearly another two weeks before the men aboard the *Beagle* learned the worst.

The enormity of the disaster began to dawn as they tacked into the Bay of Concepción, dropping Darwin off on Quiriquina Island on March 4th. There he met the *mayordomo* of the estate on the island who told him "that not a house in Concepción, or Talcuhuano [*sic*], (the port) was standing; that seventy villages were destroyed; and that a great wave had almost washed away the ruins of Talcuhuano."

Darwin was quickly convinced: "I soon saw abundant proof: the whole coast being strewed over with timber and furniture, as if a thousand great ships had been wrecked." As he explored the beach he found, "Besides chairs, tables, bookshelves, &c., in great numbers, there were several roofs of cottages, which had been drifted in an almost entire state. The storehouses at Talcuhuano had burst open, and great bags of cotton, yerba*, and other valuable merchandise, were scattered about on the shore." Later, when he himself rode into Concepción and Talcahuano, he described the scene:

---

* Yerba is the herb used in making maté, the tea-like drink that is still today an obsession in Argentina and Uruguay.

"Both towns presented the most awful yet interesting spectacle I ever beheld." He went on, "In Concepción, each house, or row of houses, stood by itself, a heap or line of ruins; but in Talcuhuano, owing to the great wave, little more than one layer of bricks, tiles, and timber, with here and there part of a wall left standing, could be distinguished." And he tried to put this earthquake in perspective: "It is generally thought that this has been the worst earthquake ever recorded in Chile; but as the very bad ones occur only after long intervals, this cannot easily be known; nor indeed would a much more severe shock have made any great difference, for the ruin is now complete."

Since its beginnings as a military outpost, founded by Pedro de Valdivia in 1550, the city of Concepción had suffered a rough time. Originally called *Concepción de María Purísima del Nuevo Extremo* (Conception of Mary the Immaculate in the New Land), the invocation of the name of the Holy Mother did little to protect the settlement from the ravages of either the town's unwilling hosts, the Mapuche, or earthquakes and tsunamis. The Mapuche destroyed the town in 1554, again in 1555, carried out a siege in 1564, and attacked the city repeatedly until they were finally more or less subdued by the Republic of Chile in the late 1800s. The settlement was razed by earthquakes and tsunamis in 1570, 1657, 1730, and 1751. The tsunami of 1751 finally did what the Mapuche had previously been unable to do, forcing it to move. The location of the city was transferred away from the coast, to its current location along the Biobío River. The original site of the city assumed a Mapuche name, Penco, and a new port, Talcahuano, grew up on the shore of the bay a few miles to the west.

◆

FitzRoy pieced together a vivid account of what had happened in the port on the day of the earthquake. "At Talcahuano the great earthquake was felt as severely on the 20th February as in the city of Concepción . . . three houses only, upon a rocky foundation, escaped the fate of all those standing upon the loose sandy soil, which lies between the sea-beach and the hills." He went on, "Nearly all the inhabitants escaped uninjured; but they had scarcely recovered from the sensations of the ruinous shocks, when an alarm was given that the sea was retiring! Penco was not forgotten; apprehensive of an overwhelming wave, they hurried to the hills as fast as possible." Remembering the fate of old Concepción, the people of Talcahuano ran for the hills. A very good idea.

"About half an hour after the shock," FitzRoy continued, "when the greater part of the population had reached the heights,—the sea having retired so much, that all the vessels at anchor, even those which had been lying in seven fathoms water [about forty feet], were aground, and every rock and shoal in the bay was visible,—an enormous wave was seen forcing its way through the western passage which separates Quiriquina Island from the mainland." Then the tsunami struck. "This terrific swell passed rapidly along the western side of the Bay of Concepción, sweeping the steep shores of every thing moveable within thirty feet (vertically) from high water-mark."

What must it have been like to be standing on a hill above Talcahuano, witnessing this wave roar down the coast? "It broke over, dashed along, and whirled about the shipping as if they had been light boats," FitzRoy wrote, "overflowed the greater part of the town, and then rushed back with such a torrent that every moveable which the earthquake had not buried under heaps of ruins was carried out to sea."

"In a few minutes," Fitzroy went on, "the vessels were again aground, and a second great wave was seen approaching, with more noise and impetuosity than the first; but though this was more powerful, its effects were not so considerable—simply because there was less to destroy. Again the sea fell, dragging away quantities of woodwork and the lighter materials of houses, and leaving the shipping aground."

But the worst was yet to come. "After some minutes of awful suspense, a third enormous swell was seen between Quiriquina and the mainland, apparently larger than either of the two former. Roaring as it dashed against every obstacle with irresistible force, it rushed—destroying and overwhelming—along the shore." Then the great wave retreated. "Quickly retiring, as if spurned by the foot of the hills, the retreating wave dragged away such quantities of household effects, fences, furniture, and other moveables, that after the tumultuous rush was over, the sea appeared to be covered with wreck. Earth and water trembled: and exhaustion appeared to follow these mighty efforts." The Bay of Concepción, separated from the Pacific Ocean by the Tumbes Peninsula, provides one of the most sheltered ports along the coast of Chile. Regrettably, the very geometry of the bay that provides protection from Pacific storms seems to amplify the waves from tsunamis. This large bay, which opens only to the north and is protected from the Pacific Ocean to the west by the Tumbes Peninsula, is the victim of resonance. The dimensions of the bay correspond with wavelength of the tsunami waves such that the waves reflected back and forth in the bay interfere constructively, building up their strength. The particular profile of water depth on the continental shelf offshore also contributes to the amplifications of the waves, just as the continental shelf off Patagonia amplifies the tides.

◆

After examining the devastation in Concepción and Talcahuano, the *Beagle* interrupted its surveys for a quick sail up the coast to Valparaíso, passing offshore of Lipimávida along the way. All but one of FitzRoy's anchors had been lost or broken. He desperately needed replacements and they were not available in Talcahuano.

On March 17, the *Beagle*, with a fresh supply of anchors—but without Darwin, who had left the ship for a second excursion to the Andes—left Valparaíso and returned south to resume her surveys along the coast of central Chile. A key objective was to survey the waters around Isla Santa María—a convenient and frequent stop for foreign ships desiring a safe harbor to rest and to replenish their supply of fresh water.

FitzRoy had been struck by the oddities he noticed along the shore of Concepción Bay after the earthquake: "beds of dead muscles [*sic*], numerous chitons and limpets, and withered seaweed, still adhering, though lifeless, to the rocks on which they had lived." These were for him, "proofs of the upheaval of the land." But what he saw upon arriving at Santa María astonished him: "It was concluded, from the visible dead shell-fish, water marks, and soundings, and from the verbal testimony of the inhabitants, that the island had been raised [in the earthquake] about eight feet. However, on returning to Concepción, doubts were raised . . ."

The proud and serious FitzRoy was not one to have his word doubted. He set out to convince the doubters: "to settle the matter beyond dispute, one of the owners of the island, Don S. Palma, accompanied us the second time. An intelligent [German] Hanoverian, whose occupation upon this island was sealing, and who had lived for two years there and knew its shores thoroughly, was also passenger in the *Beagle*."

They must have been quite a trio, the intense English naval officer with the aquiline nose, the Chilean landowner, Don Salvador Palma, likely wearing a broad, flat-brimmed Chilean straw hat with a long ribbon flowing from its band, a *chupalla*, and the apparently tall, German adventurer, the three riding their horses to examine the rocky shores of the island. "When we landed . . . Vogelborg showed me a spot from which he used formerly to gather 'choros' [a kind of mussel], by diving at low tide. At dead low tide water, standing upon the bed of 'choros,' and holding his hands up above his head, he could not reach the surface of the water: his height is six feet. On that spot, when I was there, the 'choros' were barely covered at high spring-tide."

FitzRoy with his two companions (and witnesses) rode around the island, making measurements of the elevation of the dead shellfish above the level of the tide, reckoning the amount that the island had been uplifted: "I took many measures in places where no mistakes could be made. On large steep-sided rocks, where vertical

FIGURE 1. Portrait of Charles Darwin in the late 1830s a few years after the return of the *Beagle* to England by George Richmond.

FIGURE 2. Robert Fitzroy (1805–1865). *From Schmidt, Herman John, 1872–1959: Portrait and landscape negatives, Auckland district. Ref: 1/1-001318-G. Alexander Turnbull Library, Wellington, New Zealand.*

BEAGLE LAID ASHORE, RIVER SANTA CRUZ.

FIGURE 3. The HMS *Beagle* laid on the beach at Keel Point, at the mouth of the Río Santa Cruz, Argentine Patagonia, to repair the copper sheeting on her keel, April, 1834. From a drawing by Conrad Martens included in FitzRoy's *Narrative*.

ABOVE: FIGURE 4. Cast of the skeleton of a *Megatherium americanum* from *Extinct Monsters*, 1896.
RIGHT: FIGURE 5. Reconstruction of a *Megatherium*.
BELOW: FIGURE 6. Skeleton of a *Toxodon*.

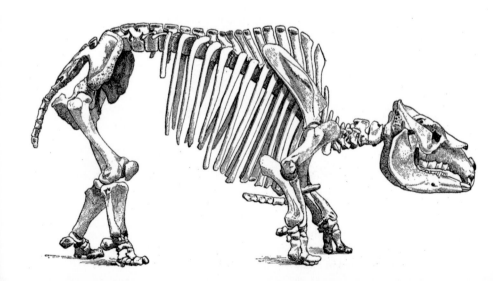

LEFT: FIGURE 7. Daniel Perea identifying fossils found by estancia owner and amateur fossil collector Javier Defferrari near the site where Darwin purchased the skull of a *Toxodon*. BELOW: FIGURE 8. Daniel Perea with the jawbone of a *Toxodon* found by Javier.

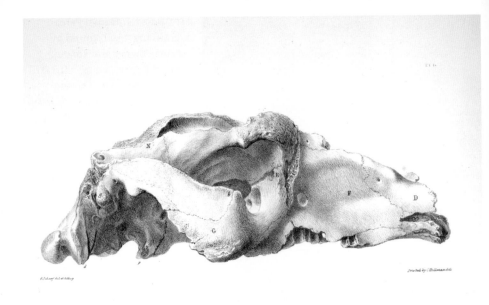

FIGURE 9. Drawing of the skull of a *Toxodon* collected by Darwin and given to Richard Owen, published in *The Zoology of the Voyage of the H.M.S. Beagle.*

LEFT: FIGURE 10. Teresa Manera cleaning a track of a *Megatherium* at Pehuén-Có, Argentina. *Image © Rolex Awards/Marc Latzel.* RIGHT: FIGURE 11. Tracks of *Megatherium* exposed on beach at Pehuén-Có, Argentina. *Image © Rolex Awards/Quentin Deville.*

SANTA CRUZ RIVER, AND DISTANT VIEW OF THE ANDES.

ABOVE: FIGURE 12. Darwin's terraces along Río Santa Cruz with the Andes in background, drawing by Conrad Martens published in FitzRoy's *Narrative*. BELOW: FIGURE 13. Our guide, Milthon, planning our next stop on the Río Santa Cruz.

ABOVE: FIGURE 14. David Catling kayaking down the Río Santa Cruz. BELOW: FIGURE 15. Conrad Martens's drawing of Basalt Glen, the valley of a tributary to the Río Santa Cruz, in 1834 from FitzRoy's *Narrative*.

BASALT GLEN — RIVER SANTA CRUZ.

ABOVE: FIGURE 16. Basalt Glen, the valley of a tributary to the Río Santa Cruz, in 2013. BELOW: FIGURE 17. Rounded erratic boulders among angular basalt boulders above Basalt Glen along the Río Santa Cruz.

ABOVE: FIGURE 18. Kayaking on the Río Santa Cruz with Darwin's terraces in background. BELOW: FIGURE 19. Darwin puzzled over this erratic granite boulder known as the Bell Stone during his boyhood in Shrewsbury. He was later much taken with the idea that this stone could have been transported to Shrewsbury from the north of England by floating on an iceberg during a time when the relative level of the land was lower. The Bell Stone was actually carried by a glacier during the Ice Ages to where it was found.

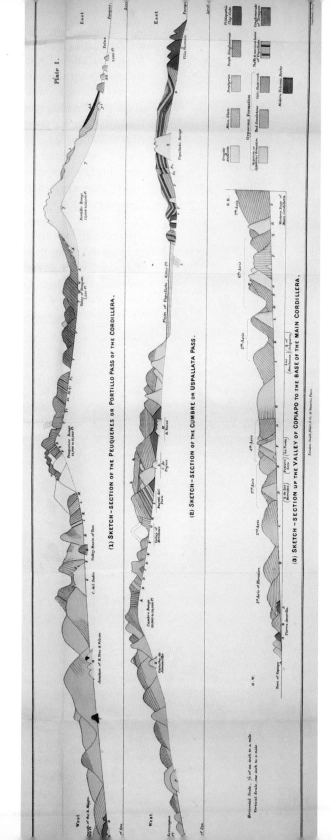

FIGURE 20. Cross sections drawn by Darwin to show the rock strata and their structure that he encountered in the Andes. The top cross section (1) is along the southern half of his loop across the range in 1835, while the middle section (2) is along the northern half of his loop on the return from Mendoza. The bottom section (3) is through the western foothills of the Andes where he visited later in 1835 just before leaving Chile.

LEFT: FIGURE 21. The author indicating his position on Darwin's section. The rocks in the cliff directly behind him are the tan-colored granitic rocks of the Portillo Range (pink in Darwin's section). Over his right shoulder (upper left of the photo) is the contact with the overlying black sedimentary rocks shown as green in Darwin's section.

BELOW: FIGURE 22. Guide Alejandro and the author crossing the Tunuyán River in the central valley of the Andes following the descent from the Portillo Pass and about to begin the climb to Piuquenes Pass.

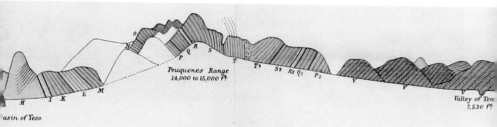

ALL THREE IMAGES: FIGURES 23, 24, 25. Darwin's cross section and panoramas from the crest of Piuquenes Pass. The upper panorama shows the view looking west to north; the lower panorama continues the view north to east. The rocky ridge in the foreground of the lower panorama is the boundary between Chile to the left and Argentina to the right. The dip, or tilting, of the rock strata in this ridge is just as shown in Darwin's section. That these rocks, now at an elevation of more than 13,000 feet, were deposited in the sea is shown by the fossils of ancient oysters (*Gryphaea*) and ammonites.

ABOVE: FIGURE 26. A boulder containing abundant fossil shells of the ancient oysters or *Gryphaea* found descending into Chile from Piuquenes Pass. These and other shells near the pass led Darwin to write "nothing . . . is so unstable as the level of the crust of this earth." BELOW: FIGURE 27. Descending from Piuquenes Pass (looking south) showing the strata dipping very steeply to the east.

ABOVE: FIGURE 28. Looking to the northwest into Chile during the descent from Piuquenes Pass. The reddish rocks to the left of the cliff in the foreground (with strata steeply dipping to the east) are the Red Sandstones that Darwin described as being a part of what he called the "Gypseous Formation." BELOW: FIGURE 29. Remains of the cathedral at Concepción after the earthquake of February 20, 1835, from a drawing by Lieutenant John Wickham of the *Beagle*, published in FitzRoy's *Narrative* of the voyage.

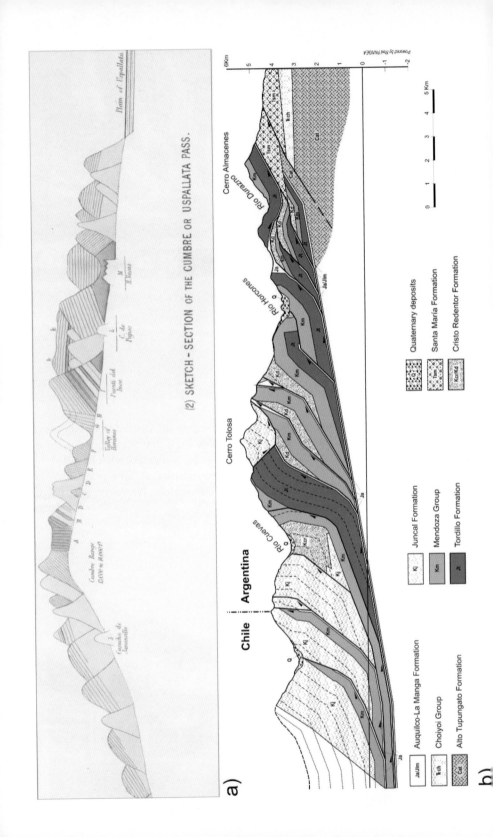

(2) SKETCH - SECTION OF THE CUMBRE OR USPALLATA PASS.

a)

b)

Legend:

- Quaternary deposits — Q
- Santa María Formation — Tsm
- Cristo Redentor Formation — Kcr/Kd
- Juncal Formation — Kj
- Mendoza Group — Km
- Tordillo Formation — Jt
- Auquilco-La Manga Formation — Jau/Jlm
- Choiyoi Group — Tch
- Alto Tupungato Formation — Cat

Chile — Argentina

Cerro Tolosa · Cerro Almacenes

Río Cuevas · Río Horcones · Río Durazno

0 1 2 3 4 5 Km

VIEW OF WHITSUNDAY ISLAND.

PREVIOUS PAGE: FIGURES 30, 31. Comparison of a portion of Darwin's cross section along the northern portion of his loop across the Andes with a modern cross section prepared by the Argentine geologist Victor Ramos and colleagues. The summit of the "Cumbre Range" in Darwin's section corresponds to the border between Chile and Argentina, and the flat lying strata marked "k k" on Darwin's section correspond to the Cerro Almacenes. The most important difference between Darwin's interpretation and the modern version is the recognition of the role of thrusting, faulting, and horizontal shortening in the uplift of the Andes.

ABOVE: FIGURE 32. The coral atoll at Whitsunday Island from Captain Frederick Beechey's *Narrative of a Voyage to the Pacific and Beering's Strait* published in 1831. This drawing, minus the boat carrying the five intrepid sailors, was used by both Charles Lyell and Darwin as an example in their writings about coral islands.

FIGURE 33. View of the South Keeling (now Cocos) Islands from space shows the classic atoll investigated by the crew of the *Beagle*. The light aqua and blue colored water represent shallow water on the margins of the interior lagoon. Soundings by FitzRoy and Darwin showed that the water deepens abruptly outside the atoll. They also found that certain species of coral formed the outermost reefs, receiving the brunt of the force of the waves, while other species enjoyed the quieter waters within the atoll.

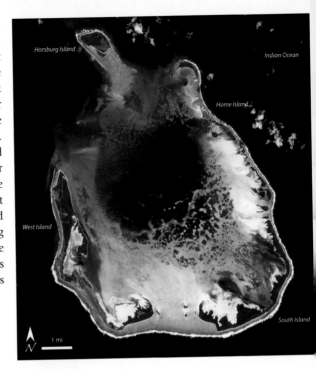

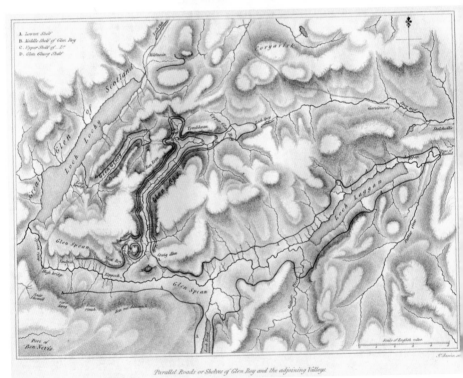

FIGURE 34. Map of the Parallel Roads of Glen Roy that Darwin used to illustrate his pape interpreting the origin of the roads (the red lines) as beaches of the sea during a time whe the level of the land was much lower.

measures could be correctly taken, beds of dead muscles [*sic*] were found ten feet above the recent high-water mark. A few inches only above what was then the spring-tide high water mark, were putrid shell-fish and seaweed, which evidently had not been wetted since the upheaval of the land."

At the northern end of the island FitzRoy described an "extensive rocky flat." Before the earthquake this was covered by water, but after the earthquake uplifted above the tides, exposed and "covered with dead shell-fish, the stench arising from which was abominable." In the years ahead, Darwin would make FitzRoy's observations famous, citing them as evidence for his theories about the geologic uplift of South America.

I wanted to see this evidence for myself.

◆

"Damn!" I cursed under my breath as I ran down the steep path through the blackberry brambles, my backpack slamming me with every step. The boat to Isla Santa María would leave at 11:00 A.M. Afraid to look at my watch, I caught a glimpse of the pier, still distant, through the houses and warehouses at the bottom of the hill. I felt my chance to visit Captain FitzRoy's island slipping away, my chance to visit the island that he and the *Beagle* crew found to have undergone an astounding uplift during the 1835 earthquake. "If only I had spent more time working on Spanish . . ."

The trip had started so well. Before coming to Chile I'd spent time searching through guidebooks, unable to find any clue about how to reach this little island, eighteen miles off the coast, or where I might stay when I arrived. I'd tried to interest Brian Atwater and Marcelo Lagos in joining me for an excursion to Santa María, but they had other things to do. Nevertheless, with Marcelo's help I'd obtained a reservation for the thrice-weekly boat to the island from the small port of Lota.

Saying goodbye to Brian and Marcelo at the bus terminal in Santiago, I'd taken the overnight bus to Concepción, then found the stop for the bus to Lota. Unfortunately the forty-five-minute ride estimated by *Lonely Planet* was apparently for an express bus, not for the local I rode, which stopped every few minutes to pick up or drop off passengers. My stomach began to knot as the time before the boat's departure evaporated. The bus finally left the main road and began to wind up a hill through the outskirts of Lota, when I realized that I was almost the last passenger on the bus. Asking the bus driver in pitiful Spanish where I should get off for the boat to Isla Santa María, I transmogrified the Spanish word *barco* for boat, into something like *barque,* and the driver, befuddled by this half-witted foreigner, reasonably enough directed me to get off at the *parque,* a mistake that I did not discover until I was two blocks down the street toward the park with the pier nowhere in sight. In response to another panicked inquiry, a policeman pointed back the way I had come.

My pace now up to a jog, I approached the edge of a hill and a dead end in the street. Fortunately, a helpful woman was able to understand where I wanted to go, and directed me through her backyard to the path down the hill through the blackberries.

The path dumped me out in a maze of narrow streets. When making the reservation, Marcelo had been told that the ticket office was "near where they eat seafood," but I figured at that point my best bet was just to find the boat. Jogging around a corner, I saw through a gate, the pier. It was covered with people: fishing, pushing carts carrying boxes and crates, sitting on bags and parcels tied with rope. I raced down the pier, looking for the boat to the island. Reaching the end, I saw what must be it . . . anchored twenty yards away from the pier. Passengers, wearing life jackets, jammed onto benches sat under a tarp stretched above the deck behind the cabin. A large wooden rowboat approached the boat, apparently with another load of passengers.

Near panic, I headed back down the pier. An official-looking man wearing a uniform was talking into a handheld radio. I tried to get his attention, but he was clearly busy. Then I saw some steps leading down to the water where the rowboats picked up their loads. A muscular young man with freshly cropped dark hair sat on the top step. I asked if this was where I could get the boat to Isla Santa María. "*Sí*," he replied, and with an urgency that I appreciated, began to hail a rowboat that had just departed from the bottom of the steps. He asked if I had a ticket. No, I replied. His face fell, and he turned away to wave off the rowboat. I asked where I could get a ticket. He pointed to the street leading away from the pier.

I raced off the pier and down the crowded street past stalls with tables heaped with fish and shellfish. Finding the ticket office at last, I asked the woman at the window for a ticket. She replied with a phrase that would become permanently etched in my brain, "*¡Se fué!*," she said, "It's gone!"

Crestfallen, I walked back to the pier.

I found my young friend still at the top of the steps. Is there another boat, I asked. Yes, he said, there will be another boat at twelve o'clock. Do I need a ticket? No, he replied. Maybe there was hope. I relaxed a little and looked around. Finally he waved to me and I scrambled down the stairs to the swaying boat. The young man rowed, an anchor tattooed on the biceps bulging from his T-shirt.

The boat that I scrambled aboard had probably been a fishing boat in a former life, with only a simple deck, no benches, no tarp. My fellow passengers and I arranged ourselves sitting on the deck along the rail, on sacks of flour, on piles of boards. The tourists chatted with excitement as they passed their bags up the ladder to a smaller deck above the boat's cabin. The captain, a short man with a quick smile, joked as he collected a hundred pesos from everyone to pay the young men in the rowboats.

Just then, a large launch, with *Armada de Chile* printed on the cabin, pulled up beside us. Here was the official looking man I had seen earlier. He boarded our boat looking extremely stern. The faces of the passengers dropped. The formerly jovial captain now looked like a chastened puppy. We were busted . . . for something. The tourists on the upper deck began passing their bags back down the ladder, their faces no longer sparkling. I asked the woman next to me what was going on. It seemed that the boat was approved to carry only twelve passengers, not the twenty or so that jammed the decks. Some of us were not going to make the trip.

The captain circulated among the passengers, shrugging his shoulders helplessly under the watchful eye of the official, searching for volunteers to disembark. It was a game of chicken. The tourists left, and a few others, but there were still too many passengers onboard. The captain negotiated intently with the husband of the woman sitting next to me. Finally I felt as if my number was up. Feeling completely stupid that I couldn't understand was going on, I stood up, heading for a rowboat at the rail. I just wasn't going to get to Santa María.

As I began to step over the rail, the woman shouted *"¡No, no!"* and motioned for me to stay onboard. Apparently one more person wasn't going to be enough to get down to the limit, and so she and her party would have to get off, leaving room for me to stay on.

The rowboats and the boarding party left. The engine shuddered to life. I sat on the deck, propped up against a pile of building supplies, the warm summer sun shining brightly, the waves sparkling, feeling a stinging dose of survivor's guilt.

◆

Arriving at Isla Santa María, two rowboats took turns ferrying the passengers and cargo to a concrete ramp at the village of Puerto Sur. Brightly painted rowboats were pulled up onto the sandy beach and floating at anchor. A farm tractor towing a trailer was being loaded at the end of the ramp.

My excitement at finally being on FitzRoy's island was tempered by the need to find a place to sleep. I set out through the dusty, unpaved streets looking for a sign indicating a willingness to accept lodgers. Eventually I encountered a large man and two young girls emerging from a shop carrying bags of fresh bread and supplies. I asked the man if he knew of a place I could stay. To my shock and glee, he responded in serviceable English, saying that I should stay with him. I followed as the two little girls chattered. We walked through a sagging gate, and past the rusting hulk of a tractor. I later discovered that it was made in Minsk, U.S.S.R., a relic from the time of Salvador Allende some years thirty-five years before. There, behind an overgrown garden, with a cracked glass window, was Paul's house.

Paul, it turned out, was a relatively recent arrival to the island, a refugee from an aluminum factory near Santiago. His passion was fishing. His dream, which he was acting out, was to start a *hostería* on the island focused on fishing, skin diving, and ecotourism. The budget available for this enterprise appeared modest. He showed me to a small, empty room with two homemade bunk beds and some other mattresses leaning against the wall. Housekeeping did not appear to be Paul's strong suit, judging by the rumpled blankets on one set of the bunk beds and the dust balls on the floor. Dinner, he explained, would be fresh fish from the island, served at the long, homemade table in the main room, where I would eat with his other guests. If the cuisine was any indicator of Paul's new endeavor, his *hostería* would do just fine.

In answer to Paul's questions of what I wanted to see on the island, I explained that I was a geologist and I wanted to see Fitz-Roy's evidence of uplift during the 1835 earthquake. No problem, he said.

The next morning Paul explained that he had arranged for me to get a ride to the north end of the island with a patrol of the *Carabineros*, Chile's national police force. Arriving at the police station, a neat building with a tall radio antenna sprouting behind it on a hill about a hundred yards from Paul's house, a policeman behind the desk offered me a seat while two young officers completed their preparations for their patrol. One unlocked a closet and took out two pistols, fitting one in his holster and handing the other to his colleague. They led me outside to a bright new crew-cab pickup, painted in the green and white colors of the Carabineros. It was one of less than a handful of vehicles on the island, and if not the newest, cleanest, sparkliest, I did not see its rival. The two young policemen, in their freshly laundered uniforms and wraparound sunglasses hopped in the truck and we bounced off to the north along the island's main road.

As part of their survey the *Beagle* crew had determined the latitude and longitude at two points on the island, one near the landing area of at Puerto Sur, and the other on a prominent headland nearly midway up the west coast. This headland was our first objective.

A mile or so up the road we turned off onto an unmarked track. As the truck lurched and pitched along the sandy ruts, I clung to the handle above the back door, struggling to keep my balance as the two policemen laughed and chatted. Then, emerging from the pine forest, the Pacific Ocean spread below us, blue-green waves crashing onto the rocks at the base of the cliff with eruptions of white spray and foam.

"*¡Llegamos!*" [We've arrived!], announced Carlos, the policeman riding shotgun, as the truck jerked to a stop. Pushing the door open, I could barely contain my excitement. The fresh wind blowing off the Pacific and the gulls squawking, I scrambled the last few steps to the top of the headland. I stood at the top of a cliff about a hundred feet high, looking out over the Pacific. This was a point that FitzRoy and his men surveyed, reporting its latitude and longitude, "Santa María Island-summit of west head." Perhaps FitzRoy had been to this exact spot. To the west, beneath me, the cliff dropped steeply to the waves crashing at its base. As the swell surged toward the cliff, patches of white spray revealed the presence of rocks, which sometimes poked above the surface. To the north, I saw a narrow beach. I imagined FitzRoy himself at this very point looking down at the newly exposed rocky flats below. Was this what he saw?

We then headed toward the northern end of the island. Once there, I thanked the Carabineros for their help and set off alone. From the top of the sea cliff I could look to the north at the prominent rock, Isla Farallón, a mile or so offshore. At the bottom of the sea cliff was a narrow beach with a rocky shelf extending outward, lying just below the surface of the water. Here and there, waves were breaking over rocks projecting upward from the shelf.

I followed a sandy track down to a wider stretch of beach, then clambered around the sea cliffs to a narrow stretch of beach wedged between the sea cliffs and the waves breaking on the rocky shelf. I was looking out at the "extensive rocky flat" FitzRoy described. But it didn't look quite as he described it. The rocky platform wasn't high and dry, it was largely below the waves. Puzzled, I sat on a rock and pulled from my backpack a copy of a recent scientific paper describing the geology of the island.

The paper, by a young Chilean geologist, Daniel Melnick, identified the fringe of current rocky platforms around the island as only the remnants of FitzRoy's platform, arguing that the level of FitzRoy's platform had been lowered by erosion. I found some flat rocks that were four or five feet above the modern beach. They looked just like a photograph in Melnick's paper, showing what he interpreted as the erosional remnant of FitzRoy's rocky flat. I pulled out my tape measure to check the elevation of the rocks. The bedrock here was soft sandstone, but was it so soft that several feet of rock over an area of several acres have been eroded away in less than two centuries? That seemed like an awful lot of erosion for such a short period.

I spent a couple of hours roaming the rocks, beaches, and sea cliffs, trying to imagine how this all looked to FitzRoy, and enjoying the wildness of the place. Concepción and Talcahuano are now urban areas, so modified by development and the works of man that one cannot really see the same beaches and landscapes that FitzRoy and Darwin saw along the coast. But here, the setting was little changed. With only a small village on the top of the hill, and a few roads and trails down to the ocean, it was easy to imagine the scene just as FitzRoy, Don Salvador, and Vogelborg—the "intelligent Hanoverian"—saw it in 1835.

Here, what FitzRoy saw and measured gave credence to Maria Graham's reports. It supplied Darwin with a concrete example of how the Andes, and indeed the whole continent of South America, rose from sea. Not only in the distant past, but within the brief span of his own experience. A *vera causa*, indeed. Isla Santa María joined the elite company of the Temple of Serapis as critical facts to be accommodated in the debate on the causes of elevation.

Nonetheless, as I gazed out at what had been FitzRoy's "extensive rocky flat covered with dead shell-fish, the stench arising from which was abominable," I began to wonder if it had really eroded to this new level beneath the waves. I wasn't satisfied.

Things didn't seem quite right.

CHAPTER EIGHT

The Andes Arising

It is an old story, but not the less wonderful, to hear
of shells, which formerly were crawling about at
the bottom of the sea, being now elevated nearly
fourteen thousand feet above its level.

—Charles Darwin

In March, 1835, as FitzRoy steered the *Beagle* south from Valparaíso to Isla Santa María in the wake of the earthquake near Concepción, Darwin instead turned to the east, to the Andes. He had long hoped for a chance to visit this great range. The end of summer was approaching, and the coming of fall meant that the high mountains would soon be covered in snow. This was his chance. He seized it.

Among the British expatriates that Darwin had met in Chile the previous autumn was Alexander Caldcleugh. Caldcleugh had traveled in South America from 1819 to 1821, first as private secretary to the British minister to the Portuguese court in Rio de Janeiro, then onward on his own through Argentina, Chile, and Peru. He was much interested in the natural sciences, and upon his return to England in 1822 he was elected a fellow of the Geological Society. He published a widely read book about his travels, a copy of which

Darwin consulted on the *Beagle*. Caldcleugh had crossed the Andes twice, by two different routes, and Darwin sought Caldcleugh's advice on how to arrange his expedition. With Caldcleugh's help he planned a large loop, first crossing "the Cordillera," as he frequently referred to the Andes, eastward by a route just to the south of Santiago, then north along the eastern flank of the range to the city of Mendoza, and finally returning by a more northerly route through Uspallata. At this time both the cities of Mendoza and Uspallata were located in the semiautonomous province of Mendoza, which earlier had been a part of Chile, but were now allied with the other newly independent territories east of the Andes that over the coming decades would become Argentina.

On Caldcleugh's transit of the southern route in the middle of March, his party encountered serious snowstorms: "The snow was so deep that it reached to my knees as I sat on the mule." Already the middle of March, autumn in the southern hemisphere was in the air. Snow in the mountains was more likely with every passing day. Darwin had no time to waste if he wished to avoid a similar fate.

With Caldcleugh's help, Darwin arranged for a guide and a muleteer with ten mules and a *madrina*. "The madrina," Darwin wrote, "is a mare with a little bell round her neck; she is a sort of step-mother to the whole troop,—It is quite curious to see how steadily the mules follow the sound of the Bell."

The expedition would prove to be rather expensive, but Darwin employed the classic ex post facto strategy of begging forgiveness upon completion, as opposed to asking permission in advance. On his return to Chile, he would write his sister, Susan, as the recipient of news for his family: "Since leaving England I have never made so successful a journey: it has however been very expensive: I am sure my Father would not regret it, if he could know how deeply I have enjoyed it."

On March 18, Darwin's little expedition set out from Santiago, heading south, then east up the valley of the Maypo (now Maipo) River.

◆

In pursuing his interest in the Andes, Darwin was following a distinguished line. The first scientists to visit the Andes, albeit more than two thousand miles north of Darwin's traverse, were not there to investigate the geology at all, but to settle an argument about the shape of the earth. A French expedition was launched to the equatorial region of the Andes in 1735 to answer the question of whether the shape of the earth was that of a sphere squashed at the top and bottom, bulging at the equator, a shape termed oblate, as theorized by Isaac Newton, or more like that of a football or watermelon, stretched toward the poles, prolate, as argued by the French astronomer, Jacques Cassini. In the end Newton, by then laid to rest in Westminster Abbey two decades before, won the argument. In any case, all the attention drawn to the debate, and the exotic accounts of the expedition to South America, opened the eyes of the emerging scientific world to the wonders of this continent and the Andes. And the future country of Ecuador got its name.

The next scientific visitor to explore this new continent was Alexander von Humboldt, Darwin's muse on the early days of the *Beagle* expedition as he lay in his hammock, wracked with seasickness. Humboldt set a very high bar. Not only did he reach the highest altitude then ever attained by a European (on an unsuccessful attempt to climb Chimborazo in 1802, then believed to be the highest mountain in the world), but in the course of his travels he measured the barometric pressure, temperature, latitude, and longitude, and he recorded all manner of detail about the flora, fauna, geology, and civilization that he met along the way. Humboldt had studied under the German geologist Abraham Gottlob Werner, propagator of the "Wernerian nonsense" that Sedgwick warned Darwin to eschew. But upon his arrival in the Andes, the ubiquitous deposits of volcanic rocks and ash, not to mention his visits to active volcanoes, quickly dissuaded Humboldt. Werner's notion that virtually all rocks were created as sediments in a primordial sea simply did the not fit the facts. Instead, Humboldt became a supporter of the idea that mountains were created by the upwelling of subterranean gases and lava. Where the lava broke through to the surface, a volcano formed. But where the lava couldn't force its way through the crust, the whole region would swell like a giant blister. This was the state of play three decades later as Darwin climbed aboard his mule, setting out to cross the central Andes from Santiago.

◆

"What are the chances the money will actually get there?" I asked.

"Not too bad," replied the tall, serious young woman with a long blonde ponytail. "Argentina is not too corrupt. We won't even accept wire transfers to some countries."

Four thousand miles from the Andes at the branch of US Bank in Evergreen, Colorado, I sat across a desk—its top swept clean except for a telephone—from Joanna, the branch manager, a keyboard and computer monitor ready at her side. A glass partition separated us from the lobby where other customers busily deposited and cashed checks.

Today, while the northern leg of Darwin's loop across the Andes largely follows a busy international highway, the southern leg remains wilderness. To follow this leg of Darwin's path, I needed to get some money to a small town in Argentina. Over the Internet I'd found an outfitter, Solo Montañas, that seemed to offer just the ticket. Their trek, Cruce de los Andes, almost exactly followed Darwin's path. My emails, painfully constructed in Spanish supplemented by the magic of Google Translate, were promptly answered by Alejandro, the firm's owner. Then a serious problem arose. How to send them the money for a deposit? This tiny enterprise, regrettably, lay beyond the global reach of Visa and MasterCard.

After a flurry of emails with Alejandro in Tunuyán, Argentina, and the representative of the Banco de la Nación Argentina in New York, I had assembled the instructions for transferring the money for a deposit. Now I had to sign the paper.

"Is there any kind of guarantee?" I asked, unsettled by her answer to my question about the odds for success.

"No," she replied, "Once our money—that is, your money—is gone, it's gone. But usually if something goes wrong, and they can't understand the instructions, it will come back." Great, I thought to myself, as I signed the paper.

Dr. Darwin's method of writing his banker in London, directing him to arrange for a letter of credit to appear in young Charles's next port of call, seemed pretty simple by comparison.

◆

As Darwin and his mules climbed up the valley of the Maypo (now Maipo) River, they left the fertile fields and orchards of Chile behind. Then higher, as they passed the last settlements, the steep sides of the valley closed in. This valley, like the others Darwin encountered in the Cordillera, was filled with an "irregularly-stratified mass of well-rounded shingle . . . to a depth of some hundred feet." "The rivers," Darwin wrote, "have removed a large part of the center [of the fill in the valley]; thus leaving a terrace of equal height, but varying width on each side." The road and any cultivation were nestled on these narrow strips of relatively level ground.

But it was the river itself, and its power, that attracted Darwin's attention, writing that the Maypo and similar rivers "should rather be called mountain torrents. Their inclination is very great, and their water the color of mud." And it wasn't just the sight, but also the sound, that elevated these rushing bodies of water beyond mere river status in Darwin's mind. "Amidst the din of rushing waters, the noise from the stones, as they rattled one over another, was most distinctly audible even at a distance. This rattling noise, night and day, may be heard along the whole course of the torrent." Then he waxed poetic: "The sound spoke eloquently to the geologist: the thousands and thousands of stones, which, striking against each other, make the one dull uniform sound, are all hurrying in one direction. . . . Each note of that wild music tells of one other step toward their destiny," the ocean.

Higher up, Darwin and his party left the main valley of the Maypo to follow the Valle del Yeso, one of the Maypo's principal tributaries. *Yeso* is the Spanish word for plaster, and also the word for gypsum, the principal ingredient of plaster of paris. "The valley takes its name of Yeso from a great bed, I should think at least two thousand feet thick, of white, and in some parts quite pure gypsum," Darwin wrote. "We slept with a party of men, who were employed in loading mules with this substance, which is used in the manufacture of wine," especially Sherry and similar wines. Decades later this thick layer of soft, easily deformable, gypsum would be used to help explain some of the spectacular geologic structures that Darwin was about to see.

Above the Valle de Yeso and an area of hot springs now called Termas del Plomo, the road steepened abruptly into a "zigzag track." The party climbed to the summit of the first of the two high passes that they must surmount to cross the Andes. "The cordillera in this part consists of two principal ranges: the passes across which attain respectively an elevation of 13,210 and 14,365 feet. The first great line . . . is called Peuquenes [now Piuquenes]. It divides the waters, and therefore likewise the republics of Chile and Mendoza [now part of Argentina]. To the eastward . . . [is] the second range (called Portillo*) overlooking the Pampas." Darwin would have been delighted to know that the name Peuquenes (or Piuquenes) is derived from the Mapundungun name for the Andean goose that inhabits this high region. Sadly, it does not seem that he encountered one.

During the vertiginous ascent to the pass across the Peuquenes, Darwin for the first time experienced difficulty in breathing. He marveled that the chilenos regarded this condition, which they referred to as puna, as some kind of disease. He laughed that "The

---

* This Portillo, now Portillo Argentino, is not to be confused with the well-known Chilean ski resort of the same name, which is now located along the northern route through Uspallata that Darwin used in returning across the Andes two weeks later.

inhabitants all recommend onions for the *puna*." But for his part, "Upon finding fossil shells on the highest ridge, I entirely forgot the puna in my delight." He had no need for onions.

Darwin found the view from the crest of the Peuquenes spectacular, and it gave an exciting picture of the geology. During his ascent he distinguished two main groups of rocks. At the bottom was a group that he called "porphyries" or "porphyritic conglomerates," rocks of largely volcanic and likely submarine origin. The thickness of this group he estimated at more than a mile. Above these he saw what he called the "Gypseous Formation," which included the beds of white gypsum, "which alternates, passes into, and is replaced by, red sandstone, conglomerates, and black calcareous clay-slate," but demurring, "I hardly dare venture to guess the thickness."

Most spectacularly, "These great piles of strata have been penetrated, upheaved, and overturned, in the most extraordinary manner, by masses of injected rock, equaling mountains in size. On the bare sides of the hills, complicated dikes, and wedges of variously-coloured porphyries and other stones, are seen traversing the strata in every possible form and direction; proving also by their intersections, successive periods of violence." Again he indulged his poetic side: "Neither plant nor bird, excepting a few condors wheeling around the higher pinnacles, distracted the attention from the inanimate mass. I felt glad I was alone: it was like watching a thunderstorm, or hearing a chorus of the Messiah in full orchestra."

And at the summit of the pass Darwin found fossils, "Even at the very crest of the Peuquenes, at the height of 13,210 feet, and above it, the black clay-slate contained numerous marine remains, among which a gryphæa is the most abundant . . . and an ammonite."

◆

The fossil that he found most abundant on this ridge was a kind of extinct oyster called *Gryphaea*, or "devils toenails" in England where it is common, as it is also in Kansas. Curiously, these fossils would become "undoubtedly the most famous putative example of

documented evolution at the species level . . . of invertebrate paleontology," according to the American paleontologist Stephen Jay
Gould, who would use his own studies of the evolution of *Gryphaea*
to propose a variation of Darwin's ideas about evolution by means
of natural selection. These oysters lived on the sea floor in shallow
water, likely in beds not unlike modern oysters. Although one rarely
encounters the words *Kansas* and *ocean* in the same sentence, during
the same geologic period, the Cretaceous, when Darwin's Andean
Gryphaea were living in a warm, shallow sea, North America was
split down the middle from northern Canada to the Gulf of Mexico
by what is now called the Western Interior Seaway. Many of the
same creatures that Darwin found, including both *Gryphaea* and
ammonites, thrived in similar conditions in the waters covering the
plains states of the U.S. The creek and riverbanks of western Kansas
thus provided a trove of fossils to explorers and paleontologists in
the late 1800s. Like trilobites, ammonites are iconic fossils, extinct
relatives of squid and cuttlefish whose spiraling curlicues resemble
a chambered nautilus.

◆

Darwin's party made camp on the slopes descending on the eastern
side of the Peuquenes under clear skies, but Darwin awoke around
midnight with alarm to a sky suddenly obscured by cloud. Fearing
that a storm was imminent, he roused the mule driver. The *arriero*,
or muleteer, assured him that "without thunder and lightning there
was no risk of a heavy snowstorm." But Darwin knew that there was
not time to waste. His party had already had "to hurry our steps more
than was convenient for geology." Now their situation was serious:
"The peril is imminent, and the difficulty of subsequent escape great,
to any one overtaken by bad weather between the two Cordillera. . . .
Mr. Caldcleugh, who crossed on this same day of the month [also
March 21], was detained . . . for some time by a heavy fall of snow."

The next day the party hurriedly descended and crossed the
bottom of the intermediate valley of the Tunuyán River, then began

the ascent of the Portillo Pass, even higher than the Peuquenes. In contrast to the Peuquenes, the core of the Portillo Range was a "coarsely-crystallized red granite." Between the two ranges Darwin saw sandstone, intruded by "immense granitic dikes" that he could trace back to see emerging from the "central mass" of red granite. The sandstone was covered by other sediments, and then by a coarse conglomerate that completely surprised him.

While some rocks are reluctant to tell their stories, a conglomerate blares its story to the world. These rocks are a cemented aggregate of chunks of the other rocks from which they are derived, their ancestry clearly on display. Although the beds of the conglomerate now dipped down away from the Portillo Range, toward the Peuquenes Range, he was astonished to find within the conglomerate "perfectly rounded [cobbles] of the black calcareous clay-slate with organic remains,—the same rock which I had just crossed in situ on the Peuquenes." But if the origin of these cobbles was in the Peuquenes Range, how could they have come to rest in beds dipping down toward the range? Rocks don't roll uphill.

Darwin's startling conclusion was that the Peuquenes Range existed prior to the formation of the Portillo Range, and that "immense quantities of shingle were accumulated on its [eastern] flank." Then as these sediments were intruded with granitic dikes from below, they were lifted up, and tilted back toward their source, "thus making the offspring [the Portillo Range] at first appear older than its parent." This insight also explained another peculiar fact. The Tunuyán River, which drained the valley between the two ranges, exited the valley through a gap in the higher range. Again, another twist. How could a river end "uphill?" This, too, Darwin realized, could also be easily explained if the existence of the river preceded the uplift of the Portillo Range. Except for this explanation, Darwin wrote, "the circumstance that rivers flowing from a less elevated chain, should penetrate one far more lofty, appears to me quite inexplicable."

After the long steep climb up to the pass across the Portillo, Darwin found himself in a cloud of freezing fog. "This was very unfortunate, as it continued the whole day and intercepted our view." The party bivouacked at the first sign of vegetation on their descent. The sky cleared and the next day they continued down as he surveyed the pink granite of the Portillo Range, a mica-slate that it appeared to have intruded, and some late arriving volcanic rocks that postdated both.

The plants and animals he found on the eastern side of the Andes were "absolutely the same, or closely allied with those of Patagonia," yet markedly different from those he found in the similar climate and conditions in Chile. "This fact," he wrote, "is in perfect accordance with the geological history of the Andes; for these mountains have existed as a great barrier, since a period so remote that whole races of animals must subsequently have perished from the face of the earth. . . . We ought not to expect any closer similarity between the organic beings on opposite sides of the Andes, than on shores separated by a broad strait of the sea."

In summing up the geology of his first traverse, he wrote, "What a history of changes of level, and of wear and tear, all since the age of the later secondary formations of Europe, does the structure of this one great mountain-chain reveal!"

◆

The next stage of my trek—where my wife, Gayle, had agreed to join me—would exactly follow the southern leg of Darwin's route. Except for one annoying detail. It was backward. Unfettered by niggling agricultural regulations, Darwin and his mules traveled to and fro over the international border from Chile to Argentina with impunity. Not so today. The border between Argentina and Chile runs along the crest of the western range, topped by Piuquenes Pass, and now creates an obstacle to the free passage of horses and mules that wasn't present in Darwin's day. While we walked, the animals would carry our equipment and supplies across the greater part of the trek, lying in Argentina, then we would backpack our gear down from the border at Paso de Piuquenes to a van waiting at the Chilean trailhead below.

Arriving at Mendoza in the heat of a summer evening, Gayle and I met Alejandro, a short, but powerfully built man, not old, but balding, dressed in a grey T-shirt, bright blue knee-length shorts, and flip-flops. To my discomfort, I discovered that without the aid of Google Translate our conversation would consist of long silences, punctuated by a few words and a smile.

Tucking Gayle's pack under his arm, he strode out of the bus station. The drive to Tunuyán in his aging car required less than an hour. He turned off into a leafy neighborhood of simple, single-story homes, and pulled to a stop in front of a large garage. Inside men were washing cars. A sign above read "Solo Montañas, Lavadero y Excursiones [Car Wash and Excursions]." Apparently the excursions didn't pay all the bills. Alejandro's two little girls in swimsuits, fresh from playing in the spray in the heat of the late afternoon, ran out and pecked our cheeks in polite greeting. As Alejandro traded his shorts and flip-flops for a pair of baggy, red-and-blue Gore-Tex mountain pants and hiking boots, Alejandro's partner, Horacio—a tall, dark young man, in shorts and a sleeveless muscle shirt—loaded our packs into a red and white Land Rover.

Before long we were jammed into the back seat of the Rover, bouncing west from Tunuyán into the setting sun, as the dark, jagged profile of the Andes rose before us. At first the paved road climbed gently, then turned to gravel, entered a canyon, and began to ascend more steeply. We arrived at the customs and immigration checkpoint after dark. The soldiers eating dinner insisted that we were too late. We couldn't be processed until the next day. After listening to half an hour of argument, I gave up, and returned to the Rover to begin unloading the packs. I had just begun when Horacio came out saying, *"Puede ser* [It could be]." I returned to the tiny office where one soldier was typing on a manual typewriter, another was drawing lines in a notebook, making a form to record our vital statistics and the serial numbers of our cameras and of my GPS.

Relieved to be finished with the bureaucracy, we piled back into the Land Rover and resumed switchbacking up the dark canyon. Spooked hares ran, zigzagging across the road in the beams of the headlights, bringing cheerful cries of *"¡La cena!* [Dinner!]" We arrived near midnight at Los Arenales, where Darwin camped to explore the geology on this side of the Portillo. Now we really were on Darwin's route.

We awoke to Alejandro's cry of *"¡Robert, arriba!"* and began to separate the things we would take in a day pack from those that would carried by the horses and mules in the corral behind our tent. After breakfast we all piled back into the Land Rover and headed up the road, as it clung to the precipitous mountainside, switchbacking upward. The bright yellowish light of the morning sun reflected off the pinkish gray granite peaks above us. When Darwin crossed the pass above us in a snowstorm, the weather finally cleared, and he was much taken by the peaks and cliffs of this beautiful reddish-pink granite. Looking back down the valley toward the flat, dusty plains of the Pampas, I could also see the cliffs of the dark mica-schist, a metamorphic rock, that Darwin concluded was much older than, and had been intruded by, the pink granite. "The colours of the red granite and the black mica-slate are so distinct, that with a bright light these rocks could be readily distinguished even from the Pampas, at a level at least 9,000 feet below."

Darwin would be horrified if he went shopping for a new stone kitchen countertop today. They are almost universally called granite, but almost none of them are. Most stone countertops are actually metamorphic rocks (rocks that have been recrystallized by temperature and pressure, but not actually melted) like Darwin's mica-slate. This inaccuracy would surely have driven him nuts.

As the Rover switchbacked up the steep track, we left all but the tiniest, hardiest traces of vegetation behind. Finally, surrounded by nothing but rock and a few snowfields that had lingered into summer, the Rover's path was blocked by a gully. The elevation was above twelve thousand feet. We piled out, continuing on foot, the notch of the Portillo Argentino visible high on the ridge above us.

After a few hours climb, we reached the notch, gasping for breath. At about 14,350 feet, Portillo Argentino is higher than all but five of the 14,000-foot peaks of Colorado. A chilling wind blew through the gap. I quickly pulled on a shell top. Despite the wind I retrieved a piece of paper from my backpack. It was one of Darwin's colored sketches of a geologic section through the Andes. A geologic section is an idealized drawing summarizing the rock formations and their structure and orientation, as if one were looking at a piece of a strange, deformed layer cake from the side, edge on. In Darwin's sketch it was as if a giant vertical cut had been made across the Andes from west to east, and I was looking at the cut from the south. On the sketch I found my position atop granite at the summit of the Portillo Range. I traced my route on the sketch, through the gap before me as the trail descended steeply into a valley surrounded by a sea of mountains. Paso de Piuquenes lay distant, hidden, on the far side of the valley before us. Darwin's sketches would be my guide to his understanding of the geology of the Andes.

We descended the steep trail as it serpentined through loose scree, passing a scalloped snowfield, glistened with little towers called *penitentes*, which for Darwin "caused some difficulty on account of the cargo mules." In fact, "on one of these columns of ice a frozen horse was exposed, sticking as on a pedestal, but with his hind legs straight up in the air." Not twenty minutes further down the trail, we passed the tan and white carcass of a modern day victim, surrounded by a scattering of bleached bones. For the beasts of burden, the rigors of Portillo Argentino remain much the same today as in 1835.

As we descended into the valley of the Tunuyán, I could see Darwin's young conglomerates, like cemented gravel, containing cobbles of the rocks from both the Piuquenes and Portillo Ranges. I was reminded of a similar conglomerate that I had met in the Rocky Mountains as a young student at geology summer field camp in Montana. The professor described it as being "deposited during the very crisis of orogeny," orogeny being geologist-speak for the process of mountain building. So too, the conglomerates of the Tunuyán valley.

After spending a day exploring the valley, we began the long climb to the Paso de Piuquenes. We stopped to camp at about eleven thousand feet, not far from where Darwin had camped descending from the pass. Here his guides blamed and cursed the pot for failing to cook the potatoes. Alejandro knew better, but we had the same problem. At this altitude, where the boiling point of water is only about 191° Fahrenheit—instead of the 212° at sea level—the lentils he stewed for dinner took absolutely forever. At last we devoured them greedily, doused in hot sauce, but still a bit crunchy.

I had struggled to level a space for our tent. But after turning in, as I lay wheezing and coughing—a cold that I had been fighting now in full bloom—I realized that the platform I had leveled wasn't long enough to stretch out. I couldn't sleep. Tossing and turning in the cramped tent, I finally woke Gayle to trade places. At last, with the foot of my sleeping bag sticking out the door, I was able to get to sleep.

The bright Andean sun greeted me as I awoke. I was pleased to have survived the night. Climbing the slopes toward Piuquenes, my excitement at seeing the spectacular geologic structure surrounding us overcame the discomfort of my cold. At first a few small green plants with yellow flowers softened the sterile landscape, but soon they gave way to a moonscape of barren, rocky slopes. Stopping to pant and catch my breath, I tried to pick out the faults that separated the rock layers jutting at odd angles. The morning sun glistened off the glaciers to the south. I was envious that Darwin had a mule.

My eyes searched the rocks at my feet for traces of Darwin's fossils, those "shells, which formerly were crawling about at the bottom of the sea." The fossil I most hoped to encounter was an ammonite. A companion's find, a rough, worn, and weathered chunk of the outermost swirl on an ammonite heightened my focus.

Eventually we crested a jutting, shaley, limestone ridge marking the pass and the boundary between Argentina and Chile. The bed of limestone sloped steeply but smoothly back into Argentina, but showed its broken, cliffy face to Chile. In the cliffs of the peaks to the north and west, I could see the strata that Darwin described and showed in his sketch—red, cream, gray, black—broken and meeting at discordant angles. But even as we shared a bottle of champagne with our Argentine companions to celebrate crossing this pass, my attention was almost totally absorbed with the hunt for a nice ammonite.

The footing was treacherous on the descent down the steep, loose, shaley, slope. Gayle found it difficult and trying on her knees. Nonetheless, imagining that Horacio's company would provide her comfort, I raced ahead, searching for fossils, oblivious to her concern. Her patience with my fossil hunting wore through quickly. Gayle maintains a list of the "life-threatening" experiences that I have dragged her through, and at that moment the descent of Paso de Piuquenes soared—at least temporarily—to the top of the list. Her fury boiled over when she found me at a small creek in a gully, searching where Horacio had told me I would find fossils. Excited at finding a fossilized, shelly mass of ancient oysters, but shamed by my indifference to her need, I took my place behind her as we continued down.

Ten minutes later, Gayle kicked a small slab of rock the size of a large chocolate chip cookie, spitting out, "Is this what you're looking for?" Sure enough, it was a beautiful, if not exactly museum-quality, ammonite, its spiral swirl, punctuated by little radial ribs. Exactly what I'd been seeking.

This little fossil, probably of the genus *Virgatosphinctes*, swam as a living creature about 150 million years ago in a shallow ocean, was now located about two and a half miles, vertically, above sea level, carried by the uplift of the Andes. The vertical travels of these fossils were, for Darwin, clear evidence that "nothing . . . is so unstable as the level of the crust of this earth."

◆

After Darwin crossed the Andes on the southern leg of his loop, he turned north, traveling a further two days along the eastern front of the cordillera to the oasis of Mendoza. There, his party bought watermelons, "most deliciously cool and well-flavored," and rested for a day. Despite the richness of their agriculture, Darwin was not impressed with the industry of the people, writing, "the happy doom of the Mendozinos is to eat, sleep, and be idle."

The party set off to the northwest, traversing a sterile plain, then finally began their climb up the Uspallata range. Just as on his southern crossing of the Andes, the mountains here are arrayed in two parallel ranges, although here the eastern range, the Uspallata, is lower and the intermediate valley wider. The Uspallata range, Darwin wrote, "consists of various kinds of submarine lava, alternating with volcanic sandstones and other remarkable sedimentary deposits." He expected that he might find some petrified wood, and in this he "was gratified in a very extraordinary manner." Indeed, on a hillside at an elevation of about seven thousand feet, he found a complete petrified forest.

Describing it in a letter to his sister, Susan, he wrote "I found a clump of petrified trees, standing upright, with the layers of fine Sandstone deposited round them, bearing the impression of their bark. These trees are covered by other Sandstones & streams of Lava to the thickness of several thousand feet." And here, too, he found an argument for ups and downs of the land: "These rocks have been deposited beneath water, yet it is clear the spot where the trees grew, must once have been above the level of the sea, so that it is certain the land must have been depressed by at least as many thousand feet, as the superincumbent subaqueous deposits are thick.—But I am afraid you will tell me, I am prosy with my geological descriptions & theories."

The word *petrified* comes from the Greek and Latin root *petra* for rock or stone. In Greek mythology anyone who gazed at the face of Medusa's severed head would be turned into stone, that is, be petrified. The way it works with forests is a little different. When a tree or log is buried by a wet sediment, or especially wet volcanic ash, and the normal biologic activities of rot and decay are denied the oxygen they need, then the organic material in the wood can be chemically replaced by silica. Usually, as warm, slightly acidic water flows through the surrounding material, it extracts the silica from the sediment or ash, and with the silica, replaces the organic material in the wood in casts down to the level of the individual cell.

Darwin's petrified forest was beauty. The trees were largely petrified standing in place, in growth position. And when Darwin found the forest he counted fifty-two fossil tree trunks, measuring about three to five feet in diameter buried in sandstones or volcanic sandstone. Some still stood as columns several feet high.

◆

Sadly, not much of Darwin's petrified wood remains.

Gayle and I returned to the now-charming city of Mendoza—surrounded by vineyards and wineries—to explore the northern leg of Darwin's loop. We started the day driving the two-lane concrete road that stretches north across the scrubby desert landscape, paralleling the mountains. It could have been New Mexico or west Texas. At last the road turned to the mountain, entered a valley, and began its climb to Villavicencio, now the site of a hot spring, resort hotel, and nature preserve.

Then the real climb began, and the paving gave out. The road switchbacked up the dusty road to an elevation of nine thousand feet before at last reaching a gently rolling plateau. There we passed a dozen guanacos grazing on the sparse plants scattered on the otherwise rocky, barren ground. It was a little hard to believe that this was the main route between Santiago and Mendoza until the 1960s.

The sky was a gorgeous bright blue. A weathered wooden cross rose at the high point on the road, a monument to the Jesuit missionaries who traveled here to evangelize the Huarpe people living to the east of the Andes, and to develop mines that Darwin perused. The snow and rock of Aconcagua, at 22,838 feet—the highest mountain outside Asia—loomed before us to the west. I couldn't help thinking that if the view was this spectacular the day Darwin passed by, perhaps he wouldn't have noticed the petrified wood at all.

Two miles farther on, a monument engraved in stone marked Darwin's passing by, but made no mention of the petrified wood. Only a few thorny bushes and cacti dotted the hillside above, where a few outcrops of laminated strata protruded. I parked the car and climbed up to take a look.

But no great logs of petrified wood. They all apparently now reside in museums around the world, or perhaps on someone's patio. Today virtually all that is left are the stumps and, in places, the casts left by the long gone trunks. In the silicified stumps I could see the outline of the trees, and even get a sense of the rings, but that was it. Darwin's petrified forest has been clear-cut to ground level.

As we were returning to our car, what appeared to be a road crew with a front-end loader stopped for a break from their workday. The crew left to roam the hillside, seemingly in search of any remaining stray chunks of the fossil wood.

◆

The town of Uspallata lies nestled in the high valley between the Uspallata Range and the main range of the Andes where the historical route that we followed from Mendoza joins the modern highway. From here Darwin's party followed a line of refuges up the Río Mendoza and its tributary, the Río Las Cuevas, then up and over the crest of the Andes at the location where the statue Cristo Redentor (Christ the Redeemer)—and the border between Argentina and Chile—are located today. Construction of the refuges, or *casuchas*, was a project proposed and then directed by Bernardo O'Higgins's father, Ambrosio, after he was almost killed trying to cross the Andes by this route during the winter of 1763. Today the road is the main route between Santiago and Mendoza—the busiest road linking the two countries—streaming with cars, trucks, and buses. A tunnel now passes beneath the crest of the range emerging near the Portillo ski area in Chile. Nonetheless, traces of the original casuchas remain, as do the magnificent exposures of geology in the mountainsides.

One of the casuchas was located at the still famous Puente del Inca or Inca Bridge, a sort of natural bridge over a mineral spring. Although Darwin was not much impressed by the Puente, what is truly impressive is the view to be had a few kilometers farther west, looking north up the valley of the Río Horcones toward the south face of Aconcagua.

◆

Darwin's last day in the high Andes was a long one. From his camp at Puente del Inca at about 9,200 feet, his party traveled up the Río de las Cuevas, over the steep ridge that forms the crest of the Cumbre or Uspallata Pass at 12,600 feet, and then down the even steeper Chilean side to the casucha at Ojos de Agua at 7,000 feet, a vertical descent of more than a mile. The distance between the two points as the crow flies is about fifteen miles, but with all the switchbacks the actual distance travel was likely at least twenty-five. Today the tunnel cuts off the top 2,000 feet or so, but an unpaved road with nearly innumerable switchbacks follows close to his actual route over the pass.

Gayle and I drove up the Argentine side, a popular excursion to the statue of Cristo Redentor, built to mark one of the successes in long-running series of border disputes between Argentina and Chile. It was a gorgeous day and the view from the summit was spectacular, the sky filled with jagged peaks. Just across the valley to the east was an enormous smooth slope of steeply tilted strata, which had been swept clean by rockslides. Chunks of rock the size of apartment houses lay in a jumble at the bottom. Glaciers draped the peak above. Nearly out of view down the valley I could just make out the red sandstones of Darwin's Gypseous Formation. Looking to the west down the Chilean side was enough to give one vertigo. The gravel road looked as if it were descending into a hole within Darwin's dark gray porphyries.

Again I turned to Darwin's beautifully colored sections that he used to illustrate his book, *Geological Observations on South America.*

In these he captured the essence and magic of what he saw in his traverses across the Andes. He was justly proud of them, explaining that although his visits were brief, that by "riding slowly and halting occasionally to ascend the mountains, there are many circumstances favourable to obtaining a more faithful sketch of [the structure of the Cordillera], than would at first be thought possible from so short an examination." These favorable circumstances included the characteristics that the mountains are "steep and absolutely bare of vegetation"; that the atmosphere is "resplendently clear"; that the stratification of the rocks is "distinct"; and that the rocks themselves are "brightly and variously coloured." In sum, he wrote, "some of the natural sections might be truly compared for distinctness to those coloured ones in geological works."

He was fully satisfied with his efforts: "Considering how little is known of the structure of this gigantic range, to which I particularly attended, most travellers having collected only specimens of the rocks, I think my sketch-sections, though necessarily imperfect, possess some interest."

Looking up a valley to the north, exposed in the bare cliffs of a ridge soaring to an elevation nearly a mile above me, to almost fourteen thousand feet, were beds of rocks standing virtually on end, just as Darwin showed on his section. Not only had these mountains been greatly uplifted, but the strata that made them up had been sorely abused in the process.

There is one particularly charming detail of Darwin's section of the Cumbre Range. Looking north, up the Valley of the Horcones— in dotted lines—Darwin shows an enormous fold, an anticline. This, although Darwin probably did not know it, was likely the hulking form of Aconcagua, the highest mountain in the western hemisphere.

◆

Overall, Darwin was thrilled with his excursion through the Andes, writing to his sister Susan, "It was something more than enjoyment:

I cannot express the delight, which I felt at such a famous winding up of all my geology in S.—America.—I literally could hardly sleep at nights for thinking over my days work.—The scenery was so new & so majestic: every thing at an elevation of 12,000 ft. bears so different an aspect, from that in a lower country.—I have seen many views more beautiful but none with so strongly marked a character. To a geologist also there are such manifest proofs of excessive violence, the strata of the highest pinnacles are tossed about like the crust of a broken pie."

CHAPTER NINE

Coral Reefs and the Sinking Bottom of the Sea

I am glad we have visited these Islands; such
formations surely rank high amongst the wonderful
objects of this world. It is not a wonder which at
first strikes the eye of the body, but rather after
reflection, the eye of reason.

—Charles Darwin

n the austral winter of 1835, FitzRoy and the *Beagle* completed surveys of the coast in Chile and Peru, turned to the Galapagos, then headed for home—the long way, across the Pacific and Indian Oceans. And as the *Beagle* left South America, Darwin already had coral reefs on his mind, even though he had never actually seen one. He was also beginning to think about the mysterious question of the origin of species, but geology was still foremost in his mind.

In his mind's eye he saw South America—and especially the Andes—clearly rising above the level of the sea. With the ever-present influence of Lyell, it began to occur to him that a wonderful test of these ideas would be a complementary proof that the bottom of the ocean was sinking. Coral reefs might provide that proof. He knew that the *Beagle* would be visiting coral islands in the months ahead. In mulling over the question of the coral reefs, he jotted in his notebook, "Is there a large proportion of these Coralls which only live near the surface.—If so, we may suppose the land sinking."

◆

For seafarers and geologists in 1835 coral reefs posed an opportunity, a hazard, and a mystery all rolled into one. Coral atolls, with their calm, protected internal lagoons—provided that the water was deep enough—afforded a safe and sheltered anchorage for a vessel in need of protection from weather or rough seas. But unexpected or unseen coral reefs could rip a hole in the bottom of an unwary or unsuspecting ship in the blink of an eye.

The most disturbing aspect of coral reefs was that they could seemingly rise up out of nowhere. The prudent sailor is always aware of the depth of water beneath the ship's keel. If there was any question about the adequacy of the depth, nineteenth-century captains would direct their crews to take continuous measurements, or soundings, of the water depth with a lead line—a weight or lead, attached to a rope—or, in very shallow water, with a long pole. They would do this especially as the ship approached land.

Comparison of the water depth determined from soundings with depths reported on a chart could also help the captain determine the position of the ship. Frequently the lead at the bottom of the sounding line was coated with tallow, or even shaped like a bell and its cavity filled with tallow. As the lead struck the bottom, pieces of the material on the bottom would stick in the tallow. Then, when the line was drawn up, the sailors would examine the tallow and report the nature of the bottom: mud, sand, rock, or shells. The nature of the bottom was reported, along with the water depth in fathoms, on many charts. Captains could use this additional information to assist in determining their positions—and alert them to the presence of coral.

But all this was predicated on the notion that the depth of water changed relatively smoothly and slowly with distance, and that the water depth became gradually shallower as the ship approached land. Coral reefs were a hazardous exception. The bottom of the sea close to a coral reef could be very, very deep, beyond the reach of the lead line—in fact, unfathomable. Consequently the lead line might give little or no warning when approaching a dangerous coral formation.

Captain Cook and the HMS *Endeavour* barely escaped the maze of treacherous reefs when first exploring the region now called the Great Barrier Reef in 1770. The *Endeavour* ran aground on coral and was severely damaged. But Cook and his crew managed to float the ship off the reef and find a safe harbor to make repairs. Two decades later, the French explorer, Lapérouse, was not so fortunate. After rounding Cape Horn, and spending more than two years criss-crossing the Pacific Ocean from Alaska to Australia and from Japan to Chile, Lapérouse and his two ships apparently ran aground on a coral reef off Vanikoro Island in the Santa Cruz Islands in 1789. Fortunately—for us—Lapérouse had sent some journals, charts, and notes home with an English prison ship that he had met in Botany Bay, Australia in January 1788. After leaving Botany Bay, no one from the expedition was ever seen again.

In 1787 the HMS *Bounty* was sent out to Tahiti with the aim of obtaining breadfruit plants to transplant in the West Indies to provide cheap food for slaves. When the Lords of the Admiralty learned of the now famous mutiny of the *Bounty*'s crew, they sent the HMS *Pandora* to the South Seas to track down the mutineers, to arrest them and to return them to England for trial, as well as to recover the ship. The *Pandora* succeeded in arresting many of the mutineers—not including Fletcher Christian and his party who had decamped and hidden on Pitcairn Island—and set sail back to England in 1791. During the search for additional mutineers, smoke signals were seen on the island of Vanikoro, then not yet recognized as the site of Lapérouse's wreck, but no attempt was made to investigate, perhaps in part because of the dangerous reefs. The

Pandora's course home led through the perilous Torres Strait. This strait, located between Australia and New Guinea and first charted by Cook, is the most direct sea-lane to eastern Australia, but the strait is narrow, laced with reefs, and the site of innumerable wrecks and losses. While passing through the strait, with fourteen of the mutinous crew locked in a box on the quarterdeck, the ship struck a reef and foundered with the loss of thirty-one of the crew and four of the prisoners who had been locked in irons. After an epic journey to Timor in small boats, the remnants of the officers and crew eventually arrived in England with the surviving ten prisoners. Of the 134 men who had left England on the *Pandora* on the hunt for the fugitive mutineers, only seventy-eight returned alive.

◆

How did these strange features, coral reefs, form and what could be learned about them? This was the challenge that Francis Beaufort gave to FitzRoy, and which FitzRoy expected Darwin, as his geologist, to meet.

Beaufort had been an honorary member of the Geological Society since 1808, notwithstanding the comment of one wag that, "Naval men have not devoted much attention to geology; perhaps . . . because it has always been their duty to avoid rocks."

In remarkable detail, probably reflecting Beaufort's attendance at meetings of the Society, he wrote in his memorandum of instructions for FitzRoy, "a very interesting inquiry might be instituted respecting the formation of these coral reefs."

A chapter of Volume 2 of Charles Lyell's *Principles of Geology*— that Darwin had received in Montevideo in November of 1832 was devoted to the formation of coral reefs. In the language of Charles Lyell's day, coral was a zoophyte, an animal that looks like a plant. Other zoophytes include sea anemones, sponges, and sea lilies. Even more telling, corals were a lithophyte, a kind of invertebrate animal with a tube-like body, a mouth at one end surrounded by tentacles, and a calcareous skeleton. The facts that Lyell had at his disposal included the results of recent explorations of the South Pacific, especially that of Captain Frederick Beechey, and an emerging understanding of the biological nature of corals.

The French naturalists Jean René Quoy and Joseph Gaimard, who participated in expeditions to the Pacific and Indian Oceans, had studied corals and written an intriguing paper, *Mémoire sur l'accroissement des Polypes Lithophytes considéré géologiquement*, on the geologic consequences of the accumulation of corals in 1825. In this paper they described the critical discovery that reef-forming corals only grow at relatively shallow depths in the oceans.

A few years before Beechey explored coral islands and in his memoir *Narrative of a Voyage to the Pacific and Beering's Strait*, he published a sketch of an atoll together with a section through it, piquing the curiosity of the English geologists. The sketch and section, titled "View of Whitsunday Island," shows four intrepid sailors pulling at the oars of a longboat through the cresting waves, while a fifth stands in the bow of the heaving boat, watching for protruding coral formations or perhaps measuring the depth of the water with a lead line. Beyond the breaking waves is the white beach of a nearly circular island, now called Pinaki, a ring of bushy trees surrounding an interior lagoon.

Above the sketch is a profile of the water depth through the entire island. And below, a more detailed profile of the seaward edge of the island that shows a bottom composed of coral and—just beyond the breakers—a very sharp increase in the depth of water quickly giving way to depths too great to be sounded. Within the lagoon, behind the tree-covered beach, the section depicts strange coral formations rising almost vertically from the bottom of the lagoon to the surface of the water.

Lyell included Beechey's sketch and sections in his chapter; although apparently to maintain a less sensational tone, he modi-fied Beechey's sketch of the atoll to remove the longboat manned by the five heroic sailors. Citing the work of Quoy and Gaimard he wrote, "In regard to the thickness of the masses of coral, MM. Quoy and Gaimard are of opinion, that the species which contribute most actively to the formation of solid masses do not grow where the water is deeper than twenty-five or thirty feet."

Despite the attention that Lyell gave to the coral reefs, his interpretation of the atolls was both rather glib and ad hoc. He opined that atolls were "nothing more than the crests of submarine volcanos, having the rims of their craters overgrown by corals." He omitted any explanation of why the craters of all these now submarine volcanoes all happened to be exactly at one elevation—sea level. In contrast, the elevations of the summits of the volcanic islands that do stand above sea level range quite widely, as do the elevations of volcanic cones on the continents. It was likely, however, that Lyell's main purpose in including the discussion of the very interesting coral reefs in his book was not so much to explain their origin, but to provide another example of perpetual change, this one at the junction of the organic and inorganic worlds, in support of his extreme version of uniformitarianism.

◆

Corals are marine invertebrate creatures that form an exoskeleton of calcium carbonate, a material that is readily preserved as a fossil. Calcium carbonate is not unique to corals. It is ubiquitous. It makes up eggshells, clam shells, snail shells, and oyster shells. In rocks it makes up limestone, chalk, marble, calcite, and aragonite. It forms a common cement in sandstones, gluing the grains of sand together. It is just plain everywhere. If you live in an area where the water is hard, the crust in your teapot is calcium carbonate. It dissolves easily in waters that are at all acidic, and it moves easily through both biologic and geologic systems.

Fossil corals first appeared in the ancient rocks that absorbed Sedgwick in Wales. In fact, Darwin thought that he had found a coral in the rocks at Cwm Idwal, although Sedgwick was unable to confirm it. While the species of coral living at the time of Sedgwick's Paleozoic are for the most part long gone, there are a wide variety of modern corals. The ones most important in building reefs live in colonies of vast numbers of individual polyps. The polyps feed on microscopic algae with which they have a symbiotic relationship. The corals provide a sheltered environment and the chemicals needed by the algae for photosynthesis. The algae provide oxygen and the products of photosynthesis. The coral polyps eat the algae obtaining the nutrients that they require for life, and that they use to produce the calcium carbonate exoskeleton. Because the algae need light for photosynthesis, this community needs clear, shallow water. Some forms of algae also secrete calcium carbonate themselves, adding their own contribution to the building of the reef.

◆

On the evening of October 20, 1835, the *Beagle*'s crew, having completed their work in the Galápagos Islands, put the ship on a course for Tahiti. The ship soon caught the southeasterly trade winds, sailing 150 to 160 miles a day. Within a few days the ship was beyond what Darwin called the "gloomy region that extends far from the coast of S. America" and entered a region where "daily the sun shines brightly in the cloudless sky." As always, better weather buoyed his spirits. He had a lot to think about as they sailed during the month that it took them to reach Tahiti. One important item on his list was the origin of coral reefs. He wrote much later "during the two previous years [I had] been incessantly attending to the effects on the shores of S. America of the intermittent elevation of the land, together with the denudation and the deposition of sediment. This necessarily led me to reflect much on the effects of subsidence, and it was easy to replace in imagination the continued deposition of sediment by the upward growth of coral."

Darwin got his first views of the reefs as FitzRoy followed a chart made by the Russian explorer Adam Johann von Krusenstern picking his way through the Low, or "Dangerous," (now Tuamotu) Archipelago. Krusenstern made the chart on the first Russian expedition to circumnavigate the globe. The first island they came to, Honden or Dog Island (now Puka-Puka), was, as FitzRoy wrote, "one of the low coral formations, only a few feet above the water, yet thickly covered with cocoa-nut trees." FitzRoy was impressed with Krusenstern's chart, writing, "Our observations corroborated the position assigned to it by Admiral Krusenstern, in his excellent chart and memoir, the only documents of any use to us while traversing the archipelago of the Low Islands." The islands were called "low" because the archipelago is a collection of atolls and sand bars atop reefs, and most rise no higher than the top of a coconut palm tree at the back of a coral sand beach.

The archipelago is made up of nearly eighty islands and atolls stretching in a streak almost seven hundred miles from northwest to southeast in what is now French Polynesia. Honden Island, offset from the main group of islands about 150 miles to the northeast, may have been first sighted by Magellan, but the name for the island, Honden, was given by Dutch explorers in the early 1600s, *honden* being the Dutch word for dogs. As the *Beagle* approached the islands, Darwin was already studying the charts of the islands and their "peculiar character."

As Darwin looked out at Dog Island, he compared it to Beechey's drawing of Whitsunday Island some three hundred miles to the south. There was some resemblance, but Darwin was not impressed. Dog Island was a little higher, but otherwise its character could be "understood by [the] drawing in Captain Beechey's work."

It took five days to cross through the archipelago, as FitzRoy would not proceed at night when the weather was murky. On the fifth day they found two islands that FitzRoy would later add to Krusenstern's chart, passing one at daylight and another at noon, which Darwin called Noon Island (now Kauehi). He was still not much impressed by the looks of the islands: "These [islands] have a very uninteresting appearance; a brilliantly white beach is capped by a low bright line of green vegetation." But climbing to the top of the mast he got a much better view. "The width of dry land is very trifling: from the masthead it was possible to see at Noon Island across the smooth lagoon to the opposite side.—This great lake of water was about 10 miles wide."

Now at last, Darwin had a clear view of a coral atoll, the origin of which would come to fascinate him.

At daylight on November 15, "Tahiti, an island which must for ever remain as classical to the Voyager in the South Sea, was in view." When the *Beagle* arrived in Tahiti the islanders determined their days of the week as if the island were located to the west of the date line, when it is actually to the east, so for the islanders it was already Monday, but for the crew of the *Beagle* it was still Sunday. On the islanders' Sunday Sabbath they were enjoined from launching their canoes, but since for them it was already Monday they provided the *Beagle* with a friendly welcome. "Crowds of men, women & children were collected on the memorable point Venus [where Captain Cook had made his astronomical measurements of the transit of Venus in 1769] ready to receive us with laughing merry faces." Darwin also quickly observed that the island was encircled at some distance by a coral reef, what he would come to call a barrier reef.

After breakfast the next day Darwin went onshore and ascended the slopes of the mountain rising on the island to an elevation that he estimated "between two and three thousand feet." He speculated that earlier the "interior mountains stood as a smaller island in the sea." From his high point Darwin could see out across the moat and barrier reef surrounding Tahiti, across to the island of Eimeo (now Mo'orea). "The island is completely encircled by a reef, with the exception of one small gateway; at this distance a narrow but well defined line of brilliant white where the waves first encountered the wall of coral, was alone visible; Within this line was included the smooth glassy water of the lagoon, out of which the mountains rose abruptly." In his diary, waxing poetic, he noted, "The effect was very pleasing & might be compared to a framed engraving, where the frame represents the breakers, the marginal paper the lagoon, & the drawing the Island itself." Here Darwin had a perfect example of what he would later describe as an island surrounded by a barrier reef.

After an even longer expedition to the interior of Tahiti extending to three days, he had a chance to hire an outrigger canoe and men to take him for a closer look at the barrier reef surrounding the island of Tahiti, where he "paddled for some time . . . admiring the pretty branching Corals." He was already forming his own ideas about the work of these creatures and their relation to the islands—and setting his bulldog—like tenacity on solving the problem. "It is my opinion, that besides the avowed ignorance concerning the tiny architects of each individual species, little is yet known, in spite of the much that has been written, of the structure & origin of the Coral Islands & reefs."

◆

The *Beagle* left Tahiti, bound for New Zealand, on November 26. On December 3 the ship passed very near to Whytootacke Island (now Aitutaki). Here Darwin immediately recognized an island setting that was partway between the simple atolls he had seen in the Low Islands—a loop of coral reefs supporting a beach and surrounding an internal lagoon—and the high islands of Tahiti and Eimeo (Mo'orea)—mountainous islands encircled first by deep lagoons, and then further out, by a barrier reef. At Whytootacke Darwin saw "a union of the two prevailing kinds of structure . . . A hilly irregular mass was surrounded by a well defined circle of reefs, which in greater part have been converted into low narrow strips of land . . . consisting merely of sand & Coral rocks heaped on the dead part of a former reef." By now his idea of the development of atolls seems to have been well established in his mind—initially an island surrounded by fringing reefs on its shallow shores, the corals continuing to grow upward as the island subsided to produce an island surrounded by a deep lagoon and barrier reef, and then the corals continuing to grow upward as the island subsided further until the island itself sinks beneath sea level, leaving only an atoll— a loop of reef and beach with a few coconut trees surrounding an internal lagoon.

◆

Before reaching the *Beagle's* next stop in New Zealand Darwin put pencil to paper and began an essay he titled "Coral Islands." He wrote quickly making many erasures and corrections, but he wanted to get the core of his idea down on paper.

While on the *Beagle* Darwin supplemented his own observations of the coral islands with the examination of the *Beagle's* collection of nautical charts of islands. In the end, he recognized three general types of coral reefs. The first type included atolls, or lagoon islands, in which an internal lagoon is ringed by a low strip of coral. The second type of reefs, barrier reefs, surrounded islands, but were separated from them by a lagoon, or as Darwin wrote, "High Islands encircled by a reef, as a picture is by a frame." The third type were fringing or shore reefs, that is, reefs formed directly on the shoreline of islands or continental coasts.

He heard of a history of earthquakes at Tahiti, but "looked in vain on the shores of Tahiti for any sort of evidence of a consequent rise." He was beginning to suspect that perhaps these earthquakes were accompanied not by the elevation of the land as he had seen in Chile, but by its subsidence. And from his observations in Tahiti and his reading, it seemed clear that the species of coral on the outer limits of the reef, most exposed to the action of waves, were different from those that grew in the sheltered interior lagoons. His hypothesis took shape. "If then the two following postulates are allowed, much of the difficulty in understanding the Coral formation, will I think, be removed.—(1st) That in certain parts of the Pacifick, a series of subsidences have taken place; of which no *one* exceeded in depth . . . [the range at which the quiet water species of coral] will flourish: & of which series, the intervals between the successive steps, were sufficiently long to allow of their growth, always bringing to the same level the upper surface of the reef.—(2nd) That those species of Lithophytes, which build the outer, solid wall, flourish best, where the sea violently breaks."

A few months later the *Beagle*'s stop at Keeling (now Cocos) Islands finally gave Darwin his chance to examine a coral atoll in significant detail. Beaufort had directed FitzRoy and the *Beagle*, that perhaps "if circumstances are favourable, she might look at the Keeling Islands, and settle their position." So after visits to New Zealand, Australia, and Van Diemen's Land (now Tasmania), FitzRoy set sail toward Cape Leeuwin, at the southwest corner of Australia, but owing to unfavorable winds, several days were required "before our little ship was sufficiently far westward of that promontory to steer for my next object, the Keeling Islands." FitzRoy had many reasons for this choice. In addition to settling their position, he intended to make a detailed survey of the islands, and to measure the tides with a tide gauge of a novel design that had recently occurred to him.

Without knowing the position of the islands, they were not all that easy to find. As FitzRoy approached what he supposed was their location, no land could be seen from the masthead. Then he got lucky. "When a number of gannets flew past the ship toward the west. We steered directly after them, and early next morning (after making but little way during a fine night) saw the Keelings right ahead, about sixteen miles distant."

From a distance the islands were not much to see. All that was visible was "a long but broken line of cocoa-palm trees, and a heavy surf breaking upon a low white beach, nowhere rising many feet above the foaming water." But then, "within five miles of the larger Keeling . . . we made out a number of low islets, nowhere more than thirty feet above the sea, covered with palm-trees, and encircling a large shallow lagoon."

They were met in the entrance channel to the lagoon at South Keeling by Mr. Liesk, an English resident of the islands, whose telling of the recent lurid history of the islands FitzRoy later recounted. "Little or no notice was taken," FitzRoy wrote, of the islands from the time of their discovery in 1608 by Captain William Keeling "till 1823, when one Alexander Hare, a British subject, established himself and a small party of Malays, upon the Southern Keeling Island, which he thought a favourable place for commerce, and for maintaining a seraglio of Malay women, whom he confined to one island,—almost to one house." This arrangement, however, proved to be a sore point with the Malay slave men—along with other aspects of Hare's harsh treatment—who deserted and sought protection from Captain John Clunies-Ross, a British merchant captain who—with his family and Liesk, his former mate—had also taken up residence on a different islet in the group. Hare then left the Keelings, and "about a year afterwards was arrested in his lawless career by death, while establishing another harem at Batavia [now Jakarta and surroundings]." Darwin seems to have been a bit embarrassed by this story, skipping the salient details in his diary, and referring to Hare only as "very worthless character."

◆

FitzRoy quickly set the officers and crew to work, and for more than a week "every one was actively occupied; our boats were sent in all directions." At first a strong wind caused difficulties making soundings, "but two moderate days were eagerly taken advantage of to go round the whole group in a boat, and get the few deep soundings." Both Darwin and FitzRoy were amazed, that, as FitzRoy wrote, "Only a mile from the southern extreme of the South Keeling, I could get no bottom with more than a thousand fathoms of line."

Both South and North Keeling met the classic definition of atolls. In FitzRoy's description, "The [southern] cluster of islets encircle a shallow lagoon, of an oval form, about nine miles long, and six wide. The islets are mere skeletons—little better than coral reefs, on which broken coral and dust have been driven by sea and wind till enough has been accumulated to afford place and nourishment for thousands of cocoa-palms." An important observation was that "The outer edges of the islands are considerably higher than the inner, but nowhere exceed about thirty feet above the mean level of the sea. The lagoon is shallow, almost filled with branching corals and coral sand."

FitzRoy noted the roles of the different varieties of corals in building and maintaining the atoll: "Among the great variety of corals forming the walls . . . and the under-water forests of the Keeling islands, there is more difference than between a lily of the valley and a gnarled oak. Some are fragile and delicate, of various colours, and just like vegetables to the eye, others are of a solid description, like petrified tropical plants; but all these grow within the outer reef, and chiefly in the lagoons."

FitzRoy was anxious to sample the outer wall of the reef, and "to ascertain if possible, to what depth the living coral extended, but my efforts were almost in vain, on account of a surf always violent, and because the outer wall is so solid that I could not detach pieces from it lower down than five fathoms [30 feet]." He tried a variety of sampling methods: small anchors, hooks, grappling irons, and chains, but none seemed to work. Although large samples eluded him, the sounding lead provided some information: "Judging however, from impressions made upon a large lead, the end of which was widened, and covered with tallow hardened with lime, and from such small fragments as we could raise, I concluded that the coral was not alive at a depth exceeding seven fathoms [42 feet] below low water."

Besides the recent salacious history, FitzRoy was impressed with several other oddities that he witnessed in the Keelings: "crabs eat cocoa-nuts, fish eat coral, dogs catch fish, men ride on turtle, and shells are dangerous man-traps . . . it must yet be said that . . . many rats make their nests at the top of high palm-trees."

Darwin spent time walking the beach and examining its structure. The atoll consisted of a "strip of dry land" only a few hundred yards wide. Inside was a lagoon surrounded by a white beach. But the outer coast was "a solid broad flat of coral rock, which serves to break the violence of the open ocean. Excepting near the lagoon where there is some sand, the land is entirely composed of rounded fragments of coral."

Darwin studied the variety of corals as they grew, and the processes by which their remains were incorporated in the reef. He recognized that certain species of coral preferred the environment of the breaking waves. These grew the fastest and built the exterior framework of the reef. He confirmed what he had read that these were also most robust on the windward side of the reef where the incoming waves were the strongest. He observed that the more fragile species of branching corals preferred the quiet waters of the interior lagoon. He found that a significant volume of the reef was actually made of the secretions of the calcareous algae, or nullil-pores, as he called them.

He talked with the residents and learned that they had excavated a channel through a portion of the reef to allow ships to come and go more easily, and he learned that the continuing growth of the coral was filling in the channel. From their observations he was able to show that the growth of coral was faster than Lyell had assumed.

As they compiled their soundings on a chart, the officers also plotted profiles through the topography of the atoll, more detailed than those of Beechey.

As the *Beagle* sailed away from Keeling Island, Darwin revised his earlier dreary assessment of the views of the atolls. He was now looking at atolls not with the "eye of the body," but with the "eye of reason," and in this manner he found them much more appealing. He expressed his amazement at the fact that the geology of the island owed so much to the tiny individual polyps of coral: "We feel surprised when travellers relate accounts of the vast piles & extent of some ancient ruins; but how insignificant are the greatest of them, when compared to the matter here accumulated by various small animals. Throughout the whole group of Islands, every single atom, even from the most minute particle to large fragments of rocks, bear the stamp of once having been subjected to the power of organic arrangement."

Then, writing in his diary, he neatly summed up his theory of the reef formation. First, he wrote, starting from FitzRoy's soundings showing that the reef rose abruptly from great depths, "Hence we must consider this Isld as the summit of a lofty mountain; to how great a depth or thickness the work of the Coral animal extends is quite uncertain." Then, he stated his hypothesis and its conse-quences, "If the opinion that the rock-making Polypi continue to build upwards, as the foundation of the Isld . . . gradually subsides, is granted to be true; then probably the Coral limestone must be of great thickness." He saw in an atoll, the trace of island—perhaps originally a barren volcano—subsequently surrounded by a bar-rier reef, finally turning into an atoll. After a long period of slow subsidence, subsidence at a rate slow enough to allow the corals to grow upward at a corresponding rate, only the reef would reach the surface of the ocean, with the original volcanic island far below. "We see certain Isds in the Pacifick, such as Tahiti & Eimeo, men-tioned in this journal, which are encircled by a Coral reef separated from the shore by channels & basins of still water. . . . Hence if we imagine such an Island, after long successive intervals to subside a few feet, in a manner similar, but with a movement opposite to the continent of S. America; the coral would be continued upwards, rising from the foundation of the encircling reef."

In Darwin's mind the sinking of the ocean floor was the compli-mentary process to the rising of the continents. In the case of a bar-rier island, "In time the central land would sink beneath the level of the sea & disappear, but the coral would have completed its circular wall. Should we not then have a Lagoon Island?—Under this view, we must look at a Lagoon Isd as a monument raised by myriads of tiny architects, to mark the spot where a former land lies buried in the depths of the ocean." One could also imagine that the fringing reefs around a large island or along a coastline, after a period of slow sub-sidence—accompanied by the upward growth of the coral—would become barrier reefs, like Australia's Great Barrier Reef.

As the *Beagle* sailed across the Indian Ocean toward Mauritius, FitzRoy must have felt a sense of satisfaction for what he and his shipmates had accomplished at the Keeling Islands. He had settled their location, made a detailed survey of their character and bathymetry and other measurements, and his volunteer geologist had come up with a new and compelling theory for their origin. Beaufort would be pleased.

The only thing that Darwin and FitzRoy had not been able to do was to drill down through the coral to discover the depth to which it extended. Darwin's theory predicted that for every foot that the island subsided, the coral should be another foot thicker, and that the coral would be up to thousands of feet thick. It would be a hundred and thirty years before the Americans, preparing to test the hydrogen bomb on the Bikini Atoll, would drill a hole and prove him correct.

◆

It actually wasn't until the *Beagle* arrived at Mauritius two weeks later that Darwin had a chance to see for himself the simplest kind of coral reefs: fringing reefs. According to his emerging theory these gently sloping reefs should grow in the shallow waters on a shore that is uplifting or remains at a constant elevation. And indeed what he found at Mauritius confirmed this view . . . Almost everything fell in to place. He did examine with some interest, guided by a retired British captain, and accompanied by John Lort Stokes, some dead uplifted corals, underscoring the obvious fact that corals could not continue to live once they were uplifted above the level of the sea. Fringing reefs, he believed, would remain relatively thin, because for them to build a great thickness required subsidence.

When he finally arrived back in England he spent many days reviewing and compiling charts of reefs and volcanoes, analyzing expedition reports, and consulting with experts to fill out the evidence for his theory of the coral reefs, but the bulk of his research was almost complete, everything save the final paperwork and analyses.

Now Darwin had—through his explanation for the development of coral reefs—a very strong argument for the sinking of the ocean basins to match his evidence for the uplifting of the continents. Just as the Andes were a vivid example of uplift on a continent, so too the atolls and barrier reefs were vivid examples of subsidence in the ocean basins. When he plotted the locations of active volcanoes on his map of the coral reefs of the world, he noted that the active volcanoes tended to be associated with regions that he believed to be rising. In contrast, active volcanoes seemed to be absent from the areas that he reasoned were subsiding. "We may finally conclude," he wrote, "that the subterranean changes which have caused some large areas to rise, and others to subside, have acted in a very similar manner." But he stops just short of telling us why these areas were subsiding, leaving us to conjecture what he was thinking.

In the back of his mind Darwin likely remembered all the way back to the early days on the *Beagle*, and his walk on the beach at St. Jago tracing the shelly, white layer of ancient beach, diving down as it passed under the volcanic crater at Signal Hill, then climbing back to its former elevation as he continued farther down the beach. Perhaps this recollection gave him the seed of an idea about the common mechanism of both subsidence and uplift: the subterranean movement of molten rock.

PART TWO

DARWIN THEORIZING

CHAPTER TEN

Faith Comforts, Facts Persuade

But the busiest time of the whole voyage has been tranquility itself to this last month.

—Charles Darwin to his cousin,
W. D. Fox, a month after the
return of the *Beagle* to England

Through the middle of 1836, the *Beagle* sailed from Mauritius to South Africa, around the Cape of Good Hope, up the Atlantic—stopping at Saint Helena and Ascension Island—and then back to Brazil. For Darwin, the days stretched endlessly. He wanted more than anything just to be home. Or at least to be on dry land. FitzRoy, however, was still a man on a mission. To Darwin's utter chagrin, the *Beagle*'s return was delayed a few weeks owing to FitzRoy's decision to divert from the most direct course to England from the South Atlantic, to make a final check on the longitudes on the east coast of South America. "This zig-zag manner of proceeding is very grievous," he wrote to his sister, Susan, "it has put the finishing stroke to my feelings. I loathe, I abhor the sea, & all ships which sail on it."

Writing to his sister Caroline, Darwin's acute and now chronic homesickness took on an almost poetic form:

I feel inclined to write about nothing else, but to tell you over & over again, how I long to be quietly seated amongst you.—How beautiful Shropshire will look . . . I am determined & feel sure, that the scenery of England is ten times more beautiful than any we have seen.—What reasonable person can wish for great ill proportioned mountains, two & three miles high? No, no; give me the Brythen or some such compact little hill.—And then as for your boundless plains & impenetrable forests, who would compare them with the green fields & oak woods of England?—People are pleased to talk of the ever smiling sky of the Tropics: must not this be precious nonsense? Who admires a lady's face who is always smiling? England is not one of your insipid beauties; she can cry, & frown, & smile, all by turns.—In short I am convinced it is a most ridiculous thing to go round the world, when by staying quietly, the world will go round with you.—

As Darwin suffered through the last months of the voyage, he planned his return to England. Even before Darwin's departure, Sedgwick had offered to propose him for membership in the Geological Society, but this task had apparently slipped the professor's mind. In July, still stuck on the *Beagle*, with nothing to do but plan on the next phase of his life—becoming a scientist and presenting his findings and interpretations to others—Darwin wrote John Henslow from Saint Helena, "My dear Henslow, I am going to ask you to do me a favor. I am very anxious to belong to the Geolog: Society. . . . Would you be good enough to take the proper preparatory steps."

Finally, in September at long last, FitzRoy was satisfied with his longitudes, and the *Beagle* headed for home.

◆

On the evening of Sunday, October 2, 1836, blown in by a storm in the Bay of Biscay, the *Beagle* anchored at Falmouth, England, having circled the globe. Within hours that same "dreadfully stormy" night, Darwin—having pined about home for months—set out on a mail coach wending his way back to Shrewsbury. The journey took two days. He arrived at The Mount in time for breakfast on Tuesday morning. FitzRoy, in contrast, still had much to do on board, but he took Monday afternoon off to have tea with the family of Robert Fox, a prominent Falmouth Quaker interested in geology and especially the earth's magnetic field.

FitzRoy examined Fox's invention of a dip needle for measuring the inclination of the magnetic field, which he seemed to think was superior to the one he had used on the *Beagle*. Fox's daughter, Caroline, a well regarded diarist, took pride in FitzRoy's reaction: "He came to see papa's dipping needle deflector and was highly delighted. He has [another dip needle] on board, but this beats it in accuracy." FitzRoy enthused about the discoveries he and Darwin had made on the voyage. Humorously describing Darwin as "fly-catcher" and "stone-pounder," he extolled Darwin's deductions about the formation coral reefs to the rapt Foxes. Caroline understood that "the coral insects do not work up from the bottom of the sea against wind and tide, but that the reef is first thrown up by a volcano, and they then surmount it, after which it gradually sinks." FitzRoy impressed them as "a most agreeable, gentleman-like young man."

Darwin and FitzRoy ended their voyage together on good terms, filled with mutual respect and affection. Shortly after arriving in Shrewsbury, Darwin wrote an effusive and chatty letter to FitzRoy describing his cheery homecoming, concluding "Good bye—God bless you—I hope you are as happy, but much wiser than your most sincere but unworthy Philos." A few weeks later, FitzRoy replied, "Dearest Philos," and writing in a similar style, described his struggles with unenlightened bureaucrats in completing the ultimate leg of the *Beagle*'s voyage back to Greenwich—for the final readings of the chronometers. And then he astonished Darwin with the news, "I am going to be married!!!!!!! to Mary O'Brien," a woman he had never mentioned to Darwin or any of the officers in the nearly five years that they were together on the *Beagle*. "Now you may know that I had decided on this step, long very long ago," he wrote. FitzRoy was riding high. His excitement was palpable. Sadly neither his euphoria, nor his warm relationship with Darwin, would survive.

On the 28th of October the ship arrived up the River Thames to Greenwich. Here FitzRoy would complete his chain of measurements of longitudes around the globe. After nearly five years of the voyage circling the earth, the difference between the time kept on his chronometers and the time kept at the Greenwich Observatory would be a measure of the accuracy of his longitudes. The *Beagle* lost a day from circling the earth by sailing west, just as the survivors of Magellan's crew had done, in contrast to Phileas Fogg, the hero of Jules Verne's *Around the World in Eighty Days*, who gained a day by traveling east. The difference between FitzRoy's time and that at Greenwich was the expected twenty-four hours plus only thirty-three seconds. This accumulated error corresponded to an error in longitude of only about eight minutes, an error in distance of about nine and one-half miles at the equator, and about six miles at the latitude of Greenwich. This was the first time a chain of longitude observations by chronometer around the planet had been accomplished, or even attempted. It was a remarkable achievement.

◆

Darwin's arrival home began a whirlwind of activity. After a quick visit to Shrewsbury, he headed off to Cambridge to seek Henslow's guidance and to begin the search for experts—he would refer to them as "the great men"—who would agree to examine his collections of rocks, fossils, insects, birds, plants, and animals. Then on to London to stay with his brother Erasmus at 43 Great Marlborough St. Most of "the great men," he found—as Henslow had warned him—were "all overwhelmed, with their own business . . . It is clear the collectors so much outnumber the real naturalists, that the latter have no time to spare." The exceptions were Robert

Owen, the anatomist and paleontologist, who was "anxious to dissect some of the animals in spirits," and his old teacher from Edinburgh, Robert Grant, who was willing to examine some of his corallines.

Darwin found his warmest welcome from Charles Lyell. Although he knew Lyell's thinking well from the *Principles of Geology*, the two had never met. Their new acquaintanceship got off to a roaring start. "Mr. Lyell," he wrote Henslow, "has entered in the most good natured manner, & almost without being asked, into all my plans." Lyell, then the president of the Geological Society, was clearly anxious to help this returned traveler. Lyell already knew, from the letters and reports read to the Society, that Darwin had gathered extensive evidence supporting his own ideas. Lyell also advised Darwin that he would have to assume the responsibility for sorting his collections himself, and suggested that he could interest experts in his specimens by taking "any odd specimen" to the different societies. Darwin slowly settled on a plan to take rooms for some months in Cambridge to sort and organize his collections—and gather his thoughts—before returning to the hurly-burly of London.

While he seemed to get on well with the geologists and the botanists, his initial reaction to the zoologists was quite negative: "I am," he wrote, "out of patience with the Zoologists, not because they are overworked, but for their mean and quarrelsome spirit. I went the other evening to the Zoological Soc. where the speakers were snarling at each other, in a manner any thing but like that of gentlemen." After a while he found a better reception, writing to his cousin W. D. Fox, "My London visit has been quite idle, as far as Nat: History goes but has passed in most exciting dissipation amongst the Dons in science." He finally even found helpful zoologists: "I find that there are plenty, who will undertake the description of whole tribes of animals, of which I know nothing." And there was also hope that he could soon get to work himself. Within in the month, he hoped "to set to work: tooth and nail at the Geology, which I shall publish by itself."

And on November 2 he attended his first meeting of the Geological Society, where he heard a paper read on "A general sketch of the geology of the western part of Asia Minor."

Amid the flurry of travel between Shrewsbury, Cambridge, and London, organizing his affairs, making obligatory visits to family and friends, Darwin even sent flowers to his old girlfriend Fanny Owen, now married to the wealthy Robert Myddelton Biddulph and ensconced in Chirk Castle. He was pulled in umpteen directions, although that was no doubt welcome after the long days of boredom at sea. The challenge now was, he wrote, trying to "settle my jolted brains into some kind of order." Nonetheless, the question of elevation was never far from his mind.

In mid November he yielded to the demands of his extended family and friends for a visit to Shrewsbury and his Wedgwood cousins at nearby Maer Hall. Darwin was anxious to see them too, but at the same time girded himself for the inevitable hubbub, writing beforehand, "It will be a most uncomfortable visit,—a mere struggle how many people can be visited, to whom I am bound, & indeed (if time, just at present was not so precious) am most anxious to see."

Darwin's first stop was at Maer Hall, the Wedgwood family home, a sprawling stone mansion and estate, in the rolling hills, woods, and fields about twenty miles northeast of Shrewsbury. The drama of his arrival was heightened by delay. The excitement among his cousins, uncles, and aunts who had not seen their world-traveling prodigal cousin and nephew for five years was already at a peak. His sister, Caroline, who had traveled from Shrewsbury, went to meet him on the expected coach. But, when he did not appear on that coach, or the next several, Caroline finally returned to Maer Hall in disappointment, empty-handed.

Then, only fifteen minutes later, Darwin appeared at Maer to the mayhem of a curious and excited assembly. "All the outlyers of the family" had been invited, reported his cousin, Emma, who a few years later would become his wife. His female cousins cooed about his appearance and manners. The party had to "distribute proper shares of his company to everybody." They "plied him with questions without mercy." Darwin's Wedgwood cousins, Harry—a barrister—and Frank, both a decade older than Darwin, were the most intense interrogators. In reply, Darwin delivered "several little geological lectures." His cousin Elizabeth reported that "the learned world are coming fast round to Mr. Lyell." And Cousin Harry "was rather anxious to know whether we are going up or going down."

Amid the howls of laughter that certainly ensued, one can only suspect that Darwin answered, "Up, but rather slowly."

He arrived on Saturday and wasn't able to pry himself away until Wednesday morning before heading on to Shrewsbury.

◆

While on his visits Darwin missed the meeting of the Geological Society at which one "Charles Darwin, Esq., A.B., Christ's College, Cambridge" was at last elected a Fellow. But the topics discussed on the Wednesday nights of the biweekly meetings of the Society that fall would have been of great interest to him. The debate about the old problem of elevation and its causes was heating up, this time in the form of raised beaches on the coast of Great Britain. A "raised beach" is a feature that shows all the signs of being a beach, such as a notch cut into a sea cliff, stratified sand or gravel, and often seashells of modern creatures, but all this above—at least in memory or lore of local residents—the reach of waves and tides. Adam Sedgwick and Rodney Murchison had found a raised beach in Barnstaple Bay in Devonshire, and as Sedgwick wrote to a friend, "Such raised beaches! They are enough to make your mouth water. The sea has changed its relative level in some places by 30 to 40 feet, and old waterworn strands may be seen up in the cliff as plain as the nose on your face."

The meeting on December 14, at which the pair's findings were reported, ran late, so President Lyell cut the discussion short. The Fellows went into revolt. As Murchison reported to Sedgwick, who had been unable to attend, Reverend William Buckland had jumped up to state, "'Mr. President, as you have enjoined us not to discuss the memoir of Messrs. Sedgwick and M[urchison], I will only declare on this occasion that I disagree . . . with the authors and do not believe one word of it'," and then sat down "evidently in a great rage." Others apparently also wanted to discuss the paper, "Then, rose immediately one or two more persons who declared that so good a memoir should not be swamped in this way & and it was moved & carried by acclamation that the subject be resumed at the next meeting." Even Greenough, the former president of the Society, who two years before had targeted Maria Graham Callcott, "was foaming to be let out, but the meeting was adjourned."

Darwin was soon to enter the fray.

◆

Over the Christmas holidays Darwin began to prepare his own first paper for presentation to the Society on January 4. The subject would be evidence for uplift in Chile. He gave a draft to Lyell to obtain his reaction and suggestions. Lyell, of course, was thrilled, writing, "Will you come up on Monday, January 2, and come to join with us at half-past five o'clock, or come at five, and I will go over the paper before dinner?" Lyell had read the paper "with greatest pleasure, and should like to point out several passages which require explanation, and must have a word or two altered." Lyell was tickled: "The idea of the Pampas going up, at the rate of an inch in a century, while the Western Coast and Andes rise many feet and unequally, has long been a dream of mine. What a splendid field you have to write upon!"

The following Wednesday Darwin made his first appearance at the Crown and Anchor, as Lyell's guest, for the pre-meeting dinner of the Geological Society Club. Here the inner circle of the Society warmed up for the meeting. By tradition the dinner began precisely at six. That evening a smallish group of only fourteen gathered, including—in addition to Lyell and Darwin—George Bellas Greenough and Leonard Horner, Lyell's father-in-law, who would also become a fast friend of Darwin. There was much to drink. The party of fourteen consumed five bottles of port, four of sherry, and one of Madeira. Darwin likely abstained, or at most drank in moderation. He had been horrified at the effects of the excesses of alcohol on the *Beagle*'s crew and even of the workmen at The Mount in celebration of his safe return from the voyage.

The conversation at the Club was always spirited. The conversation likely included a précis of the fuss at the previous meeting about the raised beaches, and perhaps a preview of the discussion to be continued. Or maybe the discussion went back to a paper presented the evening that Darwin had been elected, which argued from the study of some hills of gravel near Dublin, "that the coast around the Bay of Dublin has been elevated, though unequally, at a comparatively recent geological epoch." At about eight o'clock, coffee and tea were handed around, and then the party adjourned for the short walk to the Society's rooms in Somerset House. Despite the butterflies that would naturally have been flying in Darwin's stomach about the presentation of his first paper, he undoubtedly derived a warm feeling of acceptance into a circle that he had long wished to join. At that moment, there was no vacancy in the Club (its membership was limited to thirty-six members), but Lyell assured him that "you stand the first of those who are knocking at the door for admission."

At the center of the Society's meeting room was a long table on which new rock specimens or fossils would usually be arrayed. At one end of the table was a podium from which the president presided over the discussion of the papers read by the secretary. Rows of benches for the Fellows and guests paralleled the two long sides of the table, a bit like the layout of the House of Commons. Maps or sketches might be affixed to the wall opposite the podium.

As the participants in the dinner strolled into the meeting room, there was no less a witness than John Ruskin, who would later become a well-known critic and thinker, but was then a mere resident of Christ College, Oxford. At that time, the young Ruskin was an acolyte of the Geological Society stalwart, Reverend Buckland, and he reported the proceedings in a letter to his father:

> I was in the meeting room of the Geological Society in Somerset House on Wednesday evening last at half-past 8 o'clock precisely. The Geologicals dropped in one by one, and it greatly strengthened me in my high opinion of the science, to phrenologize upon the bumps of the observers of the bumps of the earth. Many an overhanging brow, many a lofty forehead, bore evidence to the eminence of mind which calculates the eminences of earth; many a compressed lip and dark and thoughtful eye bore witness to the fine work within the pericraniums of their owners. One finely made, gentlemanly-looking man was very busy among the fossils which lay on the table, and shook hands with most of the members as they came in. His forehead was low and not very wide, and his eyes were small, sharp, and rather ill-natured. He took the chair, however, and Mr. Charlesworth, coming in after the business of the meeting had commenced, stealing quietly into the room, and seating himself beside me, informed me that it was Mr. Lyell. I expected a finer countenance in the great geologist. Dr. Buckland was not there, which was some disappointment to me, and some disadvantage to him, inasmuch as a ground of dispute had started in the last meeting, about the elevation or non-elevation of a beach near Barnstaple bay, in which Dr. B. had taken the non-elevation, and Dr. Sedgwick the elevation, side of the question, and the decision of which had been referred to this meeting. Both the doctors being absent, two of the members rose—Mr. Greenau [Greenough] for Dr. Buckland, and Mr. Murchison for Dr. Sedgwick, Mr. Lyell being on Sedgwick's side, though, as chairman, he took no part in the debate, which soon became amusing and interesting, and very comfortable for frosty weather,

as Mr. Murchison got warm, and Mr. Greenau witty. The warmth, however, got the better of the wit, and the question, unsupported by Dr. Buckland, was decided against him. The rest of the evening was occupied by a discussion of the same nature relative to the coast of Peru and Chili [*sic*], and I was much interested and amused, as well as instructed by the conversation of the evening. They did not break up till nearly 11.

The two papers on the coast of Peru and Chile that amused and instructed Ruskin were first, one by Darwin's friend from Chile, Alexander Caldcleugh, "Some Observations on the Elevation of the Strata on the Coast of Chili," followed by his own, "Observations of proofs of recent elevation on the coast of Chili, made during the survey of His Majesty's ship *Beagle*, commanded by Capt. Fitzroy, R.N."

Caldcleugh's paper opened with a confession that he had harbored some doubts about Maria Graham's report of the uplift in the 1822 earthquake. Caldcleugh spent several months in Chile in 1821, but returned to England later that year, before returning to Chile for good in 1829. He reported that "previously to his return to South America several circumstances induced him to suspend his opinion as to the correctness of the details which had been published of the effects of the earthquake in 1822." However, since his return to Chile, and especially since the earthquake of 1835, Caldcleugh reported "his full conviction, that there have been many distinct alterations in the relative level of the land and sea." His paper then recited accounts of rocks along the coast not previously noticed by distinguished observers, "but of which seamen are now warned," and rocks now rising higher above sea level than previously reported. Finally Caldcleugh's paper "gave an account of the effects produced by the earthquake of 1835, chiefly from the observations of Capt. Fitzroy."

President Lyell then announced that he had received the translation of an article from the Chilean newspaper "*El Ara[u]ncano*, by Don Mariano Rivero; but as none except original communications were read before the Society, he could only state that Don Mariano dissents entirely from the belief, that earthquakes have produced vertical changes of level on the coast of Chili." This communication was accompanied by a letter from Colonel John Walpole, then British Consul-General in Santiago, to Foreign Secretary Lord Palmerston "strongly support[ing] Don Mariano Rivero's views."

Next up was Darwin, who took a longer view. He reported from his visit to Navidad on the Chilean coast in 1834, "Close to the mouth of the [Río] Rapel, dead barnacles adhering to rocks three to four feet above the highest tidal level; and in the neighboring country recent marine shells are scattered abundantly to the height of 100 feet" as well as several other instances of extensive beds of recent shells above or distant from the coast. Darwin discounted the possibility that these beds of shells were the remains of the ancient dinners of native people, as he had seen in Tierra del Fuego, because "the great number of shells forming extensive, horizontal beds, whereas the heaps in Tierra del Fuego collected by the inhabitants, always retain a conical shape." They were located on headlands, not near to freshwater, and contained a "large proportional number of extremely small shells."

With respect to the 1822 earthquake, Darwin jumped in with both feet, baldly stating that "he met no intelligent person who doubted the rise of land, or with any of the lower order who doubted that the sea had fallen." His own assessment of the uplift during the 1822 earthquake was just "under three feet." He also added his observations and opinions about the marine origin of the terraces near Coquimbo that had been previously described by Captain Basil Hall.

Darwin then opined that "the coast of Chili has risen, though insensibly, since 1822," and that at the Island of Chilóe, "a change effected imperceptibly is now in progress." Citing further evidence of the terraces containing recent shells along the eastern coast of Patagonia, where "earthquakes are never experienced; and it is impossible to suppose that [the] most violent of the Chilian [sic] earthquakes could produce these effects, as the shocks are scarcely transmitted to the plains at the western foot of the Cordilleras," he concluded that "the earthquakes, volcanic eruptions, and sudden elevations on the coast line of the Pacific, ought to be considered as irregularities of action in some more widely extended phenomenon."

◆

The discussions of elevation and subsidence in various forms continued at the Geological Society through the spring and into the fall, with additional contributions from Darwin. Fossil seashells on mountaintops had puzzled geologists since Steno, but there was something much more immediate about the raised beaches— and most especially about the reports of uplift during the Chilean earthquakes of 1822 and 1835.

In February, Lyell gave the annual presidential address to the Society, including a section on "Proofs of Modern Elevation and Subsidence." He reviewed the recent findings on beds of sand and gravel containing marine shells, on raised beaches, the controversy over uplift in the 1822 earthquake, FitzRoy's and Darwin's more recent observations on the 1835 earthquake, and finally, the increasingly curious and complex picture emerging in Sweden, in which "the elevatory movement . . . diminishes in intensity as we proceed southward from the North Cape to Stockholm," then seemingly, as if crossing a hinge line, farther south, shifts to "a movement . . . in an opposite direction, and thus cause the gradual sinking of Scania [the southernmost province in Sweden]."

In March another paper on raised beaches, this one by Rev. David Williams, was read that fully supported the conclusions of Sedgwick and Murchison, and "argued that their position could not be accounted for by a subsidence in the sea level, but by unequal movement of the land." By this time Sedgwick had returned from his residence as canon in Norwich, and in the discussion that followed he rose to pair off with Greenough. Present for their exchange was Charles Bunbury, who found the formal papers quite dull: "It was not til after they were finished that there was some fun, in the shape of an animated debate between Sedgwick and Greenough; the former in his glory, and very entertaining on the subject of raised beaches, which one would hardly have thought a very favourable topic for the exercise of wit or humour." On one side, he wrote, "Greenough still sticks to his notion of the sea having sunk instead of the land being raised, but I almost think he must cling to it more from caprice or habit than conviction." And on the other, "Sedgwick made a very amusing speech, in his peculiar style of rough and ready eloquence mingled with humour, on the subject of raised beaches, which one would have hardly supposed to admit much rhetorical ornament. Greenough answered him with more humour than soundness of argument, and maintained the old Wernerian notion that the sea had sunk, instead of the land being raised. Sedgwick, in his reply, justly characterized this hypothesis as wild and unjustified."

Bunbury, who had visited in Rio de Janeiro, actually attended the meeting hoping to hear about Darwin's discoveries. In fact he met and chatted with Darwin, who had, as Bunbury wrote to his brother, discovered "a kind of tiger-cat, in the immediate neighbourhood of Rio, where I did not suppose that were any wild quadrupeds at all, except rats."

The discussion of the raised beaches continued in April with one observer noting, "Dr. Buckland will never come to the scratch on elevated beaches. He has thrice spoken very audaciously of what he was to do, but has done nothing, and I suspect he feels now that the facts are against him." As to the uplifts in Chile, the same observer opined, "We have had the subject brought on by Lyell so often that it became tiresome."

On May 3, Darwin diverged from the question of elevation to describe the deposits containing the *Toxodon* that he found in Uruguay, but on May 31, Darwin described his theory of coral reefs. He argued that certain areas "of great extent" were subsiding, while other areas were being elevated. He also pointed out that active volcanoes "all fall on the areas of elevation." Lyell was entranced: "I am very full of Darwin's new theory of Coral Islands . . . I must give up my volcanic crater theory for ever, although it costs me a pang at first, for it accounted for some much." He summarized Darwin's theory, "Let any mountain be submerged gradually, and coral grow in the sea in which it is sinking, and there will be a ring of coral, and finally only a lagoon in the center. Why? For the same reason that a barrier reef of coral grows along certain coasts, Australia, &c. Coral islands are the last efforts of drowning continents to lift their heads above water. Regions of elevation and subsidence in the oceans may be traced by the state of the coral reefs. I hope that a good abstract of this theory will be published soon."

◆

In the fall of 1836, FitzRoy and Darwin discussed their plans to write descriptions of the voyage. Darwin was encouraged by his family, particularly his sisters and his cousins, the Wedgewoods, to write his own version of the trip, rather than to allow his own prose to be "mixed up [with] Capt. FitzRoy's." After some back and forth, in December, FitzRoy—happily newly married and "snowed up" at his sister's place in Bromham, Bedford, some fifty miles north of London—wrote to Darwin proposing a three-volume plan, including the provision, "the *profits* if *any*, to be divided into three equal portions." Darwin agreed. The volumes would be in the tradition of the expedition reports, or "narratives," of FitzRoy's contemporaries and predecessors stretching back to Captain Cook. The first volume, nominally authored by Captain King—who had by then settled in Australia—would largely be assembled by FitzRoy from King's journal, and would describe the first survey expedition in 1826 to 1830, including the ships *Beagle* and the earlier *Adventure*. The second volume would be authored by FitzRoy, and would describe the second expedition from 1831 to 1836. The third volume, to be authored by Darwin, would be based on his journal, and would describe his own experiences and observations on the second expedition. This volume would be his own. His prose would not be "mixed up" in FitzRoy's volume as his family had feared. The overall title would be *Narrative of the Surveying Voyages of his Majesty's Ships* Adventure *and* Beagle, *Between the Years 1826 and 1836, Describing Their Examination of the Southern Shores of South America and the* Beagle's *Circumnavigation of the Globe.*

Darwin spent the first months of 1837 mostly in Cambridge, settling his "jolted brains into some kind of order," and "arranging [his] general collection; examining minerals, reading, & writing . . . in the evenings." He made only two quick trips to London, one of which was for the reading of his paper on the elevation of the coast of Chile. Then in March, he moved to London. Once his collections were distributed—and aside from dashing off his papers for the Geological Society—Darwin's primary task was to complete his volume of the *Narrative.*

◆

FitzRoy, meanwhile, was overwhelmed. He had hundreds of charts to check, affairs of the *Beagle* to settle, the editing of King's journal, as well as his own journal to prepare for publication. FitzRoy's disposition was like a pendulum. At one extreme he could be on top of the world with no problem too difficult, no obstacle too large, no mountain too high. Nothing could stand in his way. At the other extreme, particularly when things seemed to turn against him, he would plunge into depression. He would become almost paralyzed by doubt and enveloping despair. Perhaps the most vivid example up to then was the episode in Valparaíso, when stung by his problems with the Admiralty, he withdrew from the other officers and announced his resignation as captain of the *Beagle.*

His highs, too, could border on, if not cross the line, into mania. The scene in Valparaíso in June, 1835, less than a year after his episode of depression, presented that totally different story. After the excitement of the earthquake, the discovery of the uplifts at Santa María Island, and the successful progress of his surveys, FitzRoy learned that he had been promoted to Captain. It was if FitzRoy had undergone a personality transplant.

The HMS *Challenger* ran aground on rocks during a storm on the coast south of Concepción—in country controlled by the hostile Mapuche. Her crew, camped on the shore nearby, was threatened by their advance. FitzRoy leapt at the opportunity to provide "such assistance as [he] could render." Leaving the *Beagle* in the hands of Lieutenant Wickham, he joined a rescue mission—in important respects taking charge of it, even in the presence of a senior officer. After arriving in Talcahuano, FitzRoy proceeded day and night on horseback to and from the encampment of the besieged sailors, performing his own assessment of the situation. Then, he persuaded—or bullied—the senior officer to launch a rescue attempt by sea. After some difficulties with unfavorable weather, he spearheaded the rescue of the men—leading a party of boats to shore—despite the worries and reluctance of the senior officer that his ship, too, might run aground on the rocks.

In this manic phase FitzRoy became a hero.

FitzRoy seemed to recognize this tendency for instability in his mental equilibrium, but it was beyond his control. He knew that his uncle had committed suicide. It seems likely that part of the reason he had sought out a gentleman companion for the voyage was an attempt to thwart his tendency for depression. Did this same desire for companionship play a part in his rush to get married? We will never know for sure, but many episodes in the course of FitzRoy's life suggest symptoms that psychiatrists today associate with a manic depressive, or bipolar, personality disorder. It is clear that FitzRoy was often hurting. It is also clear that with the return of the *Beagle* to England and his marriage to Mary O'Brien, the importance of his religious faith for him, which had always been high, if a little formal, soared even higher. FitzRoy was a person in need of comfort, and it would seem that he sought that comfort in marriage and in a strong turn toward fundamentalist faith.

◆

Darwin finished his volume of the *Narrative* by the end of September 1837, and by November was checking the proofs. He began to turn his attention to planning the publication of the zoology of the *Beagle* expedition, a volume that he would edit, but would be largely composed of the contributions of others. Darwin sent the draft of an acknowledgement, either for his volume of the *Narrative* or for the zoology volume, to FitzRoy for his approval. Whether it was FitzRoy's foul mood and thin skin, or Darwin's insensitivity and growing self-importance as a member of the scientific elite, and most likely a dose of each, trouble between the two flared. Whatever it was that Darwin wrote—his draft is lost—the fat hit the fire.

FitzRoy was incensed.

No "Dearest Philos" this time. Extremely hurt, FitzRoy fired an obtuse note back to Darwin, opening, "Captain FitzRoy presents his compliments to Mr. Darwin . . . and begs to say that he . . . is recommend[ing] to refer Mr. Darwin to the opinion of some third person whose personal feelings are not involved." Instantly upon receiving FitzRoy's note, Darwin understood how deeply he had offended FitzRoy. He penned a fawning apology. At least FitzRoy's reply the following day was addressed "My dear Darwin." It seems clear that FitzRoy wanted to take Darwin down a peg or two, and let him know that any success and acclaim that he was now enjoying was traceable to the invitation, courtesy, and assistance of FitzRoy, and particularly to the assistance of the other officers on the *Beagle* who had gone out their way to help Darwin. The thrust of FitzRoy's case read:

> Most people (who know anything of the subject) are aware that your going in the *Beagle* was a consequence of my original idea and suggestion—and of my offer to give up part of my own accommodations—small as they were—to a scientific gentleman who would do justice to the opportunities so afforded.—Those persons also know how much the Officers furthered your views—and gave you the preference upon all occasions—(especially

Sulivan—Usborne—Bynoe and Stokes)—and think—with me—that a plain acknowledgment—without a word of flattery—or fulsome praise—is a slight return due from you to those who held the ladder by which you mounted to a position where your industry—enterprise—and talent could be thoroughly demonstrated—and become useful to our countrymen—and—I may truly say—to the world.

FitzRoy, too, had spoken with Lyell. That conversation added insult to injury, confirming his suspicions that Darwin's volume of the *Narrative* was not being taken in the proper context: "[Lyell] does not seem to consider that the connection of your volume with mine—and mine with Captain King's—is one of feeling and fidelity—not of *expediency*."

Darwin, stung by FitzRoy's outburst, dutifully wrote suitable and gracious acknowledgements to FitzRoy, Beaufort, and the other officers of the *Beagle* in all his subsequent works. But henceforth the relationship between the two men was polite and civil, but guarded. For despite Darwin's apology—and his share of culpability in the flare-up—he too was hurt. The intimacy between the two men, always a bit fragile, was now lost. Their relationship was changed forever.

FitzRoy had so many responsibilities, so much on his plate, that what happened next was almost inevitable. By February 1838 Darwin had completely finished his volume, no doubt breathing a sigh of relief with FitzRoy's approval: "I am happy to say that there is nothing whatever in your excellent and well-filled volume, to which I have any kind of objection to offer—therefore I trust that you will entertain no further Scruple on that Subject." But FitzRoy had only finished his work on King's volume. His own volume—which eventually grew to almost 700 pages, as well as an Appendix published in a separate volume of more than 350 pages, not to mention the maps for all three volumes—was, by FitzRoy's own admission, on a "slow coach." The three volumes plus appendix finally appeared in June of 1839.

◆

Darwin and his family and scientific friends had been waiting impatiently for the appearance of the books, a bit irked by FitzRoy's tardiness. Darwin's volume III of the *Narrative* was an instant hit. It sold the best of the three, and went into a second printing only two months later. It has never been out of print since, was repeatedly revised by Darwin, and in a wide variety of editions is now known as the *The Voyage of the Beagle*.

But when all three volumes finally saw the light of day, what really shocked Darwin—and especially his scientific friends—was the final chapter of FitzRoy's Volume II. The chapter was titled "A very few Remarks with reference to the Deluge," and in this chapter FitzRoy explained the deposits along the Río Santa Cruz by reference to Scripture.

"I suffered much anxiety in former years from a disposition to doubt, if not disbelieve, the inspired History written by Moses," FitzRoy wrote, blaming in part "those of geologists who contradict . . . the authenticity of the Scriptures." Certainly referring to Darwin and the Santa Cruz River, he continued, "One of my remarks to a friend, on crossing vast plains composed of rolled stones bedded in diluvial detritus some hundred feet in depth, was 'this could never have been effected by a forty days' flood,'—an expression plainly [in] ignorance of Scripture."

Darwin told Lyell that he thought FitzRoy's chapter was outrageous. After reading the chapter himself, Lyell told Darwin, "It beats all the other nonsense [I have] ever read on the subject." Darwin wrote his sister Caroline that she would "be amused" by the chapter and added, "Although I owe much to FitzRoy, I for many reasons, am anxious to avoid seeing much of him."

As David Quammen wrote, "Faith comforts, but data persuade."

What are called data today, for Darwin were "facts." And even as FitzRoy turned to faith, Darwin continued to gather his facts.

CHAPTER ELEVEN

The Theory Comes Together

Geology of the whole world will turn out simple.

—Charles Darwin

I n early 1838, once Darwin completed his volume of the *Narrative*, and even as he waited impatiently for the arrival of FitzRoy's "slow coach"—the completion of the Captain's volumes—Darwin focused his energies on the question of elevation. His evidence that the Andes and the South American continent were slowly rising—and that the sea floor in the regions populated by coral atolls was slowly sinking—had been well received by those of his new colleagues in the Geological Society who were open to Lyell's uniformitarian ideas. The contrary point of view, arguing for a cataclysmic origin of mountain ranges—and geologic structure in general—held by Greenough, Buckland, and others, was under pressure, if not yet in retreat among those in the Society. Even Sedgwick, not a fan of Lyell's inclination to explain geology "by reference to causes now in operation," had found his own raised beaches. But simply describing the phenomena of uplift and subsidence was not enough for Darwin. He had to know why. His overpowering curiosity demanded an explanation. What was the mechanism responsible for these ups and downs of the earth's crust?

Darwin reviewed his observations and measurements of the uplifted terraces of Patagonia and Chile, of uplift in the Andes, and of the impulsive episodes of uplift in Chile during the earthquakes in 1822 and 1835. He knew that earthquakes were frequent in Chile and on the west coast of South America, but rare or absent on the east coast. He came to believe that earthquakes were only a part of the process of uplift. Perhaps the most apparent, and certainly jolting aspect, but only a part. Much if not most of the process of uplift was gradual. The uplift in earthquakes was one vera causa of the uplift, but his idea of how the terraces along the coast of Patagonia were formed implied that a large part of the process of uplift was gradual as well.

Darwin also assembled his facts about the coral atolls and islands ringed by lagoons calmed by outlying coral reefs. Here there was little evidence for earthquakes, yet subsidence was taking place nonetheless. It also seemed clear to him, from the manner in which coral and its symbiotic partner species steadily grew at shallow depths, that the subsidence was slow, gradual, and persistent.

By the late 1830s most geologists had no problem understanding that the processes of erosion were gradual and continuing. Certainly there were years of floods, and over a person's lifetime considerable erosion could take place. But it was also true that many great U-shaped valleys seemed to be far too big to have been carved by the puny streams that now flowed through them. Many still believed that there had been periods in the past during which erosion had been powerful, rapid, even cataclysmal.

And notwithstanding Lyell's arguments, most geologists did not accept that mountain ranges were the product of some unknown gradual and continuing process. Virtually all believed that mountains were the result of paroxysm.

Geologists who had worked in or visited the Alps held this view most strongly. In the cliffs and mountainsides of the Alps they saw rock strata, contorted and twisted, fractured and broken, tortured as if in a bad dream. Rock—strong enough to hold up mountains three miles high—was twisted, stretched, and bent, almost as if it were some gigantic, partially-stirred batter of stone. What must have been the magnitude and origin of the forces required to cause this kind of deformation? Something beyond the normal experience of humans, certainly.

This was the question that Darwin posed for himself. But his new friend Charles Lyell, while egging him on, had not provided him with too much to go on.

◆

Lyell, and James Hutton before him, tended to view mountains as the remnants left after an uplifted continent had been eroded. They did not directly address the question of how the long mountain chains had actually developed. As one of Lyell's contemporaries— and critics—in the Geological Society, Henry De la Beche, wrote, "one does not exactly see how those long lines of elevation were produced, that are so common on the earth's surface, and of which the most remarkable appear to be the mountain ranges of North and South America." The origin of mountain ranges was a weak spot in Lyell's theories.

De la Beche was a talented artist—and cartoonist. He used these abilities to great advantage, both seriously and in fun. De la Beche wore glasses himself, and often portrayed the characters in his cartoons wearing colored spectacles to suggest their selective consideration of the observed facts. One cartoon—aimed clearly at Lyell and his insistence on a kind of equilibrium between uplift and subsidence—shows Father Time, wearing colored spectacles, seated on a cloud holding an old-fashioned chemical balance. In place of the two trays connected by wires to the opposite ends of the balance's beam, are the continents of Africa and the Americas, alternately rising and falling as if connected to a seesaw. Beside Father Time rests a pendulum clock, the face of which bears the legend, "Millions of Centuries." The three lines of the caption below read:

The balance of power—or how to keep the sea at its proper level:
"Here we go up, up, up

Here we go down, down, down."

Although Lyell's advocacy for finding actual causes had been well received, De la Beche's cartoon reflects one of the several aspects of his theoretical framework that was much less widely accepted. Lyell insisted on a strict absolute application of his uniformitarianism, arguing that observable "actual causes" could explain all geologic phenomena. This was perhaps nowhere more of a problem than when it came to explaining the great ranges of mountains around the globe.

◆

The volcanoes of southern Chile are among the most active in the world. The sharp, conical peak of Volcán Osorno, its summit cloaked with snow and ice, rises above the clear blue waters of Lago Llanquihue and the inland waters near Chiloé. It is one of the most beautiful, and classically shaped volcanoes on the planet, rivaling Japan's Mount Fuji. Osorno put on a dazzling display for Darwin and FitzRoy only weeks before they witnessed the effects of the devastating earthquake in 1835. In Darwin's mind's eye, he saw a clear link between the two. As he had written in his diary, "The Earthquake and Volcano are parts of the greatest phenomena to which the world is subject."

Earlier, in November 1834, after Darwin regained his physical strength following his illness in Valparaíso, and after FitzRoy had regained his mental equilibrium after his apparent breakdown, the *Beagle* turned south to complete its surveys of Chiloé Island and waters to the south. From San Carlos FitzRoy dispatched a party to survey the eastern shore of the Chiloé Island. The party, led by Bartholomew Sulivan, one of FitzRoy's key lieutenants, traveled in two small, open boats, sailing south through the maze of islands along the east coast of Chiloé, and mapping the coast as far as San Pedro Island, some 120 miles to the south. Darwin joined in.

Sulivan much enjoyed the camaraderie of fieldwork. As he planned the six-week long trip he wrote about his men: "the best

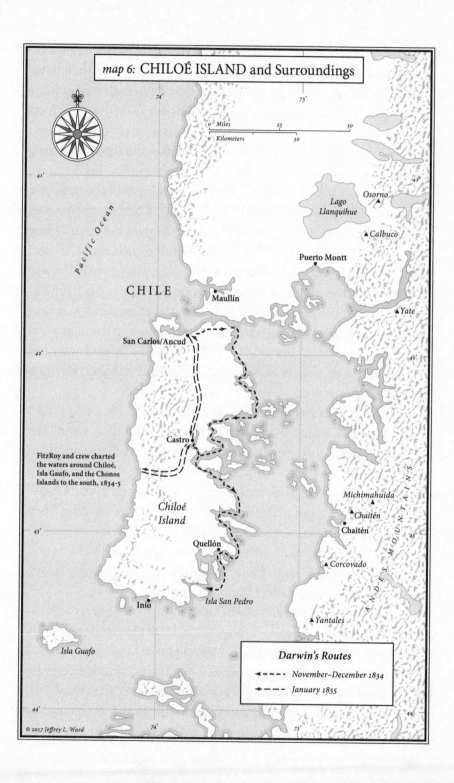

map 6: CHILOÉ ISLAND and Surroundings

Miles
0   25   50

Kilometers
0   50

Pacific Ocean

41°

CHILE

Lago Llanquihue

Osorno ▲

Calbuco ▲

Puerto Montt ●

● **Maullín**

▲ *Yate*

San Carlos/Ancud ●

42°

FitzRoy and crew charted the waters around Chiloé, Isla Gaufo, and the Chonos Islands to the south, 1834-5

Castro ●

Chiloé Island

Quellón ●

Michimahuida ▲

Chaitén ▲
Chaitén ●

43°

Corcovado ▲

Inio ●

Isla San Pedro

▲ *Yantales*

Isla Guafo

A N D E S   M O U N T A I N S

Darwin's Routes

◀--- *November–December 1834*

◀-- *January 1835*

© 2017 Jeffrey L. Ward

singers and most diverting characters in the ship are among them—and they are all of that kind, and are up to anything—we shall have, I hope, a very pleasant party." Adding that, "We shall have a large bag full of flour and raisins on purpose for a good plum-dough on Christmas day." Darwin and Sulivan—who went on to an exceedingly distinguished career in the Royal Navy—remained close long after Darwin's relationship with FitzRoy became strained.

But another member of this party would play an important role in supporting Darwin's geological theorizing. Charles Douglas was an Englishman and surveyor who had taken up residence in Chiloé and whom FitzRoy had taken on as pilot and translator. Around the campfire in the rainy evenings, Douglas shared with Darwin many of the observations he had accumulated during his time in Chiloé, about the people, their customs, and their history, as well as his observations of the activity of the volcanoes in the Cordillera just across the waters that separated Chiloé from the rainforests and mountains of the southern Andes. Darwin came to regard Douglas as a reliable observer.

Darwin was not a fan of the rain that dogged the little survey party, but on November 26 he was treated to a dazzling view. "The day rose splendidly clear. The Volcano of Osorno was spouting out volumes of smokes; this most beautiful mountain, formed like a perfect cone & white with snow, stands out in front of the Cordillera." South of Osorno, Darwin saw "another great Volcano," perhaps Volcán Calbuco, which was also emitting "from its immense crater little jets of steam or white smoke." And far to the south, he saw a third volcano, "the lofty peaked Corcobado [*sic*], well deserving of the name of 'el famoso Corcovado.'"

Only a few weeks later—after Darwin returned to San Carlos—Osorno flaunted its power. From the decks of the *Beagle* at anchor in the harbor—on the splendidly clear night on January 19, 1835—Darwin and FitzRoy were transfixed by Osorno's explosive eruption. The volcano lies about ninety miles to the northeast from their anchorage, but the view was spectacular, the fireworks resembling those for a coronation. "During this night the Volcano of Osorno was in great activity; at 12 oclock the Sentry observed something like a great star, from which it gradually increased in size till three oclock." By then most of the other officers had joined Darwin and FitzRoy on deck to watch the show. "It was a very magnificent sight; by aid of a glass, in the midst of the great red glare of light, dark objects in a constant succession might be seen to be thrown up & fall down.—The light was sufficient to cast on the water a long bright shadow."

This eruption made a powerful impression on Darwin. Later, after leaving Chiloé, after investigating the earthquake near Concepción, and after hearing reports of other erupting volcanoes—and judging his new friend Douglas to be an astute observer—Darwin wrote to Douglas back in Chiloé, requesting a report about the effects of the earthquake in Chiloé, and on the activity of the volcanoes there after his departure. Darwin, realizing that the *Beagle* was soon to embark on its passage across the Pacific and Indian Oceans, asked Douglas to write him in England. FitzRoy too seems to have been impressed with what Douglas had done for him. Besides guiding the Sulivan excursion for FitzRoy, Douglas scouted out sources of the Chilean relative of the redwoods of California, the *alerce*, upon which W.J. Hooker, the father of Darwin's friend, Joseph Hooker, later bestowed the scientific name, in FitzRoy's honor, *Fitzroya patagonica* (now *Fitzroya cupressoides*). FitzRoy also engaged Douglas to collect some more information for him about the people of Chiloé in the months after the *Beagle*'s departure from Chiloé. And Douglas used his resulting travels—"over the greatest part of the province"—to collect information about the effects of the earthquake and activity of the volcanoes for Darwin.

◆

Darwin's confidence in Douglas was not misplaced. The letter that Darwin found upon his arrival in England contained a complete and detailed report. Douglas told who felt the earthquake, and who did not. But it was Douglas's report on the activity of the volcanoes that intrigued Darwin the most. It was a missing piece to the mechanics of uplift and subsidence of the earth that he was turning over in his mind after his observations in the field and experiencing the earthquake. Along the stretch of the southern Andes extending south from Osorno to a point opposite Isla San Pedro at the southern end of Chiloé—a distance of 165 miles—there are six volcanoes rising to an elevation of 6,500 feet or more that are easily visible from Chiloé, Osorno, Calbuco, Yate, Michimahuida, Corcovado, and Yanteles. At the time of the earthquake, both Osorno and Michimahuida—both of which had been showing some activity over the previous months—kicked into high gear.

"The Volcanos," Douglas reported, "were as suddenly afected [*sic*]." Smoke curled from both Osorno and Michimahuida, and "burning stones" hurled out. But "the thundering Corcovado, showed not the least Sign of activity."

As the weeks and months passed, and as the clouds closed in, then parted again, Douglas detailed for Darwin as best he could the activity of the volcanoes. Michimahuida slowly calmed down, but the eruptions at Osorno continued, apparently reaching a peak in early December, nine months after the earthquake, when Douglas reported, "the grandest Volcanic spectacle I ever saw." He described an eruption "like an imense [*sic*] river of fire . . . an immense column of dark blue smoke, ashes & lava . . . a dense dark cloud stood high aboe [*sic*] it, and discharge forked lightning toward it." Then the eruption intensified: "I[t] appeared to be a tragic, representation of Miltons' battle of the Angels as described in *Paradise Lost.*"

Then the clouds of Chiloé moved in to obscure his view.

◆

On Wednesday March 7, 1838, the Geological Society Club gathered at the Crown and Anchor for its dinner before the biweekly meeting of the full Society. Darwin had been admitted to membership in this select inner circle the previous year as Lyell had promised, but did not join the party for dinner that night. Perhaps he was putting the finishing touches on the paper that he would read later that evening. The Reverend William Whewell—the erudite polymath, soon-to-be master of Trinity College, Cambridge, and now president of the Society—presided over the dinner. The party provided a profile of the toniest portion of the Society. Of the twenty men present, only three were born into only modest circumstance: Whewell, William Clift—the curator of the Hunterian Museum and father-in-law of the anatomist and paleontologist Robert Owen—and John Taylor. Taylor, however, who long served as the treasurer of both the Society and Society Club, had acquired great wealth as a mining engineer and investor. And while Whewell was a polymath, only Lyell, Clift, and Sedgwick's collaborator Roderick Murchison actually spent the bulk of their time engaged in the science of geology. The others, who included an earl, the Dean of Carlisle, and two baronets, all represented a considerable concentration of inherited wealth and prominence. One served in the House of Lords and another would join later. Five were serving members of the House of Commons and another would be elected a few years in the future. One of these MPs, Robert Ferguson, was well known for cuckolding Thomas Bruce—the 7th Earl of Elgin and mastermind behind the removal of the eponymous marbles from the Parthenon in Greece and their transportation to England. Like Ferguson, for whom the mineral fergusonite is named, the other aristocratic amateurs were quite serious collectors of minerals and fossils, but geology for them was not a full-time occupation.

After the dinner, the group again walked the short distance to the Society's meeting room in Somerset House to join the other Fellows and guests. Among the other Fellows were two who were also leading the trend toward the professionalization of geology, De la Beche and John Phillips. While De la Beche shared the wealthy origins of most of the members of the dinner party—he inherited a sugar plantation in Jamaica complete with a complement of slaves— his interests were dominated by geology. In 1835 he founded the Ordnance Geological Survey, the first of the national geological surveys, and later founded the Museum of Practical Geology, which together would be the focus of the remaining two decades of his life. (This turned out to be a good thing too, because the sugar business in Jamaica began to go rapidly downhill with the abolition of slavery throughout most of the British Empire, including Jamaica, in 1834.)

John Phillips was also a person who made his living at science. He was the nephew of William "Strata" Smith, maker of the first geologic map of England. Although coming from very modest circumstances, he became an acknowledged authority on fossils at the knee of his uncle. This led to his appointment as curator of a museum in York, as a professor at King's College, and to his authorship of a two-volume work, then in progress, entitled *A Treatise on Geology.*

The meeting opened with the election to Fellowship of a wealthy landowner and businessman in Dublin. Then a notice was read about some "remarkable dikes" exposed in a rocky platform at low water on the coast of Scotland.

For the main event, Darwin took the floor to read his paper, "On the connexion of certain volcanic phænomena, and on the formation of mountain-chains and volcanos, as the effects of continental elevations." As Lyell described the scene, Darwin "opened upon De la Beche, Phillips & others . . . his whole battery of the earthquakes and volcanos of the Andes."

Darwin began with a historical review of his facts. He discussed the observations of the Concepción earthquake of 1835 and the activity of the volcanoes near Chilóe, citing his communication with Charles Douglas. He summarized reports from Juan Fernandez (or Robinson Crusoe) Island, some 360 miles off the coast. The island "seems to have been more violently shaken than the opposite shore of the mainland." Also near the island, "at the same time a submarine volcano burst fourth" and "continued in action during the day and part of the following night."

"This fact," Darwin noted, "possesses a peculiar interest, inasmuch as during the earthquake of 1751, which utterly overthrew

Concepción, this island was likewise affected in a remarkable manner."

From the report by Douglas, it seemed "evident that the volcanic chain . . . was affected not only at the moment of the great shock of February 20, 1835, but remained in a very unusual activity during many subsequent months." The bottom line for Darwin was, "We see, therefore, that, in 1835,—the earthquake of Chilóe—the activity of the train of neighboring volcanos,—the elevation of the land around Concepción,—the submarine eruption at Juan Fernandez, took place simultaneously, and were parts of *one and the same* great phenomenon."

Next, he argued that the "force which elevates Continents" is the same as "that which causes volcanic outbursts." When considering the simultaneity of the earthquake and the initiation of volcanic activity, he continued, "the idea of water splashing up through holes in the ice of a frozen pool, when a person stamps on the surface, come irresistibly before my mind." He then cited calculations that the "earth's crust is not much more, and perhaps less than twenty miles in thickness; and if this be so, the crust may, indeed, be well compared with a thin sheet of ice over a frozen pool." From this he concluded that, "The facts appear to me clearly to indicate some slow, but in its effect great, change in the form of the surface of the fluid on which the land rests."

He described other "periods of increased Volcanic Action affecting large Areas," citing examples published by Humboldt in his Personal Narrative (one of which included what we now call the New Madrid earthquakes in the Mississippi Valley of the United States), and added his own example covering the years 1834 and 1835 within the Caribbean, Central, and South America. Critical for Darwin was the coincidence in time of the volcanic eruptions with the 1835 earthquake. He briefly considered, but rejected, the possibility that the nearly simultaneous occurrence of these events might be no more than a random coincidence. He was particularly impressed by the reports of an enormous eruption at Coseguina volcano in Nicaragua, the very same day as the eruption of Osorno that he and FitzRoy and watched from the deck of the *Beagle* in the harbor at San Carlos. He concluded that the "subterranean forces" remained dormant for a long period, "then bursting . . . forth throughout considerable districts with renewed vigour."

Darwin's interpretation was that "It cannot be otherwise than difficult to trace their precise origin," but he was led "to one conclusion alone—namely, that they [the earthquakes] are caused by the injection of liquified rock between masses of strata." To back up this conclusion he lists five points that "we may . . . fairly conclude, with regard the earthquakes on the west coast of South America."

Darwin's five points included, first, the primary earthquake is caused by "a violent rending of the strata," typically beneath the sea. Second, minor fractures extend upward, but reach the ocean bottom only in the "comparatively rare case of a submarine eruption." Third, the fissured area extends parallel to the nearby coastal mountains. Fourth, when "the earthquake is accompanied by an elevation of the land in mass, there is some additional cause of disturbance." And finally, "the earthquake . . . relieves the subterranean force, in the same manner as an eruption through an ordinary volcano.

Next, Darwin took aim at the catastrophist views held by Greenough—and many others—that mountain ranges were formed as a consequence of some paroxysmal convulsion in the distant geologic past. Marshaling his facts, he countered, "Therefore, we must conclude that continental elevations, one of the effects of the same motive power which keeps the volcano in action, has ordinarily gone on, since those ancient days, at the same slow rate as at present." He questioned whether there was any proof of "paroxysmal violence," stating that "the important fact which appears to me proved, is, that there is a power now in action, and which has been in action with the same average intensity (volcanic eruptions being the index) since the remotest periods, not only sufficient to produce, but which almost inevitably must have produced, unequal elevations on the lines of fracture."

He then buttressed his conclusion that "mountain-chains are formed by a long succession of small movements" with the "theoretical reasoning" of William Hopkins—a Cambridge mathematician who pioneered the discipline of mathematical geology and who would later gain fame as the tutor of the famous mathematicians and physicists William Thomson (Lord Kelvin), G. G. Stokes, and James Clerk Maxwell. Citing Hopkins's hypothetical illustrations of the fracturing caused by an uplift and swelling from beneath, he argued that observations of wedge-shaped masses of injected rock—like those he had seen in the Andes—argued for a very gradual process. Otherwise, he asked, how would it be possible "without the very bowels of the earth gushing out?"

And finally as he wrapped up his remarks, he argued that considering the vast area of South America over which the "volcanic forces either now are, or recently have been, in action . . . we shall be deeply impressed with the grandeur of the one motive force, which causing the elevation of the continent, has produced, as secondary effects, mountain chains and volcanos." It appeared to Darwin that "there is little hazard in assuming that this large portion of the earth's crust floats . . . on a sea of molten rock." And extending his conclusion still further, "When we reflect how many and wide areas in all parts of the world are certainly known, some to have been rising and others sinking . . . even to the present day," here also referring obliquely to his evidence for the ongoing subsidence of the coral islands, "and do not forget the intimate connexion which has been shown to exist between these movements and the propulsion of liquified rock to the surface in the volcano;—we are urged to include the entire globe in the foregoing hypothesis."

With a nod to the unknown, he concluded, "The furthest generalization, which the consideration of the volcanic phenomena described in this paper appears to lead to, is, that the configuration of the fluid surface of the earth's nucleus is subject to some change,—its cause completely unknown,—its action slow, intermittent, but irresistible."

In short, for Darwin, earthquakes and volcanoes were part of the same process. The crust of the earth floats on a fluid substrate of molten rock. For unknown reasons, the level of this fluid changes slowly and intermittently. When the level of the fluid is rising (presumably because of increased pressure,) it sometimes leads to a rupture, an earthquake, in the crust above. If the rupture reaches the surface, lava erupts in a volcano. Sometimes the molten rock is simply injected into the crust. The whole process leads to the uplift of the surrounding region.

◆

Following Darwin's paper, and as Lyell described it, had "done his description of the reiterated strokes of his volcanic pump," De la Beche began the discussion with "a long oration about the impossibility of strata in the Alps &c., remaining flexible for such a time as they must have done, if they were to be tilted, convoluted, or overturned by gradual or small shoves." De la Beche believed that the deformation must have occurred suddenly and catastrophically.

The discussion then turned to the Lyellian approach to geology more generally, as Phillips rose and "pronounced a panegyric upon the 'Principles of Geology.'" Phillips predicted that the dispute between those who believed in the greater intensity of the "pristine forces"—as contrasted with those like Lyell and Darwin who argued for the doctrine of the true cause and that past geologic action occurred "with the same intensity as nature now employs"—would divide geologists for centuries. William Henry Fitton, an Irish physician and one of the early and long active members of the Society, then teased Phillips for the "warmth of his eulogy," accusing him and others of Lyell's proponents of being "in the habit of admiring and quarreling with him every day, as one might do with a sister or cousin whom one would only kiss and embrace fervently after a long absence." Fitton went on, positing that this was the case with Phillips, "coming up occasionally from the provinces." Fitton then "finished his drollery by charging [Lyell] with not having done justice to Hutton, who he said was for gradual elevation." Lyell replied that, "most of the critics had attacked [him] for overrating Hutton."

President Whewell concluded the discussion saying that "we ought not to try and make out what Hutton would have taught and thought, if he had known the facts which we now know."

The discussion, while amusing, did not dwell on the specifics of Darwin's theory. Instead, it provided yet another opportunity for a skirmish between the proponents and opponents of Lyell's natural, true causes.

◆

After the meeting Darwin must have felt a bit discouraged by the tone of the discussion. At dinner with the Lyells a few days later he apparently confided that the comments of De la Beche and others constituted "a vigorous defiance" of his arguments. Lyell did his best to cheer him up. Darwin was not "able to measure the change of tone in the last 4 years." Rather than defiance, the reactions of the critics were "a diminishing fire & an almost beating of a retreat," Lyell wrote. "I restored him to an opinion of the glowing progress of the true cause."

Only a week after his paper was read at the meeting, Darwin sent it in to William Lonsdale, the curator and librarian of the Society, for publication in the Society's *Transactions*. He had spent the weekend in Cambridge and seen Sedgwick, who he knew would be the referee. "I found Sedgwick very well & in high spirits, & therefore the sooner my paper goes to him the better," he wrote.

Sedgwick supported the publication of the paper, but—in his inimitable crusty manner—challenged the author to shorten the historical part and to sharpen his arguments in the theoretical part: "The concluding or theoretical part is not all clearly brought out, & might be reconsidered by the author with some advantage: Not with any view of altering his theoretical opinions (for he only is responsible for them) but for the purpose of making them more definite & unequivocal." Nonetheless, he was quite happy with Darwin's facts, concluding, "The main facts on which the paper hinges & the immediate deductions deduced from them appear incontrovertible."

Between the time of presentation and that of publication, Darwin also made subtle, but to him important changes to the paper's title. He shifted from identifying continental elevation as the agent responsible, to including continental elevation together with mountain chains and volcanoes as effects of the "same power."

◆

From Darwin's day on the beach at St. Jago six years before, when he saw the uplifted bed of shells containing young species of shells—and observed the depression of that same bed beneath the volcanic crater of Signal Hill—he had assembled his facts, and from them he built by induction a chain of inference leading to his theory that the uplift of continents—and by implication, the subsidence of the ocean basins—was related to the movement of molten material beneath the earth's crust. From the terraces in Patagonia he concluded that the South American continent was uplifting, mostly steadily with occasional pauses, and mostly without earthquakes. In Chile and on the western side of the continent, the uplift occurred more rapidly, and in the Andes obviously to greater heights. Here earthquakes were more frequent, accompanied by increments of uplift, but here too he saw evidence that much of the uplift occurred quietly, without earthquakes. In the Andes he saw contorted and broken strata, frequently intruded with volcanic rocks. In Chile, he perceived a correlation between large earthquakes and the activity of volcanoes. He didn't yet have all the details, but it was clear to him that these phenomena were all connected through the movement of molten material at not too great a depth in the earth. He had also argued convincingly that the coral islands and islands surrounded by barrier reefs had been subsiding for some time. And this in regions of the oceans where earthquakes were rare. In explicating his theory, perhaps he was hesitant to be more direct in tying the subsidence beneath these coral islands to this same cause without more observations. Nonetheless, he had assembled, if not quite completed, a theoretical framework demonstrating and explaining the elevation of the continents, especially the mountain ranges, and the subsidence of the ocean basins.

A few months later he would write to Lyell reaffirming his conclusion: "I cannot doubt the molten matter beneath the earths crust possesses a high degree of fluidity, almost like the sea beneath the Polar ice."

Was Darwin the first to articulate this view? Darwin used the analogy of ice over a frozen pond in his diary to describe the earth when he felt the earthquake in Valdivia in February 1835. His notes after visiting St. Jago in 1832, and his correspondence with Charles Douglas after the earthquake in 1835, show that he was thinking in this direction. So it was clear that his head was filled with these thoughts at least three, if not six, years earlier. He summarized these concepts and their relation with elevation in his volume of the *Narrative*, sent to the publisher in September, 1837. But, as with the question of evolution two decades later, this idea too was in the air. John Herschel, the English astronomer and polymath, spent the years 1834 to 1838 in South Africa to study the stars of the southern sky. Herschel too was interested in geology. In February, 1836, he wrote to Charles Lyell, describing "a bit of theory." Had it ever occurred to Lyell, he wrote, supposing that the whole of the earth's crust "to float on a sea of lava," or even better "a mixture of fixed rock, liquid lava, and other masses, in various degrees of viscidity and mobility?" Darwin and Herschel met in Cape Town in June, 1836, as the *Beagle* rounded Africa on the homeward part of its voyage. Given their shared interests, it seems likely that the two discussed these ideas.

Charles Babbage also originally explained his observations of the ups and downs of the Temple of Serapis to the Geological Society at a meeting in March, 1834 in terms of thermal expansion and contraction of the earth beneath the temple, but withdrew his paper before it was published. Babbage included Herschel's letter to Lyell as a note in the second edition of his *Ninth Bridgewater Treatise*, published in 1838. Darwin was in frequent contact with Babbage, and of course Lyell himself, after his arrival back in England, so by 1838 it seems highly likely that Darwin knew that he, and at least Herschel and Babbage, were thinking along similar lines. Years later, in 1847 when Babbage finally privately published his paper on the Temple of Serapis, he added a note saying that Darwin's publications in 1838 arrived at the same set of conclusions as he had in his original 1834 paper, but close reading of Babbage's paper suggests that this is not crystal clear.

In any case, the difference was that Darwin had the data. Darwin had collected abundant "facts" to prove the uplift of South America and the subsidence of the coral reefs. He had specific examples at St. Jago and in the Andes where volcanic activity was related to vertical movements of the crust. And he had what he believed was persuasive data on the temporal correlation of earthquakes and the eruption of volcanoes.

If events had transpired differently, Darwin's theoretical musings about the connection between earthquakes, volcanoes, the movement of subterranean magma, the uplift of the continents, the sinking of the ocean basins, and the building of mountain ranges—combined with his abundant data—might have come to be called "Darwin's theory of elevation." But that's not the way things turned out.

CHAPTER TWELVE

Extending the Theory:
The Parallel Roads of Glen Roy

*The results to which I have arrived, if proved, are
of so much greater geological importance than the
mere explaining of the origin of the roads.*

—Charles Darwin

George Bellas Greenough may come across as an intimidating, if well-spoken, bully, but—out in the field—he was a sharp-eyed geologist. In 1805, two years before he and twelve others founded the Geological Society at the Freemasons Tavern on Great Queen Street, the man who later tormented William Smith, impugned the veracity of Maria Graham, and challenged Charles Lyell's uniformitarianism, set out on a geological tour of Scotland. The aim of Greenough's trip was to test the prevailing ideas about the origin of granite and basalt. Had these rocks once been molten, as argued by Hutton and Playfair, or were they originally sediments that were later compressed and cooked at depth in the earth, as argued by Werner?

But along the way he also investigated a local Highland curiosity, one that three decades later would fascinate Charles Darwin, the Parallel Roads of Glen Roy.

In the western Highlands of Scotland—in a region known as Lochaber, about nine miles northeast of Ben Nevis, the highest peak in Great Britain—lie two glens, or valleys. These glens, Glen Spean and Glen Roy, have been scrapped and battled over for centuries by the McDonalds and the Mackintoshes. The Spean River flows westward, emptying into the Great Glen of Scotland, and drains Glen Spean and Glen Roy, which branches off to the north.

The Great Glen and its chain of lochs cuts diagonally across the Highlands of Scotland, from Fort William to Inverness, as if cut by the knife of some mystical giant. Indeed, it was a giant of Gaelic mythology, Fingal, that stuck in Greenough's craw. Of the Parallel Roads, Greenough wrote, "As everything that is wonderful is attributed to Fingal they are distinguished among the common people by the name Fingal's roads." Through time the Parallel Roads have been variously referred to as lines, shelves, ridges, terraces, and grooves. As Greenough described them, "On the sides of the opposite mountains at heights precisely equal are grooves or ridges exactly parallel to the horizon and extending twenty miles. In Glen Roy, where they are most distinct, three several ridges run at different elevations, each having a corresponding ridge on the opposite side."

In the 1770s the Highlands of Scotland were still a wild and unknown land to most residents of the British Isles south of Hadrian's Wall. The Welsh gentleman and naturalist, Thomas Pennant, an early popularizer of tourism in Scotland, also alluded to the Gaelic legends attributing the Parallel Roads to Fingal. In his widely popular travelogue, *A Tour in Scotland*, Pennant described perhaps the most startling characteristic of the Roads: "They are carried along the sides of the Glen with the utmost regularity, nearly as exact as drawn with a line of rule and compass."

But Pennant's curiosity about the origin of the Roads went unsatisfied: "I cannot learn to what nation the inhabitants of the country attribute these roads." There were no causeways as might have been expected from something from the Roman times, nor were there "traces of buildings, or Druidical remains that could lead us to suspect that they were designed for economical or religious purposes." But then what can have been the purpose of these roads? Pennant tells us, "The country people think that they were designed for the cha[s]e, in order to tempt the animals into the open paths after they were rouzed, in order that they might come within reach of the bowmen."

Greenough scoffed at this idea. He rejected the theory that the Parallel Roads had been constructed by any man, mythical or physical, as "as false and absurd." Sarcasm oozed across the page as he wrote, "It is really wonderful that any man should . . . attempt to palm upon the world so glaring an absurdity. . . . let us humbly attribute to the hand of nature a work too mighty for man." Indeed, Greenough, the geologist, was the first to suggest that the Roads were shorelines of a former "immense lake."

◆

In June 1838, a year and a half after he returned home from the *Beagle*, Darwin was itching to get back to the field. He had made strong cases to the Geological Society about the elevation of South America and the subsidence of the coral atolls and islands. He had presented a theory for how this happened. But could he prove that his very own island, Great Britain, was being uplifted as well? To do so, he was even willing to risk another voyage at sea.

Darwin knew from the papers of Sedgwick and others about former beaches along the coast that were raised a few tens of feet above the contemporary strand line. Darwin vividly remembered the flights of terraces rising above the eastern coast of Patagonia, and even more spectacularly along the coast of Chile, near Coquimbo. He was convinced that these terraces and beaches were traces left as the South American continent was raised, relatively recently, from the sea. If the Parallel Roads were not the shorelines of some ancient lake, but instead ancient shorelines of the sea, he could make the elevations suggested by Sedgwick's raised beaches seem like small potatoes.

In 1817 John MacCulloch and Sir Thomas Lauder Dick independently investigated the Parallel Roads and both produced splendid papers, buoyed by spectacular maps and drawings of the Parallel Roads. Darwin studied these two papers intently. Both MacCulloch and Lauder Dick concluded that the Parallel Roads were the shorelines of ancient lakes dammed by some kind of barriers at the mouths of the glens. But their explanations had a serious flaw. There was no trace, whatsoever, of any kind of barrier that might have dammed the lakes.

Darwin had a simple answer. Imagine the islands and bays surrounding Tierra del Fuego after they had risen from the sea. After the uplift they would appear just like the mountains and glens of the Highlands. The Beagle Channel separating Tierra del Fuego from the islands to the south, when uplifted, would resemble the Great Glen.

MacCulloch and Lauder Dick both described the Parallel Roads at length. The most extensive roads actually existed at four levels. The highest road lined only the sides of Glen Gluoy, a smaller glen, adjacent to Glen Roy that opens to the Great Glen. It was separated from Glen Roy by a broad saddle at about the same level as the highest road. The next highest road, the highest in Glen Roy itself, as confirmed by a surveyor friend of Lauder Dick, was at a level twelve feet below the road in Glen Gluoy. Eighty feet below that was the intermediate road in Glen Roy. Both these roads were largely confined to Glen Roy and smaller tributary glens. In contrast, the lowest road, two hundred feet below the intermediate Road, not only lined the hills on either side of Glen Roy, but emerged from Glen Roy and ran along the sides of the Glen Spean and another tributary glen, Glen Treig. The surveys also showed that "all these different shelves are found to maintain the horizontality characterizing the surface of water, throughout all the various windings of their linear extent, and round the hollows and projections of the hills . . . and each . . . on one side of any of the glens is exactly the same level with that corresponding to it on the opposite side."

The elevation of the highest of these roads was not a mere few tens of feet above the level of the sea, like Sedgwick's raised beach, but at nearly thirteen hundred feet above sea level. If Darwin could show that the Parallel Roads were indeed ancient sea beaches, it would be spectacular evidence in support of his ideas about recent elevation and subsidence—not in faraway Patagonia or Chile, not in the distant Keeling Islands in the middle of the Indian Ocean—but right at home on the island of Great Britain.

◆

On Saturday, June 23, 1838, Darwin traveled from his rooms at 36 Great Marlborough Street through the streets of London to the docks, likely St. Katharine Dock, on the River Thames. Completion of the railroad from London to Edinburgh was still almost a decade away. There he boarded a packet steamer, perhaps the *Monarch*, operated by the General Steam Navigation Company. Any of the packet steamers would have been a far cry from the tiny cramped confines of the *Beagle*, but the *Monarch* was something special: "Descending a few steps by a commodious companion [way] you reach the entrance to various cabins, a pair of folding mahogany doors opening into the saloon, which for light, dryness and comfort is superior to all we have seen. . . . The floor is entirely covered (as are the chief cabins) with Brussels carpet." Darwin's steamer puffed down the River Thames, then up the east coast of England, to the Scottish port of Leith, serving Edinburgh, and located on his old stomping ground, the Firth of Forth.

Darwin's steam packet was much larger and more stable than the tiny, bobbing, *Beagle*. He avoided the seasickness that he had come to hate on the *Beagle*, and confessed with embarrassment, "My trip in the steam packet was absolutely pleasant, & I enjoyed the spectacle, wretch that I am, of two ladies & some small children quite sea sick, I being well."

For the trip Darwin purchased a new high tech notebook, a De la Rue & Co.'s Improved Metallic Memorandum Book. Writing in the book was done with a special stylus, and according to the manufacturer, "The point of the Pencil not being liable to break, and the writing being permanent, they will be of great advantage to Commercial Gentleman, Short-hand Writers, &c." Probably on the steam packet, Darwin jotted a few notes on the subject of the species, the subject that was beginning to consume him. But at his first geologic stop, the Salisbury Crags, steep cliffs of columnar basalt about a mile east of Edinburgh Castle, he began the geologic notes and sketches in his new book. From Edinburgh, he bounced across Scotland, "in gigs & carts, (& carts without springs as I never shall forget)."

On Saturday, June 30, a week after leaving London, he finally reached his goal. About nine miles beyond Fort William, he entered the valley of Glen Spean and the province of the Parallel Roads. His senses sharpened. His notes and sketches became more detailed. He turned up the river and examined the deposits along it, and on the slopes above. He was impressed that the deposits sloped down into the river valley, suggesting to him deposition from the sea. Finally he reached the little settlement around the bridge crossing the River Roy now known as Roybridge, and the entrance to Glen Roy itself. Just to the east, he scrambled up a steep hill, Meall Doire, leaving the forest of the valley for the open heather and rocky slopes above.

About two-thirds of the way up the hill, Darwin first set foot on one of the Parallel Roads, a shelf, several feet wide, worn into the gneiss, overlooking Glen Spean, at an elevation of about 850 feet, exactly following the contour of the hill. It immediately impressed him as a shoreline, a "beach & channel precisely as with Isld—", he jotted in his new notebook. But had this ancient island been surrounded by a lake, or by the sea? He noted too a boulder of granite—an erratic—that somehow found its way to this hilltop of gneiss, reminding him of the erratics he had seen near the Río Santa Cruz in Patagonia.

From the top of Meall Doire, Glen Roy spread before him. To the left, across the River Roy, he saw the slopes of Bohuntine Hill,

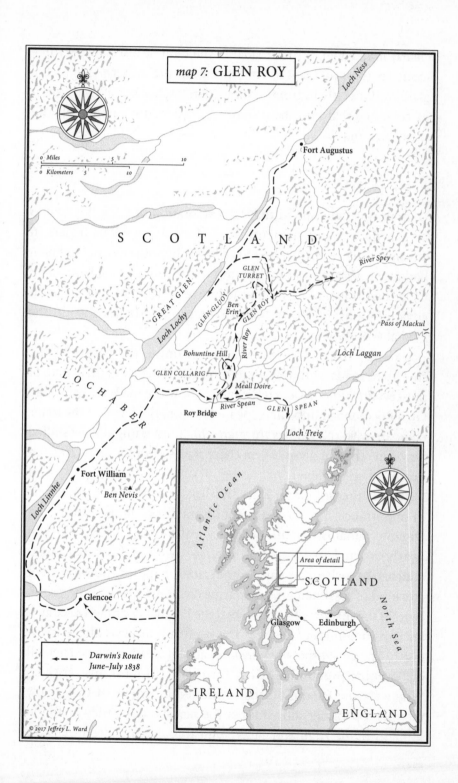

map 7: GLEN ROY

Loch Ness

Fort Augustus

S C O T L A N D

River Spey

GLEN
TURRET

GREAT GLEN

GLEN GLUOY

Ben
Erin

GLEN ROY

Pass of Mackul

Loch Lochy

River Roy

Loch Laggan

Bohuntine Hill

GLEN COLLARIG

Meall Doire

L O C H A B E R

River Spean

GLEN SPEAN

Roy Bridge

Loch Treig

Loch Linnhe

Fort William

Ben Nevis

Glencoe

- - - → Darwin's Route
June–July 1838

0 Miles 5 10
0 Kilometers 5 10

Atlantic Ocean

Area of detail

SCOTLAND

North Sea

Glasgow Edinburgh

IRELAND

ENGLAND

© 2017 Jeffrey L. Ward

clearly marked with three roads. To the right he also saw the three roads, at the same levels, contouring up the eastern side of the glen. And directly in front of him, Glen Roy with the River Roy at its center curved into the distance. He opened his notebook and sketched a map.

◆

Darwin was determined that his exploration of Glen Roy would be expeditious and efficient. He read all the previous papers in advance, and he already had his favored hypothesis, that the glen had recently risen from the sea, just as the coasts of Patagonia and Chile had done. Now he had to prove it. He had prepared—in advance—a list of fifteen "Chief Points to be Attended to" in the field at Glen Roy.

Among his tasks, he planned to examine the nature of the shelves or roads, with respect to the underlying rock, their state of preservation, and the circumstance of how the roads ended where Glen Roy opened into Glen Spean. How did the levels of the roads compare with the saddles through which water might have spilled between adjacent glens? Were there traces of other roads in Glen Roy or adjacent glens?

Two of the points on Darwin's list loomed above the others in importance. The second point on the list could be pivotal: "2. Organic remains. Balani. Serpula.—calcareous matter." Darwin dearly wanted to find some shells—weathered old barnacles attached to rock, or perhaps, some traces of marine tubeworms, or any kind of fossil remains of marine creatures. These could prove, beyond the shadow of a doubt, that the Parallel Roads were actually ancient shorelines of the sea. The twelfth point was also critical: "12th. The great problem. why lines absent in other parts. The Hill of Bountine [Bohuntine] Glen Turrit [Turret] must answer this." Why were the Parallel Roads so limited in extent? If they really represented former shorelines of the sea, shouldn't they be found in many of the Highland glens? A great problem, indeed.

For the next six days, Darwin tramped the ridges and valleys of Glen Roy and its surroundings. He carefully examined the roads on the slopes of Bohuntine Hill and the adjacent, smaller Glen Collarig, paying particular attention to the locations where the upper two roads disappeared. Near the upper roads he found deposits of well-rounded pebbles that told him he was walking on a former beach.

He found many erratic boulders, and he examined how the roads entered the ravines coming down the sides of hills. He puzzled about why the roads were present in some places and not others, and jotted in his notebook, "Even on Lauder Dicks Hypothesis [of the roads being the shorelines of a lake] impossible to explain absence of lines in certain parts."

He explored south up a tributary of the Spean River following the lowest shelf into Loch Treig where Sir Lauder Dick drew a spectacular sketch of this shelf above the outlet to the Loch. At the narrow entrance to the Loch, Darwin was duly impressed: "the gneiss is worn into smooth concave hollows, the peculiar curves of which . . . may be readily imagined by calling to mind the form of rocks washed by a water-fall." Here he concluded, "the water had remained . . . for a very long period."

"Standing on the precipitous and waterworn rocks," he wrote," it required little imagination to go back to former ages, and to behold the water eddying and splashing against the steep rocks on one side of the channel, whilst on the other it was flowing quietly over a shelving spit of sand and gravel."

Darwin used his mountain barometer to measure the heights of the shelves and of the erratic boulders that he found, some on the ridge tops, well above the highest shelf. Everywhere he made notes and sketches. He crossed the ridge of Ben Erin into Glen Guoy, made measurements on the shelf there, and examined carefully the gentle saddle that leads from Glen Guoy back toward Glen Roy. Here and at the northeast end of Glen Roy where another saddle separates it from a major drainage leading to the northeast, he studied the shelves, the deposits of gravel, measured the elevations, and found many erratics. After the first few days he felt strongly confirmed in his opinion that the shelves had a marine origin.

Late in the week, on a traverse to the northwest from Glen Roy toward the Great Glen, Darwin surprised himself, finding a new road, obscured in places by peat moss. He made several measurements with his barometer on his new road, and on terraces he found in the stream leading down into the Great Glen, but here the terraces were "so complicated that nothing can be made of them."

On his last day of fieldwork, he followed the Great Glen to the northeast, to Fort Augustus and Loch Ness. Here, along the Caledonian Canal, he was delighted to meet a lockkeeper, who, during the excavations for the locks, had seen for himself "alternating layers of coarse & fine & many Sea Shells" about sixty feet above sea level. These shells were not high in the glens, but they were something.

◆

On his return to London Darwin took a packet steamer down the west coast from Glasgow to Liverpool—this time exalting without seasickness over the miseries of "some full grown men"—then made a stopover in Shrewsbury. He took some time away from his family to poke in some sand pits where—with difficulty—he found some bits of shell. Although he didn't find any shells on the Parallel Roads, at least in Shropshire he found some evidence that argued for the land rising from the sea.

Shortly after returning to London, he wrote Lyell, "enjoyed five days of the most beautiful weather, with gorgeous sunsets, & all nature looking so happy, as I felt.—I wandered over the mountains in all directions & examined that most extraordinary district." Always enthusiastic about his most recent discovery, he wrote, "I think without exception,—not even the first volcanic island, the first elevated beach, or the passage of the Cordillera, was so interesting to me, as this week." It was "by far the most remarkable area I ever examined." And most importantly, for his theories about elevation, he became "fully convinced . . . (after some doubting at first) that the shelves are sea-beaches," indicating that this part of the Scottish Highlands had risen recently for the sea. "Glen Roy has astonished me."

◆

As Darwin set to work writing his paper on his findings at Glen Roy, he took advantage of a new resource. Thanks to Lyell and probably an influential acquaintance of his father, Darwin had gained admittance to the storied Athenaeum Club. There he had at his disposal the latest philosophical magazines, as well as a facility for comfortable contemplation and writing. The club had been founded in 1824. Gentlemen's clubs had become the rage in London, and the massive, neoclassical Athenaeum Club—with its portico guarded by four pairs of Doric columns and a statue of Athena, the goddess of wisdom, solemnly reigning above—appealed to the very uppermost class. But it also admitted the crème de la crème of the public intellectuals of the day. Lyell, Buckland, Whewell, and Owen were all members. In 1838 the club had some financial difficulties, and accepted a group of members off the waiting list in excess of its normal limitation. This group, including Darwin and Charles Dickens, came to be known as the "forty thieves."

From Darwin's rooms on Great Marlborough Street to the Athenaeum Club was about a fifteen minute walk. Coincidentally a portion of his likely route—through Golden Square—was described by his fellow recent inductee, Dickens, in his novel serialized that same year, *Nicholas Nickleby*. Although as Dickens wrote, "Golden Square . . . is not exactly in anybody's way to or from anywhere," it was precisely on Darwin's most direct walking route from Great Marlborough Street to the Athenaeum Club at the corner of Pall Mall and Waterloo Place. "It is," Dickens continued, "one of the squares that have been; a quarter of the town that has gone down in the world, and taken to letting lodgings. . . . It is a great resort of foreigners." In fact, it likely reminded Darwin of the cities he explored in South America. It was August as Darwin began to enjoy his new advantages at the Athenaeum, and as Dickens would have us imagine the scene, "On a summer's night, windows are thrown open, and groups of swarthy mustachio'd men are seen by the passer-by lounging at the casements, and smoking fearfully. Sounds of gruff voices practicing vocal music invade the evening's silence, and the fumes of choice tobacco scent the air."

Entering through the vestibule of the Athenaeum into the great hall with more columns and a grand staircase leading upstairs to the library, he needn't have worried about the foreign denizens of Golden Square. Here Darwin met his new friends from the Geological Society. On entering the door the very first night he dined there, he met William Henry Fitton, one of the Society's former presidents, who "got together quite a party."

For Darwin, the social aspects of the club were both a blessing and a curse, but the chance to visit the library, to read the latest magazines, to write at one of the desks in the Writing Room, or just to sit on one the couches and think, were an unalloyed joy. "For I am sure the first evening I sat in that great drawing room, all on a sofa by myself, I felt just like a duke," he wrote to Lyell, thanking him. "Your helping me into the Athenaeum has not been thrown away, & I enjoy it the more, because I fully expected to detest it."

❖

As Darwin put pen to paper in writing about Glen Roy, it was clear from the beginning that the paper was about much more than "the mere explaining of the origin of the roads." His results, "if proved, are of so much greater geological importance." They would provide evidence for the continuing elevation of the Scottish Highlands.

The body of the paper began with a statement that "It is admitted by everyone, that no other cause, except water acting for some period on the steep side of the mountains, could have traced these." He showed a sketch explaining how the shelves making up the roads resulted from the combination of erosion into rock on the uphill side, and the deposition of "matter in the form of a mound" on the downhill side. He pointed out the fans where the bottom of the valley rose almost to the road, or where a stream brought detritus from above "to the ancient beaches."

He described how, with the exception of the intermediate road in Glen Roy, the other roads occurred at the same levels as the saddles that would allow water to drain from one glen to another; the highest road in Glen Gluoy, at the level of the saddle to Glen Turret and Glen Roy; the highest road in Glen Roy at the level of the water gap leading to the Glen Spey and the River Spey; and the lowest road at the level of a saddle at the upper end of Glen Spean, at the Pass of Muckul, leading to another drainage eventually also reaching the River Spey. The short stretch of road that Darwin himself discovered in a small tributary of the Great Glen, was at the level of a saddle the other side of which drained to the northeast. Only the intermediate road seemed to be confined to Glen Roy. Darwin wrote, "These four cases are so remarkable, that the coincidence of level must be intimately connected with the origin of the shelves; although such relation is not absolutely necessary, in as much as the middle shelf of Glen Roy is not on a level with any watershed."

In the second section of his paper Darwin described the theories of Sir Lauder Dick, a separate lake in each valley and of Dr. Mac-Culloch, one large lake. Both of these theories, however, required barriers at the mouths of the glens to impound the water, barriers of which no trace could be found. True to the plan outlined in his "Chief Points to be Attended to," Darwin examined the shelves on near Bohuntine Hill with special attention. The extension of the two upper shelves into Glen Collarig proved that a barrier blocking Glen Roy alone was not enough. There must also have been a barrier blocking the smaller Glen Collarig as well that disappeared at the same time. He also noted that the shelves did not terminate abruptly, but rather quite gradually. Given that no traces of any barriers could be found, and that multiple, synchronized barriers seemed to have been required, Darwin felt that any hypothesis requiring barriers was fatally flawed. While admitting that "it is far easier to assert than to disprove," Darwin boldly concluded, "The conclusion is inevitable, that no hypothesis founded on the supposed existence of a sheet of water confined by barriers, that is a lake, can be admitted as solving the problematical origin of the 'parallel roads of Lochaber.'"

To Darwin the deposits of gravel, sand, and mud in the bottoms of the valleys indicated that "an expanse of slowly subsiding water did occupy these spaces." Citing the recent seashells on raised beaches around Scotland—and even in the Great Glen at the Caledonian Canal—he argued that the changes in the level of Scotland relative to the sea "is due to the rising of the land." "It can scarcely be doubted," he reflected, ". . . that the Great Glen of Scotland . . . was within the recent period an open strait; and, I may add, it must have strikingly resembled the Beagle Channel in Tierra del Fuego."

Darwin indulged in what Einstein and his German physicist colleagues would later call a "thought experiment" to imagine what "would be produced by an arm of the sea slowly retiring from inlets during an equably progressive elevation of the land." He cited his own paper presented previously to the Geological Society to argue that "There is clear evidence that the action of volcanos is intermittent; and the force which keeps the volcanos in action being absolutely the same with that which elevates continents . . . so we must suppose that the elevation of continents is likewise intermittent." Just as in the development of the terraces on the east coast of Patagonia, a pause in the elevation would provide time for the roads, or shelves, to be formed. Thus, he concluded, "I believe, then, that the hypothetical case gives the true theory of their origin."

Next Darwin answered the two principal objections that might be raised against his theory, "the non-extension of the shelves, and the absence of organic remains [seashells] at great heights." The first he dismissed by appealing to the vagaries of both the formation and preservation of the roads. He maintained that a strong fetch of wind was required to build the shelves, and that subsequent rapid vegetation of the shelves and a protected environment was required to preserve them. He then turned to the second objection, the absence of seashells on or near the shelves, realizing full well that the actual presence of shells would have sealed his case. Addressing the second point on his to-do list, he had "attentively examined, with the expectation of finding fragments of sea shells . . . but I could not discover a particle." Then—based in part on the difficulty that he had in finding bits of shell when poking in the sand pits near Shrewsbury—he argued that the preservation of shells "may be considered as a remarkable and not as an ordinary circumstance," in effect making the presence of shells a sufficient but not a necessary condition for establishing that the roads were of marine origin. He suggested that the sand and gravel in the area of Lochaber, being derived from granite and lacking carbonate, was a circumstance in which "the fragments of shells would more readily be dissolved," and asserted that "the marine origin of the shelves cannot be controverted from the absence of organic remains."

Next Darwin turned to the erratic blocks, and boldly asserted that there were only two theories for their transportation "alone worthy of consideration, namely, that of great debacles and of floating ice." The existence of the erratics high on the exposed ridges, such as those at an elevation of 2,200 feet that he found on Ben Erin, and where, under the hypothesis of the debacle, "one would have anticipated the most impetuous rush of water," convinced him that ice floating in a sea covering the entire area was responsible.

Finally, in Darwin's view, there could be no alternative but that the Parallel Roads of Glen Roy were but a symptom of the broad scale uplift of Scotland, just as South America and Sweden seemed to be slowly rising from the sea. Again he cited his previous paper, restating that the ongoing uplift of South America "cannot be attributed to any other cause than an actual *movement* in the subterranean expanse of molten rock," and therefore "it may be granted as not improbable in any high degree, that this part of Scotland when it was upraised rested on matter possessed of considerable fluidity." "We may almost venture to say," he suggested, "that as the packed ice on the Polar Sea . . . rises over the tidal wave, so did the earth's crust with its mountains and plains rise on the convex surfaces of molten rock."

Darwin emphasized several points in his conclusion. The Parallel Roads were "left . . . by the slowly retiring waters of the sea." The gradual and intermittent uplift of Scotland had been "at least 1,278 feet"—the elevation of the highest road—and probably at least 2,200 feet—the height of the highest erratic. The erratics "were transported during the quiet formation of the shelves." The elevation of the region occurred "so equably that the ancient beach-lines" were essentially undisturbed as they rose. And finally, the uplift "was effected by a slight change in the convex form of the fluid matter on which the crust rests."

◆

As Darwin toiled to finish his paper on the Parallel Roads, another important issue was on his mind: marriage.

Darwin, since taking up residence in London, had become an attractive and eligible bachelor. His friends from the Geological Society, Lyell and Leonard Horner, both took an interest in his availability. Horner's oldest daughter, Mary, was Lyell's wife, but he still had five unmarried daughters at home. Any of the eldest three of these would have been a desirable match for Darwin. He was a frequent guest in the Horners' home and enjoyed the attention of the talented, extremely well-educated, and intellectually stimulating Horner girls. In one note thanking him for a book on botany, they concluded with an invitation to come by the following Friday, adding in Maori, *"Ki te kahore hoki he mahi"* [if you have nothing else to do]. Darwin's brother, Erasmus, took to teasing Charles about Mrs. Horner being his mother-in-law. His sisters were atwitter.

Darwin, for his part, was feeling the need for a wife. He was working too hard, overwrought, and troubled by his declining health. "As for a wife," he wrote to a friend, "that most interesting specimen in the whole series of vertebrate animals, Providence only know[s] whether I shall ever capture one or be able to feed her if caught. . . . at the end of a distant view, I sometimes see a cottage & some white object like a petticoat, which always drives granite & trap out of my head in the most unphilosophical manner."

Over the spring and summer, and perhaps most intensely during his visit to Shrewsbury on his return from Scotland, he repeatedly evaluated the pros and cons of marriage. He made one list on the back of a note from a prospective father-in-law, Leonard Horner. On the negative side, if he did not marry, he would be free to make geological trips to Europe and even North America. He was initially worried about money, but it gradually dawned on him—perhaps encouraged by a talk with his father—that owing to his father's wealth, money would not be an overwhelming concern. He also fretted about how he would maintain an intellectually vigorous existence after marriage, and whether he would be better off in London, or in the countryside. He'd observed Lyell's family life in London and noted, "I have so much more pleasure in direct obser- vation, that I could not go on as Lyell does, correcting & adding up new information to old train," continually revising his books.

Finally, under the headings "Marry" and "Not Marry," he sum- marized the arguments. On the winning side he wrote:

> My God, it is intolerable to think of spending ones whole life, like a neuter bee, working, working, & nothing after all.—No, no won't do.—Imagine living all one's day soli- tarily in smoky dirty London House.—Only picture to yourself a nice soft wife on a sofa with good fire, & books & music perhaps—Compare this vision with the dingy reality of Grt. Marlbro' St. Marry—Marry—Marry Q.E.D.

In the end his attentions turned to a woman whom he had known his whole life, his cousin, Emma Wedgwood. He proposed on November 11. The choice was based on rational analysis, rather than romantic attraction, and taking the step almost finished him. But once past the trauma, he turned giddy, proclaiming it, "the day of days!"

Two weeks afterward, still dizzy with excitement, he wrote to a friend announcing his engagement, adding some lines about Glen Roy and its Parallel Roads: "I cannot get people to be half enough astonished at it.—I saw nothing in my perigrinations to the Antipodes, nearly so curious in physical geography.—I do not doubt they are old sea-beaches. . . . But just at this present time I marvel more at the prospect of having a real, live, goodly wife to myself than at all the hundred wonders of the world."

The wedding was planned for the end of January, and among the preparations for the wedding, finding and furnishing a new house in London, moving all his things—including geological specimens and "some few dozen drawers of shells, which must be carried by Hand," and a continual stream of cheery letters he exchanged with Emma, he struggled to finish his paper on the Parallel Roads.

On New Year's Day, he was almost finished, writing Emma, "I trust to be able to finish my Glen Roy Paper & enjoy my Country Holiday [and imminent wedding], with a clear conscience." Emma wrote back, thanking him for the update, and wishing that she could "have helped you & also to eat your eggs & bacon afterward." She wrote dreamily about the big event, planned for the end of the month, concluding, "I hope you will manage to finish Glen Roy now & get shut of it."

Darwin was indeed able to finish correcting the manuscript of the paper before he departed for Shrewsbury and Maer on January 11. The paper was received at the Royal Society on January 17, 1839. He was elected a Fellow of the Society on January 24, and he and Emma were married on January 29. Despite Darwin's note in his journal, "First week January correcting Glen Roy. Paper.— Did nothing during the rest of Month," it was in fact quite a busy month.

The paper was read before the Royal Society on February 7. Prior to publication in *Philosophical Transactions*, this paper too was sent to Adam Sedgwick for review. In late March, the once again supportive but curmudgeonly Sedgwick wrote back, opining that the paper "contains much original research, much ingenious speculation, &

some new and very important conclusions." However, he wrote, "I do not think that it should appear in its present form." He recommended significant condensation, sharpening, and the addition of a map. "The discussion of the erratic blocks found in the district is very good," he went on, "& has a direct bearing on Mr. Darwin's argument; & his conclusions derived from his own observations in South America, have on this subject tended to remove one of the greatest preliminary difficulties in geology." But Sedgwick was less impressed with Darwin's theorizing. The discussion of the "theory of expansion" was almost unnecessary, or "at least far too long." Nonetheless, "There are some short speculations at the end which are of doubtful value; but they are ingenious and short, and therefore I think that they should be printed."

Darwin added a map—updated from the paper of Lauder Dick—and a sketch of the Roads by his friend Albert Way, but notwithstanding any condensation that he made at Sedgwick's suggestion, the paper still contains his theorizing, and remains long, rambling, repetitive, and a challenge for the reader. Nonetheless, it was Darwin's first lengthy scientific publication, and he was on his way. And importantly, the paper allowed Darwin not only to extend the geographical extent of his proofs of recent and ongoing ups and downs of the earth's crust, but also to restate his theory of how and why these movements occurred.

◆

During the autumn of 1838, as Darwin proposed to Emma, worked on the Glen Roy paper, and as his interest in the origin of species intensified, there was one other issue that Darwin might have considered to advantage. Sometime during the late summer, likely at the Athenaeum—since the club forbade removing materials from the premises—Darwin read an article in the *Edinburgh New Philosophical Journal*. It was the translation of a paper by Louis Agassiz, arguing that the erratic blocks of the Jura area of Switzerland had not been transported to their current resting places by floating icebergs—as Lyell and Darwin supposed—but by glaciers that at some time in the not too distant past had been much more extensive. Realizing the significance of these papers to his interpretations in both Patagonia and Glen Roy, he took advantage of FitzRoy's continuing tardiness to add an addendum to his volume of the *Narrative.*

He wrote ten scathing pages, blasting the ideas of the Swiss. He agreed with Agassiz that "the presence of the boulders . . . cannot be explained by any debacle," but there his agreement came to an abrupt end. Darwin challenged Agassiz's suggestion that glaciers might have formerly been more extensive, alluding to the received wisdom at the time that the earth had previously been warmer, not cooler. If Agassiz's hypothesis was that "during the gradual cooling of the earth, there have been periods of excessive refrigeration. It is needless to state that such an hypothesis is not supported by a single fact."

One of Agassiz's arguments relied on scratches left on the bedrock that he posited were left by rocks dragged at the bottom of the glaciers. These scratches, or striations, Agassiz wrote, "never follow the direction of the slope of the mountain," but tended to be oblique or even parallel to the direction of the surrounding valley. "What explanation," wrote Darwin, fuming with skepticism, "will it be believed is offered for this fact?" Darwin countered that "these very curious facts, which we owe to M. Agassiz's observation, can be explained by the theory of floating ice."

Darwin maintained that during the gradual elevation of Switzerland the glaciers of the Alps could have descended to sea level, just as he had observed on the shores of the Straits of Magellan and the Beagle Channel. He then reasoned that the scratches parallel to the valleys could easily have been caused by rock dragged by ice flows pushed along the former shoreline by wind and currents.

One odd detail of the large angular erratic blocks reported by Agassiz from the work of another Swiss, Jean de Charpentier, was that they were sometimes found "planted vertically in the soil, in the valleys, as on the sides of a mountain, and split up throughout their whole extent from top to bottom." The Swiss contended that these odd orientations and the breaking of the blocks resulted from falls down the crevasses, or fissures, in the ancient glaciers. But Darwin preferred his own explanation of the blocks falling from floating ice, as "at least, as simple as this."

Darwin then summarizes his own explanation, floating ice, in which, "only veræ causæ are introduced, and reasons can be assigned, for the belief that these causes have been in action in these districts." He aimed his final withering blast, not at the Swiss, but at the diluvialists, maintaining that "the hypothesis of a deluge of mud and stones, fifteen hundred feet deep in Sweden, or three thousand in North America, which rushing over the country, rounded the northern fronts of the hills, and rolling by their eastern and western flanks, left them marked with oblique furrows, is to violate, as it appears to me, every rule of inductive philosophy."

◆

I too visited Glen Roy to see for myself the Parallel Roads. Of course they weren't really giant bathtub rings, but that's sure what they looked like. Three horizontal stripes wound in and out along the steep, treeless slopes on both sides of the valley, the highest at a level about 650 feet above the River Roy far below. In early October the tiny purple blooms were nearly gone from the heather, the bracken already turned yellow and brown. Sheep stared, mute, or *baa*-ed dumbly at my approach. As I climbed the steep hillside sodden from the previous night's rain, low clouds and mist threatened more. But there was no mistaking the remarkable Parallel Roads. These gently sloping shelves extended horizontally for miles along the slopes above the valley of Glen Roy almost as if some ancient bulldozer driver had run amok. But of course my views on Darwin's explanation were impacted by nearly two additional centuries of geological investigation and insight.

Darwin would later bemoan his use of the "principle of exclusion" in interpreting the Parallel Roads, and he would come to regret his harsh words for Agassiz. If only he had put more weight on the absence of seashells, looked a little harder for a water gap at the level of the intermediate road, or even been a bit more open about the ideas of the Swiss. But, as we shall see, hindsight is so easy.

PART THREE

BACK ON DARWIN'S TRAIL

CHAPTER THIRTEEN

From Uplift to Evolution

I wish with all my heart that my Geological book was out. . . . I have lately been sadly tempted to be idle, that is as far as pure geology is concerned, by the delightful number of new views, which have been coming in, thickly & steadily, on the classification & affinities & instincts of animals— bearing on the question of species.

—Charles Darwin to Lyell,
September 14, 1838

As Darwin eased into married life, and he and Emma settled into their first home at 12 Gower Street, his penchant for work did not diminish. He had a long list of things to do. After his paper on Glen Roy was sent to the Royal Society, he immediately began writing about coral reefs. He had to finish his paper on earthquakes and write a book too. But—most of all—he wanted to get to work on species.

Before Darwin immersed himself completely in species he needed to finish the book on the geology of the countries visited by the *Beagle*, the book that he promised himself to write as he sat overlooking the sea at St. Jago seven years before. He finally settled on a plan. The book would be composed of three volumes: the first on the structure and distribution of coral reefs, the second on the volcanic islands, and the third on the geology of South America.

After returning to England Darwin still employed Syms Covington, the man who had been his servant while on the *Beagle*. Covington helped Darwin to unpack and sort the specimens, and also served as a secretary and amanuensis, gathering and transcribing Darwin's notes and manuscripts. But now the man who had helped Darwin dig the bones of *Mylodon darwinii* from the sea cliff at Punta Alta and had felt the earthquake with him at Valdivia, wanted to leave.

Darwin valued Covington, but their differences in class structured their relationship. Despite his near continuous presence in Darwin's life over the previous six years, it was almost as if he were a ghost whose shadow appeared occasionally in Darwin's diaries. At the end of February 1839, Darwin wrote in his account book "Present to Covington on leaving me £2," equivalent to a little more than $300 today. He wrote letters of recommendation, and helped him find passage as a ship's cook to Australia, where Covington hoped to find his fortune.

Darwin spent the end of March and most of April on the coral reefs, but the mysterious gastric distress that would come to afflict him for life began to interfere seriously with his work. Facilitated by Captain Beaufort, still the chief hydrographer, Darwin pored over the Admiralty's collection of charts, strengthening his case about the forms and structures of the reefs, and his arguments for the process of their formation. He compiled a map of coral reefs throughout the world that would become the centerpiece of his book.

At long last, in late May the volumes of the *Narrative* were released. King's and FitzRoy's volumes were reviewed in the June 1 issue of *The Athenaeum*, and Darwin's in the issue of June 15. Darwin sent copies of his book—what we have come know in later editions as the *Voyage of the Beagle*—to geological acquaintances at home and abroad, and received many kind letters of thanks, congratulations, and interest, including from Buckland, Caldcleugh, and the French geologist, Élie de Beaumont.

Richard Owen wrote him that the book "is as full of good original wholesome food as an egg, & if what I have enjoyed has not been duly digested it is because it has been too hastily devoured. I leave it reluctantly—tired eyes compelling—at night, and greet it as a new luxury at the breakfast table." William Fitton wrote, "What I like best—however,—is the tone of kind and generous feeling that is visible in every part:—so that one sees that it is the work of a plain English gentleman—traveling for information, and not for Effect.—& viewing all things kindly." Fitton only worried that one of his sons "who has often wished to be 'a discovery man'—will now be so full of 'a voyage around the world' that he will forget, the previous discipline & labour—necessary to make it effective."

But there can be little doubt that the letter that pleased him most was from the man who had inspired his adventure in the first place, Alexander von Humboldt. In a long, gracious, and adulatory letter written in French, Humboldt heaped on praise that would have brought almost any mortal to tears. He noted passages that he particularly liked, and commented on several scientific points related to plants, animals, volcanoes, elevation, and climate. Although he doubted Darwin's view that erratic blocks were carried on ice rafts, he complimented Darwin for his "strength of talent, solid and wide knowledge, and a felicitous literary disposition." Humboldt's words cannot but have inspired Darwin to redouble his efforts, writing, "You told me in your kind letter that, when you were young, the manner in which I studied and depicted nature in the torrid zones contributed toward exciting in you the ardour and desire to travel in distant lands. Considering the importance of your work, Sir, this may be the greatest success that my humble work could bring. Works are of value only if they give rise to better ones."

◆

Despite the joy that the birth of his first child, William, in December, brought to Darwin, the latter half of 1839 was largely lost to illness, as was much of 1840. He did however finally finish his "earthquake paper." But the fall of 1840 also brought the appearance of the first dark clouds over his geological theories, and the storm that they foretold was in the person of none other than Louis Agassiz.

In September 1840, Louis Agassiz published the result of his studies of glaciers in a beautifully illustrated book, *Études sur les Glaciers*. He traveled and lived among the ice and rocks of glaciers in the very heart of the Swiss and French Alps—at Grimsel, Grindelwald, Zermatt, and Chamonix—and in the now ice-free Rhône and other great valleys descending from them, and in the Jura Mountains. He detailed the characteristics of glaciers, their moraines, the landscape, and other traces—including striations—left in their wake, and the story of their changes during historical times. He concluded by arguing that the former extent of glaciers went far beyond the Alps—based largely on the distribution of moraines extending along the sides of former glaciers. The moraines extended, for example to Lake Geneva, Switzerland, and the Italian lakes Maggiore and Como, and as much as sixty miles or more from existing glaciers. He described erratic boulders too, such as the enormous Pierre-à-Bot erratic near Neuchâtel with a volume of 50,000 cubic feet, at a similar distance from the mountains. The impact of Agassiz's book was immense, although he lost many readers with his final conclusion that the earth had been swaddled in glaciers preceding the catastrophic uplift of the Alps, which broke through the ice. Sedgwick summarized the general reaction, writing, "I have read his Ice-book. It is excellent, but in the last chapter he loses his balance and runs away with the bit in his mouth."

Later that autumn Agassiz visited Great Britain. In the company of Reverend Buckland, he toured Northern England and Scotland, then journeyed on to Ireland, evangelizing all the way on the subject of glaciers and their former extent. Most troubling for Darwin was their visit to Glen Roy. On November 4, a paper by Agassiz on Glen Roy was read at the Geological Society. To Agassiz, the explanation of the Parallel Roads was obvious. An enormous set of glaciers had once covered the Scottish Highlands. As the climate warmed and the glaciers shrank, there came a time when glaciers had blocked the mouths of Glen Roy and the other glens, impounding lakes. As the warming continued and the glaciers retreated, the ice impounded lakes at successively lower elevations. At each level, governed by the elevation of the lowest saddle holding back the lake, a shoreline, or road, was formed. After all the ice melted, no trace would remain of the barriers that dammed the lakes—except to the practiced eye of someone who could read the spoor of glaciers. If one could accept the idea that extensive glaciers had once covered the area—a high hurdle at the time—the Parallel Roads were perfectly explained as shorelines, at different levels, of a lake dammed by the glacier.

Darwin was intrigued by the new explanation, but dug in his heels in favor of his own. He wrote Buckland, "I should much like to hear your's & Agassiz's opinion on the parallel roads, though I believe I know its outline." He continued, "I cannot give up the sea, after thinking over many points of minor detail in that country." Then he added, a bit strangely, perhaps out of politeness, "Though I am very sure, if your theory had occurred to me, during the first two days of my examination, I should have given up their marine . . . origin at once."

Darwin studied, and was much impressed by Agassiz's book. Realizing that he had been a bit hard on him in the addendum to the *Narrative*, Darwin sent a copy to the Swiss, writing, "I have lately enjoyed the pleasure of reading your work on Glaciers, which has filled me with admiration. As I have briefly treated of the boulders of S. America in the accompanying volume I thought you possibly might like to possess a copy; and sending it you, is the only means I have of expressing the regret I feel at the manner in which I have alluded to (although probably the fact is unknown and quite indifferent to you) your most valuable labours on the action of Glaciers."

Over the winter Darwin stewed about glaciers and Glen Roy. He parsed and reparsed his facts. Despite the appeal of Agassiz's glacial theory for the origin of the Parallel Roads, he stuck to his explanation of ongoing uplift from the sea. There was one fact that Agassiz's theory did not explain at all. There was no saddle corresponding to the level of the middle road in Glen Roy. "The difficulty," he wrote Lyell, "about ice-barrier of middle Glen Roy shelf keeping so long at exactly the same level does certainly appear to me insuperable." It may have been a thin reed, but he clung to it fiercely.

Darwin also went on the offensive, presenting a paper to the Geological Society in April on his observations on erratic blocks in Patagonia and Chilóe, restating and expanding on his arguments for their transportation by icebergs.

That summer, Buckland—armed with the new ability to see traces of vanished glaciers passed to him by Agassiz—visited one of the places that Darwin thought he knew well, Cwm Idwal in north Wales. There Buckland saw moraines, striations left by the rocks dragged by ice, and erratic boulders dumped in along the way. The Cwm itself, Buckland concluded, had been sculpted by a glacier. After Buckland presented his views to the Geological Society, Darwin knew that he had to see for himself.

In June 1842, Darwin combined an excursion to North Wales together with a visit to families in Shrewsbury and Maer. For him this was a period of intense intellectual ferment. His mind was now filled with questions and ideas about species, but nonetheless, he was drawn back to Snowdonia. He wrote to William Fitton from Capel Curig, a few miles from Cwm Idwal, and where he had stayed during his traverse after leaving Sedgwick a decade before: "Yesterday (and the previous day) I had some most interesting work in examining the marks left by extinct glaciers—I assure you that no extinct volcano could hardly leave more evident traces of its activity and vast powers." The valley at Capel Curig, he concluded "must once have been covered by at least 800 or 1,000 ft in thickness of solid Ice!—Eleven years ago, I spent a whole day in the valley, where yesterday everything but the Ice of the Glacier was palpably clear to me, and then I saw nothing but plain water, and bare Rock." Darwin too had achieved the power of Agassiz's glacier vision, if not to its fullest extent, for notwithstanding what he saw: "it convinces me that my views, of the distribution of the boulders on the S. American plains having been effected by floating Ice, are correct. I am also more convinced that the valleys of Glen Roy & neighboring parts of Scotland have been occupied by arms of the Sea, & very likely . . . by glaciers also."

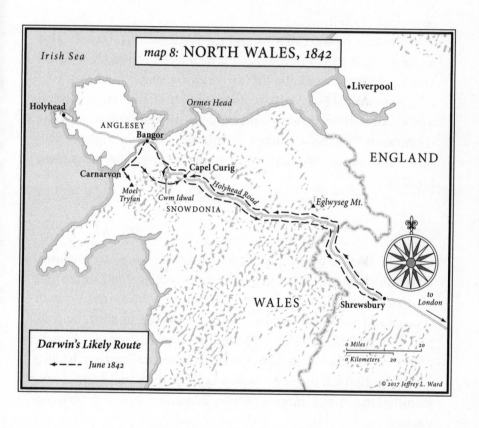

map 8: NORTH WALES, 1842

Irish Sea

• Liverpool

Holyhead

Ormes Head

ANGLESEY
Bangor

ENGLAND

Carnarvon

Capel Curig

*Moel
Tryfan*

Holyhead Road

Cwm Idwal
SNOWDONIA

▲ *Eglwyseg Mt.*

WALES

Shrewsbury

to
London

Darwin's Likely Route

◄— — — *June 1842*

0 Miles   20
0 Kilometers 20

© 2017 Jeffrey L. Ward

About eight miles from Cwm Idwal, a hill called Moel Tryfan rises to an elevation of more than thousand directly from the Irish Sea. Darwin also visited Moel Tryfan, hoping to confirm for himself what another geologist, Joshua Trimmer, had found there. Although Darwin couldn't find it himself, this evidence provided some of the strongest support for Darwin's conclusion that North Wales, like Glen Roy, had risen—relatively recently—from the sea. Some bits of shell.

While visiting the Darwins and the Wedgewoods, Darwin also wrote a pencil sketch of his species theory. His trip to Cwm Idwal and Moel Tryfan would be his last serious geologic fieldwork.

In September 1842, Darwin moved with his family from London to the quiet of Down House and eighteen surrounding acres in the countryside of Kent, a journey by cab, the Croydon Railway, and phaeton of about two and a half hours southeast of downtown London.

◆

Over the decade following the return of the *Beagle* to England, Darwin slowly shifted his primary focus from geologic topics and his theories about the vertical movements of the earth's crust to species and natural selection. And as he made this shift, he took with him a geologist's understanding of the ancient age of the earth, and of the changes in its past. On this intellectual journey, he also took with him his mentor, Charles Lyell, and his ideas about uniformitarianism, the immense power of gradual, but never-ending, change.

Darwin's ideas about how species evolved grew from the first sketches in his notebook in 1837, to a penciled essay of his ideas for himself in 1842, to a more formal essay in 1844 (which he put in an envelope and still kept to himself), to his nearly full-time focus on barnacles as a case study in species beginning in 1846. During these years Darwin finally completed his three volumes on the geology of the *Beagle*. He published the first volume on coral reefs in 1842, the second on volcanic islands in 1844, and finally, the third on the geology of South America in 1846.

The question of species crossed the boundaries between geology and biology. Geology provided the historical record—through fossils—of how the complexity of life grew from almost nothing, through the simplest forms to the complex forms, finally including the mammals that occupy the earth today.

When Darwin completed his essay on natural selection in 1844, he wrote an odd memorandum to his wife Emma, directing her—"in case of my sudden death"—to find someone to edit and publish his ideas. The memo included several specific instructions, but clearly indicates his understanding of how his ideas spanned natural science. "The Editor," he wrote, "must be a geologist, as well as a naturalist . . ." Darwin, by his own admission, had "dabbled" in several branches of the natural sciences, and he was stung when his friend the botanist, Joseph Hooker, wrote to him saying, "I am not inclined to take much for granted from any one who treats the subject [of species] in his way & does not know what it takes to be a specific naturalist." Darwin realized that if his ideas about natural selection were to be taken seriously, he would need to build a reputation in biology or botany in addition to his growing standing in geology. Ultimately he chose barnacles as his vehicle to become a "specific naturalist."

◆

By October, 1846, Darwin had set his course. It wasn't exactly crossing an intellectual Rubicon. He wasn't risking everything, or shouting to the rooftops that he was now going to work on proving the validity of his ideas about natural selection, but he started to drop hints to influential people within Britain's scientific community that he was working on something new—and potentially revolutionary.

He had been "slaving," he had written Lyell in August, "to finish my S. American Geolog." This would be his third and final volume of his geological trilogy. He had to finish it, as he desperately wanted to get on to his new project, the one that beckoned, tantalizing him. Nonetheless, he thought that Lyell would find "the collection of facts on the elevation of the land & on the formation of terraces pretty good."

Finally in early October as the publisher, Smith, Elder & Co., prepared to release the final product, Darwin wrote to his friend Hooker, letting him know in Darwin's peculiarly oblique yet incorrigibly determined way, that he was taking Hooker's comment to heart, as he set his course ahead. "I am going to begin some papers on the lower marine animals, which will last me some months, perhaps a year, and then I shall begin looking over my ten-year-long accumulation of notes on species & varieties which with writing, I daresay will take me five years." Darwin's estimate of the time required was way off the mark. He would spend about eight years working on the barnacles—becoming the world expert on both the living and fossil species—and then another five years of thinking, illness, personal tragedy—and dithering—before he finally rushed the *Origin* into print.

But his course was now set. As he wrote to one correspondent, "I have lately been busy with my geology & shall [now] be for some time employed on Invertebrate zoology."

◆

As Darwin's consuming interest moved from his focus on the ups and downs of continents, islands, and the sea, toward the changes in the forms of life, how did his experience as a geologist inform his thinking and potentially even inspire this new intellectual trajectory? What principal "facts," as he would have said, from geology did his theory have to explain? What "facts" from geology provided clues to how the process of change took place?

Fact #1: The earth is quite old. He did not know how old, but millions of years at least.

Fact #2: The surface of the earth, and therefore the environment for life, had changed dramatically over the eons. Much of what was once sea was now land, even to the top of the Andes.

Fact #3: He knew that the processes responsible for uplift and subsidence were still going on and that the shifting of oceans and continents, and the uplifting of mountains, in turn affected species and their distribution.

Fact #4: The uniqueness of some groups of species to a particular region, such as the armadillos and tree sloths peculiar to the Americas, extended back in time as shown by the related fossils of the extinct glyptodonts and ground sloths.

Fact #5: In the oldest rocks, there were few if any signs of life, shown by fossils or "organic remains" as Darwin would have said. The earliest fossils represented relatively simple creatures. Then as time progressed, the fossils in younger and younger rocks represented increasingly complex forms of life. First came invertebrates, then reptiles, then mammals, and finally, at some late date, the mammal Homo sapiens.

Fact #6: The fossil record is incomplete. What remains for the geologist to find depends on the vagaries of preservation. Indeed, preservation is commonly the exception. Destruction, decay, and disappearance are the rule.

Fact #7: Once a species goes extinct, it is gone. It doesn't come back.

Fact #8: Changes can happen in small, abrupt increments, or they can occur slowly, continuously, and "insensibly," but over the vast expanse of the geologic time the accumulated change is extraordinarily breathtaking. Just as mountains and continents rise, islands sink, changing almost imperceptibly, but ultimately producing dramatic change, so too, he concluded, could life.

So as Darwin brooded on his questions about the species, his thinking was strongly conditioned by what he knew about the history of the earth, the changes it had experienced, and the earlier life forms that preceded those inhabiting the earth today. All these facts he knew from geology, and he carried them with him as he sought to explain the development and evolution of life on earth.

◆

But even as Darwin wrote to colleagues around the world requesting specimens of living and fossil barnacles, as he dissected them, as he studied their organs, the functioning and purpose of their body parts, and their ethology (their behavior), as he absorbed himself in barnacles . . . the problem of Glen Roy would not go away.

The leaders of the Geological Society, among them Lyell, Murchison, and Horner, had largely dismissed the theory for the formation of the Roads offered by Agassiz and Buckland as being fanciful, and largely supported Darwin's explanation. Darwin, for his part, wrote Horner in 1846 that "there never was a more futile theory" than the one offered by Agassiz.

David Milne (later Milne-Home), a wealthy Scot and geologist, made his own investigation of the Parallel Roads in 1847. Milne, who most famously studied earthquakes of Scotland—and gave us the word "seismometer"—found something in the field that Darwin and his other predecessors had missed, a saddle just at the elevation of the middle shelf in Glen Roy, tucked away at the head of a twisting tributary glen. And, on the slopes of this saddle leading away from Glen Roy, he also found deposits of sand and gravel suggesting that this had once been the site of a river draining a lake.

He described similar deposits suggesting former rivers, indeed, overflow channels, on the slopes of the saddles leading away from Glen Roy to the River Spey at the level of the highest shelf, and at the level of the lowest shelf, at the head of Glen Spean, at the Pass of Muckul. Milne, however, was no fan of Agassiz's glacial hypothesis, and argued that his newfound discoveries supported the earlier ideas of lakes dammed by barriers that had somehow eroded away.

Darwin was once again rattled by Milne's findings, but refused to admit defeat. He wrote Milne, "You will, I fear, think me very obstinate when I say that I am not in the least convinced about the Barriers: they remain to me as improbable as ever." Milne's discovery of the missing saddle, however, "staggered [him] in favour of the ice-lake theory," although, he wrote, "I am not, however, as yet a believer in the ice-lake theory, but I tremble for the result."

Darwin dashed off a letter to *The Scotsman* defending his earlier paper in the light of this new attack, but it was never published.

◆

Darwin's work with the barnacles was arduous, tedious, and nearly all consuming. He complained. One of his young sons is alleged to have asked a friend, "Where does your father do his barnacles?" But the results of Darwin's years of sitting in his study, staring down his microscope, dissecting and studying his barnacles, ultimately gave him the same level of credibility in biology that he had already gained in geology from his explorations on the *Beagle* expedition.

While Darwin's geologic fieldwork was outside, in the elements, on a horse, a mule, or on foot, his studies of the barnacles were mostly in the comfort of his study in Down House. Although many of his geological studies were descriptive, he quickly rose from a descriptive plane to a theoretical one, constantly seeing explanations of how and why. His publications about the barnacle were also intensely descriptive; however, his attention to variations within the species and search for evidence of natural selection were motivated by his theorizing.

The two phases of Darwin's career also differed in another intriguing character. In studying geology one almost never sees the whole picture. As a geologist Darwin could see what was on the surface of the earth, or exposed on hillsides, or in cliffs in the mountains and along the coast, and in the banks of rivers and streams, but the opportunities to even glimpse the third dimension were rare. He would even write a brief paper in 1863 about some wells that bored through the Pampas to reveal the nature of the bedrock beneath. In contrast, with his barnacles, once he was able to get his specimens, he was usually able to dissect them to his heart's content and get to know the creature from every possible dimension. Even though the dissection was exacting, and sometimes frustrating work, he was able to control the extent and nature of his observations of the barnacles in ways that were impossible with his geologizing in the field.

What initially had grabbed his interest was a most curious parasitic barnacle that Darwin called "Mr. Arthrobalanus." With this he began eight years of dedication to the living and fossil barnacles, or Cirripedia as they are more formally called. His work culminated in two thick volumes on the living barnacles totaling more than one thousand pages, and two slimmer volumes on the fossil barnacles. The first volumes on these two sets were devoted to the stalked or gooseneck barnacles, and the second to the barnacles which attach their bodies directly to rocks, logs, or other creatures, of which the most common are the familiar acorn barnacles.

Darwin brought to his studies of the barnacles, and particularly the fossil barnacle, his geologic appreciation for the element of time. He knew the relative ages of his fossil barnacles from the stratigraphic positions in which they were found, so he could watch—at least in this rich, but tiny part of the animal kingdom—both the variation and succession of species. What he saw solidified his convictions that natural selection was the means of evolution.

◆

While Darwin dissected his barnacles, he struggled on through his own mysterious illness, periodically seeking relief from water treatment. Even as his own ill-health ebbed and flowed, in 1851 Annie, his ten-year-old daughter, and perhaps his favorite child, died. She had suffered from scarlet fever in 1849, and never completely regained her strength. Nothing seemed to help. In desperation, Darwin sent her to the establishment where he received hydrotherapy. When, after a few weeks, her condition deteriorated with fever and vomiting, he joined her. Despite seeming to turn a corner, she died. It was a bitter blow, and perhaps the low point in Darwin's life. He seemed to accept some blame for her death, thinking that she had inherited his "wretched digestion." He was so overwhelmed that he left the funeral arrangements to his sister-in-law, who had joined him, and returned to Emma and the rest of his family in Down.

In 1853 Darwin received the Royal Medal from the Royal Society. In 1854 he finally finished his work on the barnacles and came face-to-face with the prospect of writing up his ideas about evolution. In his journal for September 9, he wrote, "Finished packing up all my cirripedes. Preparing Fossil Balanidae distributing copies of my work &c &c.—I have yet a few proofs for Fossil Balanidae for Pal: Soc: to complete, perhaps a week's more work. Began Oct. 1 1846. On Oct. 1st. it will be 8 years since I began! But then I have lost 1 or 2 years by illness. . . Began sorting notes for Species Theory."

He dithered. He did more reading. He experimented with seeds, to see if they could germinate after long exposure to salt water or after passage through a bird's digestive system. (Answer to both: Yes.) He wrote yet another paper about the power of icebergs to make scratches on exposed bedrock beneath the sea. His children, now seven of them, would get sick but thankfully recover. He went on family vacations. He traveled to Scotland for the meeting of the British Association for the Advancement of Science. He wrote letters to the *Gardeners' Chronicle* seeking information about lizards' eggs, the vitality of seeds, cross breeding, the color of ponies, and sharing advice with fellow gardeners about how a fancy wire rope helped him to overcome the problems of a deep well, and inquiring whether any readers had experience with a gutta-percha, or hard rubber bucket.

By 1856 Lyell—who didn't necessarily agree with his hypothesis—nonetheless urged him to get on with it. Write it up. Slowly he began turning out the chapters about natural variations, geographic distribution, the struggle for existence. In December, 1856, Emma, at age forty-eight, bore another child, Charles Waring Darwin. But baby Charles was not quite normal. Darwin kept writing—and battling his illness. By June of 1858, still fighting his illness and his writing time at Down interrupted by trips for water treatment, he had about ten chapters complete.

Then, within ten days, the world closed in around him.

About June 18 Darwin received a letter that shook him to the core. Alfred Russel Wallace, a self-educated collector of rare creatures with whom Darwin had been corresponding, wrote from a distant island in the Malay Archipelago asking for Darwin's opinion on an enclosed twenty-page manuscript titled, "On the Tendency of Variations to Depart Indefinitely from the Original Type."

The ideas spelled out in Wallace's paper clearly overlapped Darwin's own. Darwin, after agonizing and brooding about how variations and natural selection led to the evolution of species for two decades, was about to be scooped. It was a catastrophe. He wrote to Lyell and to Hooker, pleading for help. But his youngest son, the baby, was quite ill. Darwin couldn't leave Down.

Lyell and Hooker, corresponding with a rapidity that would challenge the Internet, quickly engineered a plan. A meeting of the Linnean Society was scheduled in London for July 1. They would arrange for a reading of both Wallace's paper and an abstract of Darwin's work at that meeting. In this way, they reasoned they could give credit to Wallace, but also preserve recognition of Darwin's independent work. The arrangement pushed the envelope in a very sensitive area of ethics in science, the allocation of credit for discovery, but they must have felt that they owed it to their friend.

Then on June 28, baby Charles, barely eighteen months old, died. Darwin must have felt as if he were surrounded by darkness.

The meeting of the Linnean Society took place as planned in London three days later. Both papers were read. Neither of the two authors was present. After the readings, nobody stood up and cheered, or objected, as David Quammen wrote, "The most remarkable thing . . . is how little immediately came of it." But it was as if adjoining mining claims had been staked out. Now it remained to be seen who would develop the mine and strike gold.

Darwin was distraught, but rebounded. By the end of July he began writing with a renewed purpose and newfound sense of urgency, an "abstract of [his] notions about Species & Varieties." This would become *The Origin of Species*. He finished checking the proofs of the book on October 1, 1859. By his careful reckoning, it had taken "13 months & 10 days." Twelve hundred and fifty copies were printed. All were sold the first day, November 24. The book was a sensation. He immediately began correcting and revising for a second edition of three thousand copies.

Regarding the controversy that his book triggered, Darwin recorded in his spare journal "Multitude of Letters." Darwin was immensely relieved and pleased to have his book completed. In fact, he was rallying support for his theory, sending copies of his book off to those who he thought might become allies in the struggle to have his ideas accepted. He knew that he would draw fire from many directions. The publication of the *Origin* vaulted Darwin from being relatively well known to achieving both fame and infamy in one step.

◆

The year 1860 brought out the second edition in the UK, and the first edition in the United States. The third UK edition of another two thousand copies was published in April, 1861.

Then, with the excitement over the publication of the *Origin* still ringing, on September 3, 1861, Darwin received another troubling letter. This letter was not about the *Origin*, nor was it antagonistic. Its tone was almost apologetic. The letter was about Glen Roy.

Thomas Jamieson, a Scottish geologist twenty years Darwin's junior had earlier written Darwin that he was planning field work in the area around Glen Roy and had sought Darwin's guidance, which Darwin had graciously provided. Now Jamieson was reporting the results of his fieldwork. His findings were devastating for Darwin's hypothesis.

He preceded a lengthy recitation of his findings with a summary of the bottom line. He hastened to inform Darwin that it was now his conviction "that these parallel roads have been formed along the margin of freshwater lakes and finding the marks of ice action so plain over the whole district I cannot help thinking that Agassiz hit upon the true solution of the problem when he pronounced these marks to be the effect of glacier-lakes."

Darwin was crushed. After studying the points in Jamieson's letter, he wrote back, "I thank you sincerely for your long & very interesting letter. Your arguments seem to me conclusive. I give up the ghost. My paper is one long gigantic blunder." He was less restrained in describing his disappointment to Charles Lyell: "I am smashed to atoms about Glen Roy." The Parallel Roads of Glen Roy were created not by uplift, but by a lake dammed by a glacier.

◆

It was a very wet November when I visited Snowdonia. At breakfast in the hotel where Darwin stayed in Capel Curig, now the Plas y Brenin Mountain Sports Center, I commented to a young man that I supposed that this kind of weather was to be expected in Wales in November. "Sadly," he replied, "you could see the same in July." After breakfast I asked to see the pane of glass, formerly in the bar, on which Darwin had scratched, "C. Darwin." But since the pane also bore—among the autographs of other visitors—the etching, "Victoria R.," it had been stored away. Apparently it was judged too valuable to be exposed to potentially boisterous mountain adventurers and visitors from former colonies.

As I drove the narrow A5 leading from Capel Curig, past Llyn Ogwen, and toward the trail leading up into Cwm Idwal, I focused on staying on the left hand side of the road and shifting with my left hand—counter to all my well-trained impulses—and at keeping my eyes on the road rather than scanning the countryside for telltale signs of recent glaciers. The rains paused briefly, but the sky remained overcast, the clouds thick, the rivers overflowing their channels.

On the way to Wales I had had the special treat to meet Hilda Padel, Darwin's great-granddaughter, in Oxford. Hilda, then ninety, was a tall woman, with large round glasses and curly hair. Neither her appearance nor sharp mind gave any hint of her advancing years. She joked that her grandfather, Horace, was the slow one of Darwin's children. In fact he was elected to the Royal Society, and eventually knighted. Learning that I was on my way to Cwm Idwal, she recounted how as a college student she had been evacuated from London to Bangor, Wales, during the Blitz. She vividly recalled accompanying a group of bleary-eyed students directly from their exams on a rainy camping trip to Cwm Idwal, and washing dishes in the mud. The weather in North Wales has not improved.

From the parking lot, the trail led a half a mile to the top of a classic terminal moraine damming a lovely lake, Llyn Idwal. Standing on the blocky moraine and looking up over the cold dark waters of the lake at the fog-shrouded cliffs of the cwm above, the glacial origin of this amphitheatre seemed obvious. While I felt as if I had been looking at glacial features like this one almost my entire life, there is no doubt that this was a very good spot for Darwin's epiphany.

A few steps to the left lay a set of automobile-size boulders. Darwin recognized that these angular blocks were just like those described by de Charpentier and Agassiz, and posited, like the Swiss, that these blocks in Cwm Idwal had fallen down a crevasse in a now-vanished glacier. Nearby, as the rain began in earnest, I found clear striations—long clear scratches on the dark gray rock cutting across the bedding—just as Darwin had found.

Back in the car I continued on down the A5 toward Bangor through a beautiful, glacially sculpted valley. What was it, I wondered, that stopped Darwin from going all the way to accept the glacial hypothesis then and there? A few hours later, I began to understand.

While Darwin and many of his geologic colleagues in the middle of the nineteenth century were prepared to accept the idea of small glaciers in the steep mountains of Scotland and Wales, they choked on the notion of large glaciers blanketing an entire relatively flat-lying landscape. They ascribed many of the deposits of sand, gravel, and erratic boulders—not only across western and northern Britain, but across northern Europe as well—to a high stand of the sea. Darwin had seen glaciers descending from the clouds in Tierra del Fuego, but he had never seen the immense glaciers in the heart of the western Alps, not to mention those of Greenland and Antarctica, then still terra incognita. So he and his colleagues could not visualize a sheet of ice that could blanket a continent. It would also be decades before they fully appreciated the enormous power of glaciers to excavate solid rock. But as I drove up through the lanes that rose above the town of Caernarfon, I began to understand what really stuck in their craw.

The hill of Moel Tryfan rises fourteen hundred feet above the sea. My objective was an abandoned quarry at an elevation just over a thousand feet. It was here that Darwin searched for bits of broken shell, mixed in with the young sand and gravel, similar to those found by Joshua Trimmer.

As I looked out over the dark gray Irish Sea from high on Moel Tryfan, through the wind and spitting rain, I began to comprehend how difficult it must have been to accept the notion that those bits of shell had been scooped from the bed of that sea, and dragged up this hillside by a hypothetical glacier. Darwin visited here in 1842. Only in the early 1860s, after admitting defeat on his paper about Glen Roy, did he became a full convert to the ideas then advocated by Sir A. C. Ramsay about the power of glaciers to cover and sculpt the landscape. In 1862 when Ramsay, then the president of the Geological Society, wrote Darwin asking if he would review Jamieson's paper on Glen Roy, Darwin replied, "With respect to Glen Roy paper, I shall be happy to undertake it. It will in fact be no trouble, as I have corresponded with Mr J. on subject & know most of his facts & arguments, & know, alas, too well that I am everlastingly smashed."

It was not until near the end of Darwin's life that the bits of shell on Moel Tryfan came to be understood by British geologists—compelled by overwhelming evidence—as debris pushed there by an icy bulldozer from the basin of the Irish Sea.

Just as Darwin was himself forcing a paradigm shift with *The Origin of Species*, in Glen Roy he had found himself largely on the wrong side of a paradigm shift about a relatively recent widespread glaciation affecting the crust and oceans of the planet. It did not invalidate his evidence for uplift in South America and subsidence beneath the ocean's atolls, but he had put a lot of chips on Glen Roy being another example of uplift. His bet did not go well.

CHAPTER FOURTEEN

From Natural Selection to Plate Tectonics

I found, on landing [at Guafo Island], that the formation of the island, like that of [Guamblin] and [Ipun] Islands, is a soft sandstone, which can be cut with a knife as easily as a cake of chocolate.

—Robert FitzRoy

F*aro Guafo, faro Guafo, cambio,*" Daniel barked into the walkie-talkie, imitating the monotone and cadence of military speech, a mischievous smile covering his face. Receiving no response, he repeated, "Guafo lighthouse, Guafo lighthouse, over." Six of us sat around an aluminum camp table in an orange-and-blue dome tent sheltering us from the mist on the beach outside. As I finished a Chilean breakfast of mashed avocado on toast while tottering on a three-legged campstool sinking into the sand, I contemplated the day's task. Carry a couple of hundred pounds of GPS equipment, tools, and batteries five miles down the shore and up a 500-foot cliff to the lighthouse, home of the only permanent inhabitants on Isla Guafo.

The *Beagle* visited here twice in 1834–5. On the first visit—while Darwin was with Sulivan surveying the eastern shore of

Chilóe—FitzRoy thought the geology of the island interesting and unusual. He brought Darwin back for a look, but Darwin was not too impressed. It would be a century and a half before the real importance of this island emerged.

On May 22, 1960, the sailors in the lighthouse on Isla Guafo had the ride of their lives during the greatest earthquake of the twentieth century—and a ringside seat for what was to follow. About ten minutes after the earthquake, the sailors watched in awe as the sea withdrew about a third of a mile to the west. Farther out to sea they could see a great wave forming, then moving rapidly toward them. The first megawave was followed by three more, reaching more than thirty feet above the beach.

Two days later the sailors discovered that their dock, a flat shelf carved in the rock, was useless, far above the water. The island had been uplifted about ten feet.

The understanding of this great earthquake represents a watershed in the development of what is now called plate tectonics—the concept that the geology and physiography of Planet Earth are determined by the movements of great plates on the surface of the planet above a viscous mantle. And also marked another success in the energetic and ultimately fruitful, but sometimes disharmonious back and forth between the geologists, like Darwin, who base their observations on what the they see for themselves in the field, and their more mathematically inclined brethren who argue from a basis in physical theory. Our expedition to Isla Guafo aimed to push that understanding even further, to advance understanding of the processes going on in the time period between great earthquakes.

◆

The branch of geology that deals with the large-scale processes that shape the earth's crust—how seemingly solid layers of rock are folded and faulted to form mountain ranges, how continents are formed, and including how and why an island might bob up and down—has its own name: tectonics. And the ripening of the ideas that make up modern tectonics, plate tectonics—between the time that Darwin abandoned his explanation for Glen Roy and our adventures on Isla Guafo—is a tangled skein of story lines, a heady mix of personality and institutions, of observation and theory, and of curiosity versus pragmatism. The story involves not only purely geology, but advances in physics, chemistry, and . . . evolution.

One key thread through these snarls is the maturation of Darwin's idea of a fluid beneath the earth's crust, an idea that he shared with his contemporaries Charles Babbage and John Herschel. This thread twists through many corners of science, including the study of that most basic physical force, gravity.

To back up just a little, the force of gravity is proportional to mass, and for a planet, the gravity on its surface is related to its density, and the variation of density within it. Isaac Newton, writing in *Principia* in 1687, made an intuitive estimate of the density of the earth. Newton guessed the density was "five or six times greater than if it consisted all of water." Not too far off from the modern value of 5.51 gm/cm3 (with water at 1.0). Pierre Bouguer, the real brains of the French expedition to South America in the 1700s, tried to test another of Newton's ideas. Newton suggested that a plumb bob suspended at a point near a large mountain would be deflected slightly toward the mountain by its gravitational attraction. Bouguer tried this experiment near the large volcano Chimborazo, which was then thought to be the highest mountain on earth. He calculated that the plumb bob should be deflected about 103 seconds of arc from the vertical, but was disappointed that the measured value was only about 7½ seconds. "One can say," he concluded, ". . . [that] mountains act at a distance, but that their action is much less considerable than the scale of their volume." Since Chimborazo was a volcano, he thought that it might be partially hollow, and that perhaps this was the problem.

Measurement techniques improved, but understanding of how and why the measurements of gravity varied upon the surface of the earth was elusive. When Sir George Everest, the surveyor general of India, discovered that his plumb bobs were deflected toward the mass of the Himalayas—but again not as much as was predicted by theory—a whole new area of inquiry, later named isostasy, began. Was the attraction reduced because the rocks of the Himalayas were less dense than those of the surrounding region, or because the mountains had the same density, but had a root of low-density rocks that penetrated into the higher density of the earth's mantle below? Many geologists were a little confused by the debate and why it was important. But once again, those curious raised beaches and that pesky ice played a role.

Thomas Jamieson—to whom Darwin had finally yielded on the question of Glen Roy—first pieced together a history sea level during the ice ages in Scotland. Jamieson argued that Scotland had been significantly depressed by the weight of the thousands of feet of ice piled up on it, but then gradually rose after the melting of the ice. This, he concluded, was the best explanation for the raised beaches of England and Scotland and the slow rising of Sweden. Jamieson also noted one peculiar fact that he did not explain. Parts of Scotland experienced a period of submergence just after the ice had melted. Jamieson did not anticipate that the melting of the glaciers led to a rapid rise in sea level throughout the oceans, inundating low-lying land. In contrast, the rebound of the land after the melting of the ice was a much slower process—one that is still going on today.

In the late 1880s as Jamieson's ideas were finding their way to America, Grove Karl Gilbert, a pioneering geologist of the U.S. Geological Survey, found the Glen Roy of the United States—on a massive scale—in the region surrounding the Great Salt Lake. Gilbert found and mapped four ancient shorelines of the lake, the highest about one thousand feet about the current level. The ancient lake—he named it Lake Bonneville—covered about ten times the area of the current Great Salt Lake. But unlike Glen Roy, Lake Bonneville filled a closed depression. There was no mystery about what dammed this lake.

From careful measurements of the elevations along the shorelines of the former lake, Gilbert discovered something amazing.

The shorelines were not flat, as the shorelines at Glen Roy were perceived to be. Shorelines on mountains that were once islands and peninsulas near the center of the lake were nearly two hundred feet higher than those on the periphery. The level of the shorelines was a mirror image, albeit subdued, of the depth of water in the ancient lake. Gilbert hypothesized that the earth's crust was like an elastic plate—overlying a highly viscous, yielding substrate—and had been bowed down by the extra weight of two thousand cubic miles of water. Gilbert was an exceptional geologist, combining skill in the field with an appreciation for mathematical approaches and physical understanding. At the time, he was unique.

What geologists saw led them to believe that the crust of the earth had had the ability to move around, to be mobile. This belief was anathema for many scientists grounded in physics and mathematics. If the common sense-interpretation of a geologic observation didn't agree with the physics as understood at that time, it couldn't be correct. This erroneous way of thinking extended even into the twentieth century. The leader of this group was William Thomson, the British physicist later ennobled as Lord Kelvin, and not only did Thomson oppose mobility, but his views about the age of the earth also came to vex Darwin, endangering the acceptance of natural selection. Fortunately, eventually, physics would catch up.

◆

Geologists who adopted Lyell's point of view—that the geology of the earth was shaped by ongoing processes—quickly realized that the earth had to be very old. Darwin estimated at least three hundred million years. Lord Kelvin estimated the age of the earth by imagining it as a cooling iron sphere, and settled on an estimate between twenty and one hundred million years, a number that Darwin believed was inadequate to allow the progress of evolution by natural selection.

The discovery of radioactivity in the 1890s began a whole new ball game. First, heat generated by the decay of the naturally occurring radioactive elements completely invalidated Kelvin's calculations. Second, as understanding of the radioactive isotopes resulting from these same decay processes progressed through the twentieth century, this understanding provided a clock to measure the age of the earth and to date events in its history. Now, with the earth's age reliably determined at about 4.6 billion years, Darwin has all the time he needed, as well as a clock to time evolution's progress.

But Kelvin's conceptual model of the earth as a rigid iron sphere also impeded other aspects of thinking about the earth. Even Darwin's son, George Henry Darwin, who became a colleague of Kelvin's at Cambridge, and a pioneering mathematical geophysicist, saw no room for some kind of fluid layer. Following Kelvin, he too was convinced that the earth was "throughout a solid of great rigidity."

In the first decade of the twentieth century the new science of seismology produced two discoveries that complicated the picture. From the beginning, seismologists identified the two types of waves that propagated through the earth, as predicted by the theory of the French mathematician Pierre-Simon Laplace. The first to arrive was called the primary, or P-wave, and the second to arrive, the secondary, or S-wave. The faster P-waves are analogous to the sound waves that carry our voices through the air and the shrieks of dolphins through the ocean. These waves compress the medium as they pass through it. S-waves are different. When they pass through the earth, or indeed any solid, they jerk it from side to side or up or down, with a motion that is transverse to the direction that the wave is going. Only solids can transmit S-waves, not fluids like air, water, or the molten rock that Darwin imagined beneath the earth's crust. This was highlighted by the discovery of the earth's core by Robert Oldham in 1906 and confirmed by Sir Harold Jeffreys, a professor at Cambridge, about twenty years later. Oldham discovered that beginning at a depth of about eighteen hundred miles, nearly halfway to the center of the earth, there is a fluid core through which S-waves do not pass. This was far too deep to provide the kind of ice-over-a-pond situation that Darwin imagined.

In 1909 a Croatian with a name that twists the tongue of an English speaker, Andrija Mohorovičić, studied the waves from an earthquake near Zagreb on seismograms recorded at new seismological observatories across central Europe. To his surprise, on the seismograms recorded at distances greater than about two hundred miles from the earthquake, he found two groups of primary waves. Mohorovičić concluded that the faster group had propagated within the earth's mantle. What he discovered, and was named in his honor, the Mohorovičić discontinuity, or Moho for short, marks the bottom of the crust and top of the mantle. There was no evidence for anything like a fluid layer. This was another big problem for Darwin's idea of the crust floating on lava, like ice upon on a pond.

In 1906 the famous earthquake shook San Francisco and gave the strongest indication to date that earthquakes were generated by movement on "faults." While fractures and offsets in rocks and mountainsides had been long noted by geologists, the possibility that the largest of these, now referred to as faults, could continue for hundreds, even thousands, of miles, emerged slowly. The 1906 earthquake, however, provided spectacular evidence of the horizontal offset and the immense distance over which it could extend along a fault. Something that would have warmed Lyell's heart.

In the months after the 1906 earthquake G. K. Gilbert and others documented horizontal offsets of up to twenty feet or more along a three-hundred-mile stretch of the San Andreas fault. Before the earthquake, this "line or narrow zone characterized by peculiar geomorphic features" running for six hundred miles through the coastal mountains of California was recognized by geologists and local residents alike. Thus arose the understanding of the San Andreas fault as a modern day source of large earthquakes. The paradigm of an "earthquake fault" was born. But one puzzle remained. Land surveys showed that the displacements after 1906 decreased with distance from the fault. Far from the fault there was no change. Fortunately for the future of earthquake science, the commission charged with investigating the event reached out to an Easterner and non-geologist, Harry Fielding Reid.

Reid, a professor at Johns Hopkins and member of the upper crust in Baltimore, was trained as a physicist, spending time in Germany and England with the likes of Lord Kelvin. But increasingly Reid chose to study geologic phenomena. Reid was asked to look at the pattern of these movements of the land, and his insight has proved long-lasting. Reid hypothesized that the process leading up to and during the earthquake was much like what happens when you take a stick in your hands to break it. As you bend the stick, elastic strain is stored in the stick. When the stick finally breaks, the elastic strain is released, and the two pieces of the stick fling apart. But in the case of an earthquake, what was bending the stick?

◆

Also about the time that Darwin was giving up on Glen Roy, a young Austrian, Eduard Suess, was also thinking about both natural selection and tectonics. Suess would argue that the presence of fossils of an ancient and extinct fern called *Glossopteris* in South America, Africa, India, and Australia (where Darwin himself had found these fossils) indicated that these land masses had once been joined in a single continent that he called Gondwanaland. Darwin noted that certain species represented by the fossils that he had collected in South America, such as the giant sloths, as well as their descendents the modern tree sloths, the guanaco and llama, and armadillo, were all unique to the Americas and Caribbean. Again, somehow, they had evolved in isolation from the other continents. But the rocks containing the fossil ferns *Glossopteris* were much older. The evidence from the fossils suggested that South America had been joined to the other continents at the time of Gondwanaland, then later isolated, before finally rejoining with North America.

Curious minds had long noticed the striking congruence of the coastlines of eastern South America and western Africa, appearing like the outlines of two giant pieces of a puzzle, longing to be put back together. In 1912 the German meteorologist Alfred Wegener took up the cause where Suess left off, and with a vengeance. He began publicizing his own theory of tectonics that he called "continental drift," hypothesizing that the continents had moved around, drifting to their current positions. He matched not only the coastlines and Suess's fossil ferns, but also a variety of other rock formations on continents that were now separated by thousands of miles of ocean. Wegener picked up a few important followers for his ideas, especially Alexander du Toit in South Africa, Arthur Holmes in the UK, and Harvard professor Reginald Daly, but most geologists—particularly in the United States—either thought the idea was absurd, or at least too speculative to be useful or taken seriously. And just as Darwin criticized Agassiz for being unscientific in his hypothesizing the previous extent of glaciers, so critics attacked Wegener.

Sadly, Wegener had no Hooker to gently explain to him—as Hooker had indirectly intimated to Darwin back in the 1840s—that to gain credibility in a field of science different from one's own, it was necessary to labor in those trenches too. Wegener didn't spend his eight years with barnacles. And a bit like the anonymous author of the speculative *Vestiges of the Natural History of Creation**—who preceded Darwin in publicizing in 1844 the then-scandalous idea of evolution of the species, but without a satisfactory, much less persuasive, mechanism to explain the process—Wegener was smashed by his critics. He was only able to attract a small following among geologists as he, tragically, died a relatively young man during an ill-fated expedition on the Greenland ice sheet.

◆

Over the next several decades the debate about continental drift sputtered along between the drifters and anti-drifters. The drifters didn't go away, but neither did they produce compelling evidence. The paleontologists, most notably the South African du Toit,

---

* The author was later revealed to be Robert Chambers, a Scottish journalist.

continued to see patterns in the distribution of plant and animal fossils that supported the idea. Later, geochemists who studied the ages and composition of the oldest rocks on the continental shields also saw matches across the presumed connections, and joined in. Geologists in the alpine countries, just as a century before, were impressed by the complexity of the structure of the Alps and increasingly saw evidence for large-scale horizontal displacement and mobility. In Britain, Arthur Holmes promoted the idea that continental drift was a consequence of a process called convection in which giant currents of hotter rock somehow rise in the mantle, then sink when the rock cools, in an overturning motion as in a giant pot of simmering soup. Opposing Holmes was the seismologist Sir Harold Jeffreys, an implacable foe of continental drift, who argued that the mantle was too strong and that flow in the mantle was impossible, notwithstanding the evidence for isostasy and postglacial uplift. Most American geologists regarded continental drift as if it were science fiction. Many likened the movement of the continents to battleships plowing through solid rock. How could it possibly work?

◆

While some seismologists, like Jeffreys, focused their attention on the structure of the earth, others were intrigued by the location and distribution of earthquakes and how they might be related to geology. One leading group was the Seismological Laboratory at Caltech, known simply as the Seismo Lab. Originally founded near Caltech as a part of the Carnegie Institution of Washington in 1921, it was transferred to Caltech in 1937. The group's leaders, Beno Gutenberg, Charles Richter, and Hugo Benioff, although all trained in math and physics, devoted considerable energy to the descriptive aspects of earthquake studies. Richter and Gutenberg pioneered earthquake magnitude, but also a description of where earthquakes occurred, the seismicity of the earth. Most of the earth's earthquakes (and volcanoes as well) occur around the margins of the Pacific Ocean, the so-called Ring of Fire. But they also documented some important subtleties, like narrow bands of earthquakes running down through the middle of the Atlantic, Indian, and Pacific Oceans. While most earthquakes occur within about twenty miles of the surface of the earth, Benioff—and independently, Kiyoo Wadati, a Japanese seismologist—recognized that some earthquakes around the Pacific Rim occur in planar zones that dip landward beneath the coasts down to depths as much as four hundred miles. These zones are now called Benioff, or Wadati-Benioff zones, in their honor.

Meanwhile other seismologists showed that the earth's crust was indeed thicker beneath most mountain ranges, partially, if not completely, supporting the iceberg-like model of isostasy. In contrast seismologists investigating the crust beneath the oceans discovered that it was fundamentally different, thinner and more dense, than the crust beneath the continents.

◆

After Chile was battered by the great earthquake in 1960, no one found a fault. The geologic origin of the earthquake presented a huge puzzle. Then, on Good Friday, March 27, 1964, another great earthquake struck—half a world away from Isla Guafo—in Prince William Sound, Alaska. Although not quite as large as the Chilean earthquake, it too was a giant earthquake, and still holds the record for second place since score has been kept. And studies of this earthquake in Alaska, half a world away, held the keys to understanding the earthquake in Chile.

Between 1960 and 1964 seismographic instrumentation had improved significantly—as had interest in seismology generally—owing to the possibility of using seismic waves to detect nuclear explosions and to differentiate them from natural earthquakes. Both the geologists and newly energized seismologists were on top of this earthquake.

Geologists flew to Alaska and began their investigations in the field. The seismologists studied the records on their permanent instruments, and some took portable instruments to the field to determine more precise locations and depths of the thousands of aftershocks spread across a region 250 by 500 miles, from the Aleutian trench landward beneath the coastal waters and mountains of Alaska. Instrumentation and techniques available at the time did not allow a very precise depth of the main shock, but most of the aftershocks were shallower than about twenty-five miles.

George Plafker—a tall, brash, USGS field geologist with dark, wavy, slicked-back hair and a degree from Brooklyn College—was one of the first on the ground. Already an Alaskan veteran, he hopped rides with a bush pilot, visiting small fishing villages around Prince William Sound, and as far to the west as Kodiak Island. The tsunami following the earthquake damaged many coastal towns and villages, obliterated some. Initially George was looking for a fault that broke the surface of the ground, something like G. K. Gilbert found after the 1906 San Francisco earthquake, a gash offsetting fences and roads by several feet. Except for a relatively short fault on one island, it wasn't there.

Instead, what he found over much of the area were barnacles newly elevated above the high tide line, dying. Species that usually lived in the zone between the limits of the tides were now high and dry, just what Maria Graham, FitzRoy, and Darwin had seen in Chile. Elsewhere he found evidence for subsidence. He collected measurements from tide gauges, land surveys, and interviews. Fishermen, he would later say often, have a keen sense of where sea level is and when it changes. As his fieldwork continued over the summer of 1964, he discovered that new, young barnacles began to grow and colonize just below the new, post-earthquake high tide line. The difference between the upper limits of the pre- and post-earthquake barnacles provided a measure of the change in the level of the land during the earthquake. Eventually he cataloged more than eight hundred measurements of vertical changes over one hundred thousand square miles.

George shared some of his preliminary measurements with a prominent seismologist, Frank Press, then the director of the

Seismo Lab at Caltech. Press would later go on to MIT, and subsequently to be President Jimmy Carter's science advisor, and then president of the National Academy of Sciences. But in 1964 he was trying to use George's data to understand the Alaska earthquake. He and a bright young Caltech senior, David Jackson, plotted twenty-six of George's data on a profile perpendicular to the coast and to the trend of the aftershocks of the main shock. Press and Jackson compared these data to some simple calculations from the emerging theory of elastic dislocations. Their conclusion, published in *Science* magazine less than a year after the earthquake, was that the earthquake occurred on a vertical fault with the Pacific side up and the Alaska side down. The hypothesized fault, located along the hinge line separating the zone of uplift from the zone of subsidence, extended from within a few miles of the surface to a depth of 60 to 120 miles.

George Plafker might not have been on quite such an august academic track, but he was not so sure that Press and Jackson were right. For one thing, their hypothesized fault didn't fit with the geology at all. The geology of the islands and coastal mountains was characterized by folds and shallowly dipping thrust faults suggesting compression perpendicular to the Aleutian Arc. The surface fault that George and his colleagues did find showed the landward side was up and the seaward side down, opposite to the sense suggested by Press and Jackson. A few months after Press and Jackson's paper appeared, George published his more complete data set and an array of arguments for why the 1964 earthquake had not occurred on a deep vertical fault, but rather on a shallow thrust fault, extending landward from the Aleutian Trench and gently dipping to the north, beneath the Alaskan coast.

Over the preceding decade seismologists had been struggling to use their observations to determine orientation of the fault responsible for an earthquake and the direction of slip upon it. It was still a work in progress. There were many problems, both practical and theoretical. But what had not been possible for the 1960 earthquake seemed almost within the reach of the seismologists for the 1964 earthquake. The theory available at the time allowed them to determine the orientation of two perpendicular planes: one, the plane of the fault, the second, a plane perpendicular to the direction of fault motion, but they could not distinguish which plane was which. For the Alaska earthquake one plane indicated a vertical fault with the seaward side up, as Press and Jackson suggested, the other a very shallowly dipping, almost flat, thrust fault, with the landward side thrusting up and out over the ocean as George contended.

◆

In publications and at scientific meetings the debate about the 1964 earthquake raged on. After George made a presentation at a meeting in 1967, another well-known earthquake geologist from Caltech, Clarence Allen, approached him with a suggestion: why didn't he go to Chile to see what level changes had occurred during the 1960 earthquake? With Allen's help, George got a small grant, took leave from the USGS, and went to Chile.

Just as in Alaska, George traveled the length and breadth of the area affected by the earthquake. He chartered a fishing boat to visit the remote Chonos Islands far to the south. Where he couldn't visit, including Isla Guafo, he obtained reports from credible observers. For Guafo he had two independent measurements made by Chilean naval officers. He obtained data from Chilean surveys made before and after the earthquake. In all he collected vertical change data at 155 sites. The resulting picture was quite similar to Alaska, although in Chile most locations along the coast subsided during the earthquake, while uplift was largely limited to the offshore islands, with Islas Guamblin and Guafo reporting the two largest values. This time, in writing up his results, George teamed with another USGS scientist, Jim Savage, an expert in elastic dislocation calculations himself. George's data was well fit by a shallow dipping thrust fault. The case was closed. Both these great earthquakes occurred on shallow dipping thrust faults, with the continent thrusting up and out over the ocean.

◆

During this time the old debate about continental drift warmed up to a fever pitch. In 1963 two geologists at Cambridge University, F. J. Vine and D. H. Matthews, had an amazing insight. Oceanographers had mapped an extremely puzzling pattern of magnetic anomalies across the oceans around the world. One clue seemed to be that the anomalies were vaguely symmetrical about the mid-ocean ridges. The Cambridge pair realized that this pattern was consistent with a recently proposed idea called seafloor spreading. The idea was that the seafloor was formed by the solidification of volcanic rocks at the mid-ocean ridges, then the new rocks were split in two and moved away from the ridge by upward and outward circulating convection currents in the earth's mantle below. Vine and Matthews knew that as the rocks cooled they would be magnetized by the earth's magnetic field as it existed at that instant. Further, it had just been shown that the earth's magnetic field alternated polarity over the last three and a half million years. They posited that the ocean floor on each side—as it moved away from the ridge—acted as if it were a humongous piece of magnetic tape, faithfully recording the history of the magnetic field as it shifted back and forth from one polarity to the other, over the millennia.

Gradually the data and techniques available to the observational seismologists allowed them to make reliable determinations of the fault orientation and slip direction, if not the essential ambiguity between them. Lynn Sykes and a group at Columbia University in particular, made these determinations for earthquakes along the mid-ocean ridges, elsewhere beneath the oceans, and in the Benioff zones. What they found exactly fit the models proposed for sea floor spreading at the mid-ocean ridges. In the upper parts of Benioff zones they agreed with George Plafker's conclusions from the great Alaskan and Chilean earthquakes. Piece by piece the picture came into focus. The mid-ocean ridges were the consequences of the up-going limb of convection currents, while the ocean trenches, their associated Benioff zones, and giant thrust earthquakes were all consequences of the corresponding down-going limb.

By the early 1970s, the revolution was complete. The rebels were in control of the castle. The geological sciences had a new paradigm combining continental drift, seafloor spreading, and convection currents. It was called plate tectonics. Plates form at the mid-ocean ridges. They move away at rates of a few inches per year, carried by thermal convection currents in the earth's mantle below. Some plates are purely oceanic, others carry continents upon them. The deep trenches of the oceans, including those around the margins of the Pacific Ocean, are the results of one plate, typically the denser oceanic plate, sliding beneath another, typically the continental plate. This is the newly named process of subduction, which also leads to melting above the down-going plate and to the formation of volcanoes. The San Andreas and other great faults are the boundaries between plates sliding past one another. When two plates bearing less dense, buoyant continents collide, neither is willing to subduct, and the resulting crash gives rise to the largest mountains on earth, including the Himalayas.

◆

Once, as I stood in my waders, sinking into the mud on the bank of Alaganik Slough in the delta of Alaska's Copper River, George Plafker stood above, retelling the story of his battle with Frank Press and the geophysicists about the 1964 Alaskan earthquake. It was mid May, early spring in Alaska. Minutes before I had watched a pair of trumpeter swans paddling in a pond, then take off gracefully, just skimming the water until they gained altitude and lifted into the sky.

George, now in his late eighties, is slowing down just a touch, but still riding his bike, and still sharp as a tack. As a graduate student I had attended a seminar that George gave on the 1960 Chilean earthquake. A few years later I learned not to try to keep up with an Alaskan field geologist when he drank me under the table one night after an earthquake meeting. Here on the banks of the Alaganik in the 1980s, George had found a sequence of buried layers that told the history of repeated great earthquakes in Prince William Sound prior to 1964. We were revisiting the site as part of a group on a field trip to see the deposits at his classic site. He loves to tell the stories of his Alaskan and Chilean adventures, and with only a bit of prodding he retold this one. I asked George what he thought Frank Press had been thinking during their face-off. George said that Press had told him later that he had only made one mistake, and that George had caught him in it. Despite working closely with geophysicists over many years, he delights in needling them. Hanging in a frame on the wall of his office is an inscription of what George calls "Plafker's Law." The law states, "When the geology and geophysics clash, throw the geophysics in the trash." One can easily imagine that—in retrospect—this epigram would have given Darwin a chuckle, remembering his difficulties with Lord Kelvin. Notwithstanding the value of George's experience, I'm not quite ready to accept his law. Progress, it seems to me, benefits from multiple views. Nonetheless, George stands with Maria Graham, FitzRoy, and Darwin on the list of top geologic contributors to the understanding of great subduction earthquakes.

◆

One place affected by the great Chilean earthquake that George Plafker had not been able to visit himself was Isla Guafo. I was tickled to be there. From the point of view of plate tectonics, Isla Guafo is a very special place. It rides the leading edge of the South American Plate, as the oceanic Nazca Plate slides beneath it. Guafo's proximity to the edge of the plate, only thirty-five miles to the west, explains the large uplift during the 1960 earthquake. Few places on the planet—above sea level at least—share such an intimate view of subduction.

The year before my visit to Guafo—even as I was looking for deposits from ancient tsunamis in central Chile with Brian

Atwater—Daniel Melnick and Marco Cisternas were astonished to discover that Isla Guafo was sinking. Now, a year later, in 2008, we were back to fill in the story.

When Daniel and Marco first walked on the rocky platform extending seaward from the beach at low tide, they were shocked to find something sandwiched between the sand of the beach and the rock below, something completely out of place. Lying on the platform, above the tide pools, was a layer of peaty organic soil with the roots of bushes still visible, a freshwater soil now being actively eroded and washed away.

As they looked more closely, the plot thickened. The erosion of the soil exposed the holes bored into the soft mudstone by a small mollusk, holes still occupied by their empty shells. These creatures—with the charming name of *Petricola patagonica*, given them by Darwin's French contemporary, Alcide d'Orbigny—spend the happy days of their lives below the limits of the tides. Now their empty shells were many feet above that level. It seemed likely that for these poor *Petricolas*, May 22, 1960, was a very bad day indeed.

But that wasn't all. As Marco—a master of observation—poked around in the eroding soil, he made another remarkable find, a piece of plastic that he recognized from his school days. In the early 1970s a Pinochet-era government program distributed a protein supplement powder, Fortesan, to Chilean schoolchildren to improve their nutrition. Now, three and a half decades later, Marco recognized the remainder of a plastic bag, clearly labeled, that once contained the Fortesan that he had been given as a child.

With these seemingly disparate discoveries, this episode of the geologic story of the island slowly came into focus. At the time of the 1960 earthquake, as Guafo rose about ten feet, it lifted the rocky platform—with the *Petricolas* sadly attached—from below the level of the low tide to above the level of the high tide. There, as the years passed with the platform above the reach of salt water, a freshwater soil rapidly developed upon it. Then at some point the island began to sink, allowing the waves and tide to erode the new soil away.

What was going on and why?

◆

Three days before, in front of the church on the town square in Puerto Montt, I had finally met Marco Cisternas. Marco, a tall bespectacled man with short salt and pepper hair and a quick sense of humor would serve as the adult-in-residence for our team. With him was Marcelo Lagos, who had been the silver-tongued guide for our team the previous year.

After a few minutes, Cesar Vera, a small, but powerfully built young man, joined us. "Our professional digger," Marco introduced him. Marco knew Cesar and his family well from many years of studying tsunami deposits on their land near Maullín, on the coast west of Puerto Montt.

Later that afternoon I also met Daniel Melnick, our leader, who I previously only knew through his publications on Isla San María. Many Chileans might mistake Daniel, a tall young geologist with receding blond hair and bright blue eyes, for a gringo. Until he opens his mouth. Though currently living and working at the University of Potsdam in Germany, Daniel is Chilean, and switches effortlessly back and forth between refined *castellano* and native slang. With him was Kako Rodríguez, a witty, gnomish mountain guide hired by Daniel to handle our logistics.

After Daniel and Kako finished rushing around town, gathering the last of our supplies, we drove south from Puerto Montt, crossed by ferry to the large island of Chiloé, then drove a hundred miles further to Quellón. There, despite our arrival after midnight in pouring rain, Daniel had arranged for dinner. The next morning we met Kapi, the captain of the converted fishing boat, the *Tirana*, that would take us on to Isla Guafo.

After a delay for the weather to improve, we set off around the southern end of Chiloé to the outpost of Inío, then across the twenty-mile stretch of open sea to Isla Guafo. Crossing the heaving open water, with swells rolling in from the Pacific, my sympathy for Darwin and his propensity toward seasickness found new dimensions.

◆

On the morning of our planned hike to the lighthouse, Daniel, Marcelo, Kako, and I shouldered our heavy loads. Most annoying was an awkward set of long orange plastic pipes that we carried by turns. Eventually the mist lifted, and the sunshine sparkled off the waves crashing on the rocks along the beach. Sea birds strode the beach and soared above, piercing the air with the familiar squawk of gulls, and occasionally the metallic, duck-like honk of the *bandurrias*. Oystercatchers prowled the tide pools. Graceful, long-necked cormorants, unfazed by the waves sloshing at their feet, surveyed the scene.

One bird, the *caranca*, always appeared in pairs, a bright white male and a dull black female, constantly walking the beach within a few feet of one another. "They mate for life," Daniel explained with a smile. "No divorce, no infidelity, no messing around with other birds." Later I learned that in English we call the caranca by a considerably less romantic name, the beach goose.

Most of the hike was along a narrow, rocky beach crammed between a high cliff and the sea. After about four miles the cliffs were broken by a large landslide. Here, aided by a long rope anchored to trees, we pulled ourselves and our loads, including the vexatious orange pipes, to the top of the cliff. Following a faint trail through the dense rain forest, we emerged on a windswept headland overlooking the Pacific. The lighthouse loomed into view.

Daniel had two objectives. First, he planned to install a permanent GPS station that would record continuously, receiving power and nominal oversight from the lighthouse and its staff. Our troublesome orange plastic pipes would protect the cables carrying power to, and data from, the receiver antenna. He also wanted to repeat, with a portable GPS instrument, a survey done fifteen years before to learn how Guafo had moved in the intervening years. Both of these instruments were much more accurate than the ordinary GPS unit in a car, or a handheld GPS taken on a hike, yielding measurements accurate to about a tenth of an inch.

Leaving Kako to begin the installation of the permanent station, Daniel, Marcelo, and I set off to find the markers left by the previous surveyors. One site at the top of the cliff immediately seemed problematic. A step off the trail the grass and bushes were knee high. Even worse, it appeared that active landslides had been eating back into the cliff where we thought the site should have been. After a brief, fruitless search for some kind of monument, we abandoned the effort.

We hiked down a steep switchbacking trail to the water and the rock surface that had been uplifted ten feet or more in 1960. A concrete path led to a simple pier on the rocks. A small white shrine, topped by a cross, and containing a small statue of the Holy Virgin stood on a ledge. Somewhere near here a small brass monument, about the size of a dime, should be fixed in the rock. We began to search.

Daniel found it first, a tiny disc of metal, fixed on the surface in the expanse of rock. Here we carefully set up the antenna exactly over the indentation in the center of the disc, and connected the receiver and batteries, which we placed in weatherproof boxes behind the shrine.

Returning to the lighthouse, we had more work to do installing the permanent station. The sailors insisted we stay for dinner, and we still had a five-mile hike back to our camp. As I projected our estimated time of arrival back at camp, I realized that I had committed a blunder beyond foolish. I had failed to bring a headlamp.

Though long and challenging, our return was otherwise uneventful. I found my way along the rocky beach in the soft light of my colleagues' headlamps, arriving at our tents after midnight.

◆

The next two weeks sped by in a flash. Most of our work was in Cesar's pits in the magical rain forest that rose behind the beach. To reach them one would first pass through thorny nalca plants with giant leaves and dense bushes of wild fuchsia. Farther back the first red-barked arrayán trees appeared with dense bushes at ground level. A hundred yards from the beach rose a Tarzan-esque jungle with vines and delicate air ferns hanging from giant trees. Sunlight does not reach the ground here and neither did my feet. I swung from fallen log to root.

I often paired with Marco taking notes as he stood in a pit almost out of sight below me describing the deposits. Even during breaks from the rain—with spectacular sunshine high above the forest canopy—the water table in the forest was so high that the pits quickly filled with water. My notes would go along fairly cleanly, until Marco would pass me a sample of sand or mud for my opinion. Notwithstanding the waterproof paper of my notebook, its pages frequently turned into a muddy mess.

What we found were buried soils like the one on the beach created after the 1960 earthquake, but older and likely indicating previous episodes of uplift and subsidence. We searched for samples of organic material we could use to date with carbon-14. In one pit we dug into the remains of an ancient *curanto*, the still locally popular method of cooking shellfish and potatoes by burying them in the sand with the coals of a fire.

In the evenings our dinner conversations in the dome tent were dominated by what we were finding at Guafo and the seemingly clear evidence that Isla Guafo, once uplifted in the 1960 earthquake was now sinking rapidly. One night the conversation expanded from the ups and downs of Isla Guafo to include Isla Santa María, some 450 miles to the north. Daniel and I shared our experiences and views about what was happening there. I told Daniel about the chart that I had found of the *Beagle*'s survey at Isla Santa María made shortly after the uplift of the island in the 1835 earthquake. He too was now suspicious that Santa María was sinking. It would be simple enough, we both thought, to get some kind of echo sounder with a GPS to repeat the *Beagle* survey, to see what we might find.

We resolved to give it a try.

◆

After two weeks of fieldwork at Isla Guafo, finishing the installation of the permanent GPS station, and retrieving the portable station and its data, we awoke to see Kapi's *Tirana* bobbing off the beach. We were on our way home. Over the two weeks we had accumulated many interesting "facts," as Darwin would have described them. But did we have enough to unravel the history of ups and downs at Isla Guafo?

Back in Germany in the months that followed, Daniel processed the GPS data and submitted our samples for carbon-14 dating. The results confirmed some tentative correlations of the buried soils in the different pits that we had made in the field. It appeared that we had evidence for six distinct episodes of vertical change ranging from about 1315 B.C.E. to about 1360 A.D.

And the GPS data? Isla Guafo was moving to the east at the rate of almost two inches per year, and sinking at almost half an inch per year. A very good collection of "facts" was beginning to emerge, facts to illuminate the movements of the island in the periods between earthquakes.

CHAPTER FIFTEEN

In the *Beagle's* Wake

> *The most remarkable effect (or perhaps speaking*
> *more correctly, cause) of this earthquake [of 1835]*
> *was the permanent elevation of the land.*
>
> —Charles Darwin

Two days after feeling the earthquake in Valdivia of February 20, 1835, the *Beagle* set sail to the north to continue her surveys onward toward Concepción. Beaufort had instructed FitzRoy to pay attention to this section of the coast: "It will require no great expenditure of time to correct the outline, and to fix the positions of all the salient points. Mocha Island is supposed to be erroneously placed: and the depth, breadth, and safety of its channel are not known."

It wasn't quite so easy. Late summer along the coast of south-central Chile is characterized by a strong ocean swell from the southwest—now prized by surfers—but for FitzRoy it presented a real hazard. "A dangerous coast," Darwin described it.

As FitzRoy struggled to land on Isla Mocha and complete a survey of the surrounding waters, the crew felt frequent, strong aftershocks. Some led FitzRoy to think "that the anchor had been accidentally let go, and the chain was running out."

And as the officers completed their survey, they fixed the position of one point on the mainland, opposite Isla Mocha, Cape Tirúa. Tucked beside the cape, probably hidden from FitzRoy's inquiring gaze by a bar, lay the mouth of a small river and a settlement of the native Mapuche, Tirúa.

The relationship between the Mapuche and the Spanish was not a happy one. Indeed, despite efforts to subdue the Mapuche, beginning with the conquistador Pedro de Valdivia in the 1500s, the Spanish had failed, just as the Incas had failed before them. No less a personage than Sir Francis Drake was wounded and lost two men in a battle with the Mapuche on Isla Mocha in 1578. In the 1830s, except for the lonely outpost at Valdivia and the offshore islands of Mocha and Santa María—where the Spanish had largely cleared the native people—the Mapuche controlled the coast of Chile more or less from the Biobío River at Concepción to the Island of Chiloé. Europeans were not welcome, and those who tried to visit risked their lives. At Isla Mocha and Isla Santa María, pirates, smugglers, whalers, and sealers rested and took on fresh water[*], but visitors ventured onto the mainland only at their peril.

No one from the *Beagle* took the risk.

FitzRoy was sympathetic to their plight, but knew enough not to tangle without good cause. "I reflected on the multiplied sufferings undergone by their ancestors—the numbers that perished in mines—or in trying to defend their country—and the insidious attempts made to thin their numbers by frequent intoxication, if not by introducing deadly disease." Nonetheless he regretted their claim on the fertile lands of south-central Chile: "Probably the finest district in all South America, is still kept by the brave Araucanians [as FitzRoy referred to the native people]." After independence the Chileans had little more success, until finally mostly subduing the Mapuche in the 1880s. Nonetheless, the Mapuche community

---

[*] A white whale, Mocha Dick, known to whalers who used Isla Mocha as a base around the time of the *Beagle*'s visit, was the inspiration for the whale in Herman Melville's novel, *Moby-Dick*.

ABOVE: FIGURE 35. A modern view of the three principal Parallel Roads within Glen Roy. The highest road here is at an elevation of about 1,148 feet and the lowest at 853 feet. BELOW: FIGURE 36. The author examining the largest of Darwin's boulders in Cwm Idwal in Snowdonia, Wales. Behind the boulder and to the left is a glacial moraine that dams the lake. The moraine was deposited by a glacier at the end of the Pleistocene, or possibly during the Younger Dryas.

FIGURE 37. This outcrop of bedrock below the moraine at the lip of Cwm Idwal shows numerou
scratches, or striations, made by rocks dragged across it by the ice at the bed of a glacier. On
prominent striation is indicated by the white arrows. These striations, along with the moraine
and other signs, are what convinced Darwin that this valley had been filled by a glacier. Th
valley in the background displays the classic U-shape now associated with valleys that ar
carved by a glacier.

ABOVE: FIGURE 38. The lighthouse on the western headland of Isla Guafo looking out over the Pacific Ocean. Sailors manning this lighthouse experienced the Chilean earthquake of 1960, which uplifted the island around ten feet, and witnessed the accompanying tsunami rushing toward the island. BELOW (LEFT AND RIGHT): FIGURES 39, 40. Daniel Melnick and Marco Cisternas were astonished to find this rapidly eroding soil below the beach on Isla Guafo. They were even more astonished to find, embedded within the soil, this plastic wrapper (Figure 40) of the food supplement Fortesan that Marco and other Chilean school children had been given in the 1970s. These discoveries established that the soil had formed after the 1960 earthquake, and that the island is currently subsiding.

ABOVE (LEFT AND RIGHT): FIGURES 41, 42. Marco Cisternas examining shells of *Petricola patagonica*, a rock-boring mollusk that normally lives below the low tide line. These creatures were killed when the rock platform was uplifted with Isla Guafo in the 1960 Chilean earthquake, and are now covered by the post-1960 soil. BELOW: FIGURE 43. Marco Cisternas logging a pit in the rain forest on Isla Guafo. The sequence of sands and soils exposed in the walls of the pits studied by the team indicate a series of vertical changes over the last few thousand years.

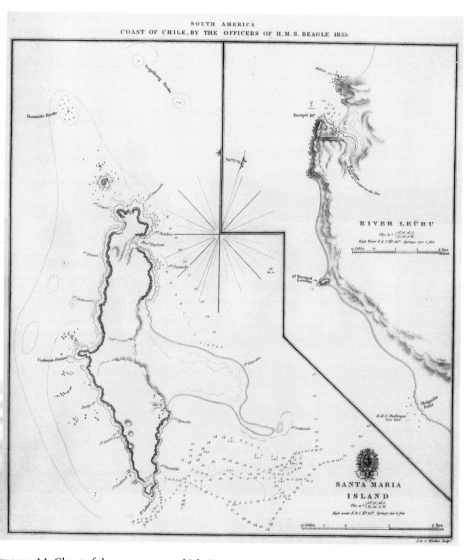

FIGURE 44. Chart of the waters around Isla Santa María made by the officers of the *Beagle* a few weeks after the earthquake of February 20, 1835. Making charts like this one was a principal task of FitzRoy. Upon arriving at the island in late March, 1835, FitzRoy realized that the island had been uplifted eight to ten feet in the earthquake. The author and colleagues compared the water depths shown by the soundings on this chart (the small numbers) with their own survey in January 2010 to show that the island had subsided almost five feet in the intervening 175 years.

FIGURE 45. Lisa Ely receiving a lift from Don Toti and his horse Ruperto during field work on Isla Santa María in January, 2010, six weeks before the magnitude 8.8 earthquake on February 27 that uplifted the island about five to six feet.

ABOVE: FIGURE 46. Daniel Melnick looking out over part of what was FitzRoy's rocky platform at the north end of Isla Santa María in January, 2010. This platform had been uplifted and covered with dead sea life when examined by FitzRoy in 1835. The platform became submerged as the island subsided between 1835 and 2010, then partially reemerged in the earthquake of February 27, 2010. BELOW: FIGURE 47. Javier, a local shellfish diver, piloting his boat in January 2010 during the resurvey by the author and colleagues of the water depths in the bay southeast of Isla Santa María. Comparison with the soundings made by the crew of the Beagle 175 years before showed that the bay was about five feet deeper in January 2010.

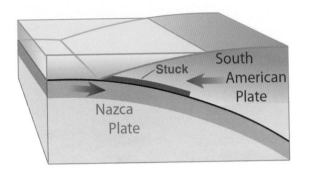

FIGURE 48. A SLICE THROUGH THE SUBDUCTION ZONE
The Nazca Plate descends, or "subducts," beneath the South American Plate. This kind of plate boundary is called a "subduction zone." When the plates move suddenly in an area where they have been stuck, an earthquake happens.

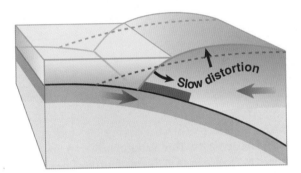

FIGURE 49. BETWEEN EARTHQUAKES
Stuck to the Nazca Plate, the South American Plate gets squeezed. Its leading edge is dragged down, including islands like Santa María and Guafo, while an area behind bulges upward. This movement goes on for decades or centuries.

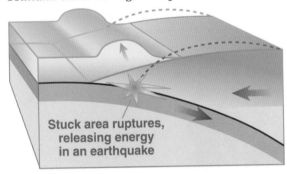

Tsunami starts during earthquake

Stuck area ruptures, releasing energy in an earthquake

FIGURE 50. DURING AN EARTHQUAKE
An earthquake happens when the leading edge of the South American Plate breaks free and springs seaward, raising the sea floor, islands, and the water above it. This uplift starts a tsunami. Meanwhile the bulge behind the leading edge collapses, commonly lowering coastal areas mainly onshore.

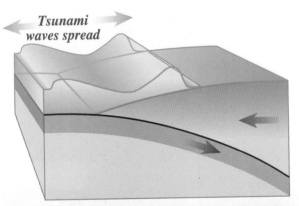

Tsunami waves spread

FIGURE 51. MINUTES LATER
Part of the tsunami races toward nearby land, growing taller as it comes ashore. Another part heads across the ocean toward distant shores.

ABOVE: FIGURE 52. A street in downtown Talcahuano after the earthquake and tsunami of February 27, 2010. *Image © AP Photo/ Natacha Pisarenko.* BELOW: FIGURE 53. Debris on the beach in the seaside town of Dichato, Chile, after the destruction wrought by the tsunami of February 27, 2010. *Image © Ric Francis/ZUMA Press/Almay Stock Photo.*

FIGURE 54. Don José Luis in the yard of his home in Lipimávida describing his experience escaping the tsunami of February 27, 2010 to Marco Cisternas.

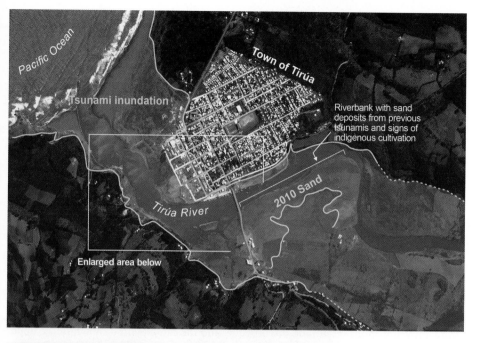

FIGURE 55. Post-tsunami image of Tirúa showing the locations of people whose stories of the tsunami in 2010 are related in the text. Prior to this tsunami, the author and colleagues recognized the deposits of at least three prior tsunamis in the riverbank and field. The 2010 tsunami created a new deposit of sand, but this time the team had the stories of the people affected by the tsunami (and their own) to go with the sand. *Base image: Google, Digital Globe.*

ABOVE AND BELOW: FIGURES 56, 57. In Tirúa in April 2009 the author and colleagues stayed with Don Pedro in his restaurant, discotheque, and the cabin behind shown in the upper photograph. They tried to stay here again in January 2010. In May 2010 Lisa Ely and the author encountered the scene shown in the lower photograph. Both buildings were completely destroyed in the tsunami of February 27.

Upstream →

2010 tsunami sand

1960 sand

1751 sand

1575 sand

FIGURE 58. Lisa Ely and Tina Dura examining the deposits left by tsunamis in the riverbank at Tirúa. The sand layers deposited by the tsunamis are more easily eroded than the silty layers deposited by river floods leading to the prominent notches in the riverbank.

FIGURE 59. Marco Cisternas and his assistant surveying elevations of dead intertidal sea life on Isla Santa María after the uplift accompanying the February 27, 2010 earthquake. Marco is surveying the same rocky platform where FitzRoy encountered dead sea life after the island uplifted in the 1838 earthquake. In the intervening 175 years the platform had again disappeared beneath the waves as the island subsided, dragged down by the subduction of the Nazca Plate.

ABOVE AND BELOW: FIGURES 60, 61. A small inlet, Puerto Ingles, on the northeast corner of Isla Santa María six weeks before (upper photograph) and one year after (lower photograph) the February 27, 2010 earthquake. The uplift of about five and a half feet during the earthquake exposed the rocky platform.

FIGURE 62. The author in January 2010 examining trees flattened by the explosion of Chaitén volcano, about twenty months. The dome of the volcano is still steaming in the background. The light gray sandy material covering the ground is volcanic ash—fragments of volcanic glass and rock—left by the pyroclastic flows accompanying the explosion.

has not been particularly well treated since, and resentment—and occasional violent episodes—continue to the present day.

◆

April 2009 again found me jammed in a camioneta, this time headed south from Concepción, through modern Mapuche country, toward the town of Tirúa. With me again were Marco and Marcelo, but this time we were joined by Lisa Ely, from Central Washington University, and her student, Caitlin Orem. A canoe was fixed to the top of our truck.

Lisa, a tall, energetic woman with a quick sense of humor and a mop of tight curly hair, had been dubbed *la geóloga crespa*, the curly geologist. This was Lisa's first trip to Chile. She too had fallen under Brian Atwater's spell and been entrained in the search for evidence of ancient tsunamis. She'd spent the previous two months on a sabbatical at Marco's university in summery Valparaíso, and had a small grant from the National Geographic Society to see if she could find any evidence of "Darwin's tsunami" from the 1835 earthquake.

We began our search for evidence of the 1835 tsunami in marshes on the coast around Concepción. We had found some possibilities, but now we were headed a hundred miles south to try our luck in Tirúa. We passed through Cañete, near where Pedro de Valdivia met his end at the hands of the Mapuche. Finally, the road emerged at the coast. As it crossed a terrace, high above the sea, we got our first glimpse of Isla Mocha.

Marco, two years before, had done a reconnaissance of the entire coast of south-central Chile, living in his small camper truck, searching for evidence of historic and earlier tsunamis. In Tirúa he had seen something promising, and we were on our way to investigate.

It was April, early fall in Chile, and the afternoon was overcast, threatening rain. My first impression of Tirúa was of a gritty, soggy little town of dreary, wooden houses with corrugated iron roofs, spreading down off a hill toward the beach. Chileans seem to have a relatively relaxed approach to hotel reservations. We had none. We had hoped to stay at a small *hostal* known by Marco and Marcelo, but upon arriving there we found no one home, except for a large, fierce-looking, reddish-brown dog who growled at us from behind an iron gate.

After a fruitless search for the proprietress, Doña Clemedia, we drove down the hill to the main street and turned left three short blocks to the river. A concrete seawall protected the town from the river, but on the opposite side, a low bank rose to a flat grassy field. This bank was our objective.

At low tide, from the bridge crossing the river, Marco had seen three horizontal and parallel notches in the muddy riverbank, extending at least a few hundred yards up the river. He suspected that each notch marked a layer of sand. Sand would be more easily eroded from the bank that was mostly composed of dense silt. And a layer of sand might easily have been carried up the river by a tsunami only a mile from the sea. Later, I couldn't help imagining that Darwin—had he braved the Mapuche and rowed in a whaleboat up this river—would have noticed these notches, and wondered about their origin.

Eventually when the tide fell and exposed most of the bank, we parked the truck and everyone donned waders or rubber boots. To get a really good look at the riverbank, wading in the river would be required. We shouldered our packs and shovels, and slithered through a wire fence. Atop the bank were piles of fishing nets and gear. Local fishermen tied up here to clean and dry their nets. A flock of ugly black vultures stood nearby. We picked our way through the fishermen's ropes, buoys, and nets.

"Ooh!" Lisa exclaimed. A penguin caught and drowned in a fisherman's net lay near her booted foot. The penguin's carcass was intact, but was missing its eyes. "Yuck, let's get out of here!" she said as we hurried past. The vultures had backed away at our approach, but now we understood what they were after. When we returned the next day all that remained of the *pobrecito* (poor little) *pinguino* was a shiny, curvy skeleton, picked totally clean of flesh, looking like something ready for a museum of natural history.

◆

About 150 yards upstream from the bridge, and a respectable distance from the penguin, we unshouldered our shovels, took off our packs, and began to dig. Using a flat bladed shovel that Marco had nicknamed the "geoslicer"—after a very fancy piece of Japanese equipment with the same purpose—we cut back the bank to make a clean, flat, vertical surface, the better to see and delineate the deposits. The bank was everything that Marco had advertised and more. Indeed the notches seen from a distance did mark the erosion of sandy layers. I sketched a cross-section in my notebook, labeled "Tirúa #1."

My notes were improving, but I still had a long way to go to match the real pros, like Marco and Lisa, who could distinguish between silty sand and sandy silt, between clayey silt and silty clay. Lisa felt the material by scraping it with her mason's trowel (which she preferred to a nejiri gama). Then she rubbed small bits between her thumb and forefinger to judge the texture and amount of grit. One of our key challenges would be to prove that these deposits had come up the river with tsunamis, rather than down the river in floods.

I listened as Lisa and Marco discussed this challenge and the possible environments for the deposition of each deposit. Was this clean sand from a beach, or was it mixed with finer material like silt or clay indicating a source up the river? Shell fragments would be a giveaway of a beach or shallow marine environment, but these fragments are easily dissolved by the acid in shallow ground water and do not survive for long. Deposits from the river would tend to be a mix of sand and silt, more likely to appear brownish. This sand definitely looked like it came upstream, from the beach.

◆

As we dug in the riverbank, describing and tracing the layers of sand, darkness began to fall and the tide began to rise. I had not noticed that Marcelo had left, but he now returned with the good news that he had found a place for us to stay. Not Doña Clemedia's, but a place with a few rooms and a small cabin just across the road from the field and the gas station. Don Pedro, the proprietor, would also provide dinner and breakfast. Tired, hungry, grubby, and a little cold and wet, we thought it sounded great.

Don Pedro, a short, older man with thinning gray hair and large aviator-style glasses that constantly threatened to slip off of his nose, had established what appeared to be a discotheque. The large front room had a bar, some tables, a dance floor, a rotating mirror light, and a booming sound system. Behind the front room was a kitchen and four bedrooms. He lived in one and had a tenant in the second. Marco and Marcelo shared the third, and Lisa and Caitlin, the fourth. I drew a room in a small cabin behind the main building. Although closed in, the cabin was still being finished inside. I lugged my belongings past a saw, other tools, and piles of lumber, to find my bed. The shower was cold.

As his only customers, we implored Don Pedro to turn down the music, gobbled our dinner, and slaked our thirst with *cervezas* as dappled reflections from the rotating mirror ball flitted across the walls.

The next morning we rose before dawn to catch the low tide, crossing the road and field by the light of our headlamps to continue our work at the bank. With daylight, came the vultures. They maintained a wary distance from us as we worked, but I did not expect sympathy from them should things go badly.

◆

In the following days we found that we could trace the deposits up the bank about half a mile. Later we canoed down the last few miles of the river but found nothing similar in the river above. A flood deposit would likely extend a long way up the river. In contrast, a tsunami deposit would be confined to the lower reach of the river. The deposits of clean sand were confined to the area near the field.

After demonstrating that the sandy layers originated from the sea, our second task was to determine the times of their deposition. Marco, as a professor of both geography and history, had—with colleagues—been combing the records in Chile, in Lima, Peru, and at the Archivo General de Indias in Seville, Spain, to find records of earthquakes and tsunamis in Chile. He aimed to increase the understanding of their times, location, and the nature and distribution of damage.

Written records of earthquakes and tsunamis in Chile began with the arrival of Diego de Almagro and his band of conquistadors from Peru in 1535. The records were spotty at first. Then their quality rises and falls, especially south of the Biobío River, with the strength of the colonists in their battles with the Mapuche. Although the Mapuche lacked written records, their oral mythology is rich with images suggesting both earthquakes and tsunamis, even if dates are lacking.

One day while we dug in the bank, Marcelo left to interview one of Tirúa's older residents, a man who had served as the town's mayor. He lived there at the time of the 1960 earthquake. He described the severe shaking that terrified the townspeople. More importantly for us, he also described the tsunami and how it inundated the low-lying portions of town and had left a deposit of sand in the field behind the riverbank that we now occupied. That our topmost layer of sand was associated with the 1960 earthquake and tsunami seemed to be a slam dunk. But for the lower layers, things got tricky. We hoped to use a variety of techniques, especially carbon-14 dating, to determine their dates of deposition.

◆

The naturally occurring carbon in all life forms, including our own bodies, contains a tiny amount of the radioactive isotope, carbon-14. It is at least a little bit creepy (or perhaps tickling) to think about these naturally radioactive atoms decaying within our own bodies, but as we breathe and metabolize food, the ratio of carbon-14 to ordinary carbon-12 maintains equilibrium in our bodies with the atmosphere (or ocean in the case of a sea creature). When a life form dies, this equilibrium is no longer maintained. Carbon from the atmospere or ocean is no longer being introduced through metabolism. As the carbon-14 atoms decay spontaneously to the stable isotope of ordinary nitrogen by emitting chunks called beta particles, they are not replaced, and the ratio of carbon-14 to carbon-12 decreases in a systematic way, following the curve of a decaying exponential.

After 5,730 years, half the carbon-14 present at the time when a plant or animal stops growing will have decayed to nitrogen. This is the half-life of carbon-14. At that time, the ratio of carbon-14 to carbon-12 in a seed, or a piece of wood or bone will also be reduced by one half.

We searched for bits of wood or the rhizomes of marsh plants that died or stopped growing close to the time of becoming part of the deposit. Roots are to be avoided since they grow down vertically through layers of older deposits thus giving ages that are too young. Charcoal survives long after it was burned and commonly yields dates much too old. In contrast, a rhizome that extends laterally from a plant provides ideal material for dating.

After several days digging in the bank, describing the sand layers and convincing ourselves that their origin was overwhelmingly likely to be from a tsunami coming upstream, rather than a downstream flood, we packed our samples and gear and headed back to Concepción.

◆

Marco maneuvered the truck into a parking place on the main square in front of the green municipal building in Talcahuano, the port city for Concepción. The canoe strapped to the top of our truck provided a novelty for passersby. For once we didn't have on our grubby field clothes. No boots. No waders. No floppy hats. Marco and Marcelo sported white shirts and ties. Marcelo's sometimes wild long hair and beard were neatly trimmed. Both looking quite professorial. Lisa and Caitlin wore skirts and sandals. I didn't have a tie, but at least I had on a shirt with buttons down the front, and my pants bore the hint of a crease. We climbed up the broad stairs familiar to visitors of local government buildings almost everywhere, and were shown into a small conference room. Marcelo and the director of emergency affairs for the city had arranged this meeting, a chance for senior staff in the city administration, including the mayor, to discuss the risks that the city faced from tsunamis.

The officials began to jam the small room, bantering as we waited for the mayor. At last he entered: a tall man, with carefully coiffed white hair, a light brown coat and tie, Gastón Saavedra. Mayor Saavedra, a socialist who worked his way up through the business, unions, and politics of the port, had been elected the year before, promising a program called *El Salto Gigante* [The Great Leap], with 46% of the vote.

The banter stopped, the director of emergency operations introduced Marcelo, and he explained that we had come to Talcahuano to research the tsunami that struck the town just before Darwin's visit in 1835. The director of emergency operations introduced the mayor and the other officials. The discussion began. The rapid Chilean Spanish was almost impossible for me to follow but I did my best to hang on. Marco and Marcelo tried to explain that if anything a similar earthquake and accompanying tsunami were overdue. After only about ten minutes, the mayor announced that he wasn't really worried about a possible tsunami. He seemed to think that if such a large earthquake occurred, all the buildings in the city would fall down anyway. But it seemed that he had another engagement and that he had to leave. Marcelo and Marco each got in a few more words before he left.

We finished the briefing. Marcelo described the anticipated run-up and inundation zones. Much of the new development in Talcahuano, both industrial and residential, had been located in the zone likely to be flooded by a future tsunami. It was not going to be a pretty picture.

Later, we met with some reporters. They scribbled notes, took photos, and shot video. Reports appeared on television and regional newspapers about the Darwin team and the need to prepare for tsunamis. Marco was quoted in the paper *El Sur* as saying, "No one can deny that a tsunami will occur in this area. The problem is that we don't know when."

But Mayor Saavedra didn't have time to hear about tsunamis. Then.

◆

In January, 2010, nine months later, Marco, Lisa, and I again sat in a rented truck. We were half a mile away from the modern air terminal in Concepción, Chile, down a quiet road parallel to the runway. Beyond a fence was the hangar of the Club Aéreo de Concepción. I would again be traveling to Isla Santa María, but his time in style.

A visit to Santa María hadn't been Marco and Lisa's top priority, but when we sat in Marco's office in Valparaíso finalizing the plans for our fieldwork a few weeks before, I had pressed the case for a visit to the island. I argued that we knew from FitzRoy's descriptions that the tsunami in 1835 attacked the island. FitzRoy reported that the wave had come around both ends of the island causing great damage to the small town now known as Puerto Sur. I described what I thought might be prospective sites to search for tsunami deposits. I also explained how I hoped to repeat the *Beagle*'s survey with the small echo sounder I had in my duffel bag, following the plan that Daniel and I had hatched the previous year in the dome tent on Isla Guafo.

We would repeat the *Beagle*'s survey in the shallow bay off the southeast coast of the island. The officers of the *Beagle* surveyed this bay, Rada Santa María, a few weeks after the earthquake in 1835. If the elevation of the island had remained constant since then, the water would still be the same depth as shown by the soundings on the *Beagle*'s chart. But if Isla Santa María was being dragged down by the subduction of the Nazca plate accumulating elastic strain, in effect the bending of a stick—as predicted by the elastic rebound theory of Harry Fielding Reid—the water would be deeper. Indeed, this was what Daniel believed was happening at Isla Guafo. These two small islands, we hypothesized, were being dragged down in the time period between earthquakes, only to be flung upward when the faults beneath them gave way, giving rise to giant earthquakes.

We reached Daniel who was in Chile preparing for his own field-work. He was enthusiastic to join us. Eventually Marco and Lisa were persuaded. Daniel, being well-financed by a German science foundation, urged us to make arrangements to fly to the island, in preference to the slow boat that I had taken two years before. We agreed to meet at the Club Aéreo.

Daniel arrived and we carried our equipment to the front of the hangar. Even after trimming down our gear, we still had coring tools, a couple of shovels, and personal things. It added up to a good deal of paraphernalia. We would need two of the Club's Cessnas.

Looking at our pile of gear, one of the pilots murmured, "¡Muchas cosas!" [Many things!], and the two pilots began the mysterious, but solemn ritual of lifting each item, judging its weight, and stowing it in some part of the plane to give the desired distribution of weight.

With the gear stowed, we loaded the planes and taxied to the end of the runway. I was hoping that everything had been worked out so that a 737 wouldn't be landing as we took off.

With a roar, our little plane gained speed, and in a few moments we were quickly gaining altitude. Concepción and the Biobío River spread before us to the left, to the right the Tetas del Biobío, two mound-shaped hills that served as key land marks for FitzRoy and other early sailors as they navigated this coast. Behind us was the port of Talcahuano, its yards filled with stacks of shipping containers. Directly ahead loomed the Gulf of Arauco and in the distance Isla Santa María.

Approaching the island it was easy to see the rocks at the northern end of the island that marked FitzRoy's platform. We made a loop over Puerto Sur for the pilot to gauge the wind. Then suddenly before us was the tiny runway on the island. With a few bumps and a roar of the engine, we were there. Twenty-five minutes in all. What a difference from my previous epic of several hours on a rolling boat.

Within a few minutes we had unloaded the gear, bid *"¡Ciao!"* to the pilots, and the planes lifted into the air. Shortly after, a stout man arrived in a creaking cart pulled by an indifferent white horse. *"¡Don Toti! ¡Hola!"* Daniel called out. Daniel had made arrangements for us, waved to the man in greeting, and introduced us. Don Toti wore a rakish flat-brimmed felt hat with a braided leather band. It was held in place by a leather thong pulled tight under his chin. His dark black hair had begun to gray in his sideburns and mustache. We loaded our gear onto the cart. Lisa climbed onto the bench beside him, and Don Toti slapped the horse on its rump with the old white ropes that served as reins. The cart jerked into motion. The rest of us trailed behind along the sandy road.

To my surprise the house where Daniel had arranged for us to stay was only about fifty yards from the house where I'd stayed two years before. But this time we were hosted by Señora Ruth, a woman whom Daniel knew well. He had been raving about her seafood cooking ever since I had met him. Lisa was fascinated by the scene. While Ruth prepared lunch, she and I set out to explore. We stopped at the house where I had stayed before, but there was no sign of my previous host. I showed her the remains of the rusting Soviet tractor. We walked on to the quaint harbor of Puerto Sur, a beach no more than fifty yards wide, covered with the colorfully painted boats of the fisherman with names like *Nautilius*, *Fortuna*, and *Mi Jesús*.

To the right of the beach was a concrete ramp running down into the water, and to the left an old wooden pier, no longer used and missing many planks. Many boats were pulled up onto the beach, others floated at anchor in the waters just offshore. A farm tractor pulled a trailer across the beach. The crew of the *Beagle* had landed on this very beach to set up their instruments to measure the latitude and longitude, and they noted their location with a cross on their chart.

To our amazement we found a rusty white-and-blue sign showing the small figure of a man running to escape a giant wave that towered over and threatened to engulf him, together with an arrow showing the *Vía de Evacuación*, a version of the universal sign showing the route to high ground in case of tsunami.

"Wow," Lisa said, "good for them!" as we both shot photos of the sign.

When we returned to Señora Ruth's for lunch, we quickly realized that Daniel had not been exaggerating at all. It was delicious.

◆

That afternoon we all piled into the cart to take a look at the north end of the island. Daniel, who had investigated the geology of the island for his PhD thesis was our guide. The one real road on the island, extending between the island's two villages, Puerto Sur and Puerto Norte—sandy ruts during my visit two years before—was now paved with concrete paving stones. On the flat, Don Toti's horse Ruperto had no trouble pulling us along at a trot, retracing the route I had taken with the Carabineros two years before. But as we climbed onto the low ridge that forms the spine of the island, Ruperto slowed to a crawl. I was ready to jump off the cart to lessen poor Ruperto's load, but Don Toti flicked the reins, made loud kissing noises, and called out, "*¡Vamos* [Let's go], *Ruperto!*"

Don Toti loved to talk, and he and Daniel bantered easily. With Don Toti's thick country accent, I was completely in the dark.

Ruperto, however, knew exactly what was going on. Whenever Don Toti fell deep into conversation, the horse would slow his gait, until suddenly Don Toti would realize that we were hardly moving at all, smack his lips, call out "*¡Vamos, Ruperto!*" and snap the reins, urging the horse back into action.

We rode past the lighthouse to the bluff at the north end of the island. From there we could look down on FitzRoy's reef. A half mile off the shore waves crashed the rocks surrounding the islet called the Farallón. As the blue-green swell off the Pacific approached the island itself, breakers would rise up and crash into streams of white foam a few hundred yards from the base of the cliff beneath us, at the outer edge of the rocky reef. Between the line of breakers and the narrow beach below us, bits of rock peeking through the water and streams of foam revealed the shallow reef lying just below the surface. This was the reef that FitzRoy described as being above water even at high tide, giving rise the "abominable stench" of dead and dying shellfish.

Marco asked Don Toti if he knew any older people in the village who might have been living here at the time of the 1960 earthquake. A recollection of the effects of the tsunami, especially of sand deposited in a field, could give us clues about where to look. Don Toti asked some passersby. They directed us through the winding lanes of the village to the house of Don Gabriel.

Don Gabriel was a short man with a weathered face and neatly trimmed, graying hair and mustache, who wore a two-tone zippered sweater. Sitting on a pile of lumber, he agreed to answer Marco's questions. He was born in 1930 and had come to Puerto Norte in 1950, Marco wrote in his yellow notebook. Don Gabriel did not show his eighty years. "It was already a village even then?" Marco asked. Yes, Don Gabriel replied. He had worked with an old-fashioned diving helmet collecting *locos* [Chilean abalone] and sea urchins.

A smile crossed Daniel's face. For many Chileans, and especially Daniel, there is almost nothing better to eat than locos. Marco asked if he remembered the 1960 earthquake. Don Gabriel did, and Marco asked the day and time. Marco, a skilled interviewer, asks questions to which he knows the answer well, to warm up an interviewee, and also to test the quality of the interviewee's recollections. Don Gabriel passed the test with flying colors, giving the times of the two large earthquakes of which the second led to the tsunami.

He recalled that at the time of the earthquake he was sitting with his mother near the top of the cliff. The section of the cliff where they were sitting began to slide, carrying them down with it, but they were not injured. A girl was buried, but they were able to rescue her. The wave from the tsunami was not too high, only about twelve feet, coming from the northeast. They did not lose any boats. It did, however, carry kelp into the little lake in the valley south of town.

Marco asked whether the rock platform at the base of the cliffs had changed since Don Gabriel arrived in Puerto Norte. To our amazement he replied that, yes, it had changed. Formerly there was a marsh at the base of the cliff with cattails and rushes. Geese and goats used to graze there. A former resident of the village, Don Oscar Saavedra, who been born in 1910, but was no longer alive, told Don Gabriel that earlier there had been plowed fields at the base of the cliffs. Marco asked Don Gabriel why he thought this was happening. Don Gabriel said that the scientists who sometimes came to the island (most likely a friend and colleague of Daniel's) said that the island was sinking, but he didn't believe it. He thought that the level of the sea was rising. Daniel, in a serious breach of interview protocol, began to argue with him, earning a scowl from Marco.

As we rode back to Puerto Sur, I was barely able to contain my excitement. The descriptions of the disappearing land at the base of the cliff is exactly what one would expect if the island was, in fact, sinking, just as at Isla Guafo.

Again, just as Daniel had advertised, Ruth's dinner was spectacular. His face lit up as she brought out plates of locos, freshly collected from the island, and fresh potato salad. Daniel had tucked a few bottles of good Chilean wine in his duffel and we were all set. In the city a person could pay big bucks for seafood not half as good. After dinner, Ruth's nephew, Javier, a shellfish diver, appeared. Javier had a boat and was willing to help us with our survey the following morning.

◆

Daniel and I awoke early and gathered our gear. I had entered a route into the GPS that would cover the smooth, flat parts of the bottom of the bay where we had the best chance of comparing our measurements with the *Beagle*'s soundings. In areas where the bottom was rough or rocky, errors in matching the *Beagle*'s chart with the GPS would result in depth differences unrelated to whether the island had moved up or down. Fearing seasickness I had only a bite to eat before we set off to meet Javier and his brother Eduardo, trusting that the tasty meal from the night before would sustain me. The sun was just coming up and the skies were gray.

Javier's boat, made of wooden planks, was powered by an outboard motor that had seen better days. In the center of the boat was the compressor that Javier used when diving for shellfish, with coils of yellow air hose draped beside it. We pushed the boat off the beach. Once afloat I began to set up the echo sounder. I'd mounted the transducer for the echo sounder on a piece of pine board with the idea of clamping it to the transom of the boat, but the clamps did not provide a firm grip. Seeing my difficulty, Javier produced two rusty nails, and—as the boat rolled gently in the waves of the harbor—with an old pair of pliers, nailed the board to the transom.

As Javier started the motor and we steered toward the bay, I gradually figured out how to direct Javier to follow the route that I'd programmed into the GPS. The plan was to begin with a line for about three miles long along the shore of the island, then to make a sequence of similar lines, each about five hundred yards farther off the shore.

I watched as the water depths popped up on the little screen of the GPS. In the anchorage at Puerto Sur the water was only about six to eight feet deep. As we steered a course to along the shore of the island, avoiding the beds of kelp that could foul the propeller, the water depth slowly increased. We wanted to go as slowly as possible, only a couple of miles per hour, but Javier's motor was really made to go fast, not to putter along, so we had to come to a compromise, to find a speed at which the echo sounder gave stable results and at which the motor was not laboring too badly.

We reached the end of the first survey line, and made a boxy U-turn, turning to head northeast. The wind was coming out of the southwest. The wind had been in our face on the first leg, and having the wind to our back was much more comfortable. But when we finished the second line and turned again to the southwest, not only were we headed into the wind, but we were no longer protected from the ocean swell by the shore of the island. Each time the boat hit a line of the rolling waves with a smack, we were doused with spray. I had not noticed until now that Javier and his brother had brought along their waterproof fisherman's pants and jackets. The brothers pulled on their waterproofs. Daniel and I began to get wet. As we continued the survey, every minute or two on the southwest legs, the boat would hit another large wave and the spray would hit us again, drenching us with cold water. We both wore Gore-Tex jackets, but compared to the waterproofs that the Javier and his brother wore, these were nothing. We were getting totally soaked.

Finally after about two hours, Daniel said to me, "These guys are really tough. They will go on forever. Maybe we should call it a day." I agreed. Daniel signaled Javier to head back to Puerto Sur. My respect for the crew of the *Beagle* soared. They carried out surveys in conditions far worse, pulling at the oars by hand.

When Daniel and I reached the cottage where we were staying, both of us cold and dripping wet, we gave each other a high five. Our first look at the data suggested that the island had sunk about five feet since 1835. Not exactly what Darwin anticipated, but in line with elastic rebound theory.

◆

The basic idea of elastic rebound is easy to think about on the San Andreas fault in California, which like the giant fault beneath Isla Santa María, forms the boundary between two of the earth's plates. The San Andreas is a vertical fault. Along it geologic formations and topographic features are offset. All these offsets, as well as geodetic measurements, show that the Pacific Plate on the west side of the fault is moving northwestward, relative to the North American Plate on the east side.

Imagine a giant, straight stick, sixty miles long or more, fastened on the surface of the earth, perpendicular to and centered on the fault. Geodetic data show that the movement of the plates would bend the stick, moving the western end about one and a half inches per year relative to the eastern end. Along most of the San Andreas—except in places in central California where even pipes and curbs are broken by continuing slow movement—the fault is locked or stuck at the surface. Nonetheless, higher temperatures at depths greater than ten or twelve miles allow the rocks to slide easily. After a century our stick would be bent into a gentle curve that mathematicians call an arctangent, with the west end about twelve and a half feet or so north of the east end. A large, 1906-type earthquake would break our stick into two right at the fault. In another minute or two, the two halves of the stick would straighten out, surging, as the parts of the two plates that had been stuck together broke apart—driven by the release of the elastic strain built up within them. After the dust cleared, the two ends of the stick would be separated by about twelve and a half feet for every century since the previous earthquake. This is elastic rebound.

In the case of Isla Santa María the basic idea is the same, but the geometry is quite different. Instead of a vertical fault like the San Andreas, the huge fault beneath Santa María is nearly flat, dipping at a gentle angle down from the Chile Trench, fifty miles west of Santa María, down beneath Santa María, and then further down beneath the coast of Chile, as the Nazca Plate is subducted beneath the South American Plate. Eventually at depth of a hundred miles or more beneath Argentine side of the Andes, the Nazca Plate is reassimilated into the mantle.

Near the trench and shallower than about ten miles, the soft rocks are weak, and they slide. The higher temperatures at depths greater than twenty-five miles also weaken the rocks allowing the plates to slide silently past one another. But at depths between about ten and twenty-five miles, the two plates are stuck together, or locked, along the fault. This stuck or locked region—when it finally breaks—is the source of the great earthquake and is termed the "seismogenic zone."

The seaward edge of the locked seismogenic zone is just about under Isla Santa María. As the Nazca Plate slides downward and toward the coast, because of the locked seismogenic zone, it drags Isla Santa María down with it. This is just what Daniel and I were seeing. The theory also predicts that when a great earthquake occurs, the edge of the South American Plate will fling upward and westward, uplifting Santa María, just as FitzRoy saw in 1835. And when this earthquake occurs, it has one other very dangerous consequence. The upward fling of the edge of the plate raises not only islands and parts of the coast, but also the seafloor and the water above it, triggering perhaps the most dreaded effect of a giant subduction earthquake, the tsunami.

◆

The next day the wind was as strong from the southwest as it had been the day before. Not a good day to finish the survey in the most exposed part of the bay. Daniel had to fly back to the mainland that afternoon, but he arranged for Javier to meet me the following morning if the wind was not quite so strong.

The next morning was calm and the survey went smoothly. Although the swell was large, the wind was not too strong, and there was no spray. I finally began to relax and look around. It was a bit like a low-tech amusement park ride as the boat rode up onto the crests of the swell—from where I could see the Punta Lavapié and the southern coast of the Gulf of Arauco eight miles away—then fell down into the troughs where I would face nothing but a ridge of blue-gray water. Javier's ancient outboard motor would roar for a moment as the propeller was exposed as the boat went over the crest, then gurgle as the propeller again submerged.

I couldn't help thinking how much more difficult the survey would have been in a whaleboat, sailors pulling at the oars trying to keep the boat pointed into the swell, one man sounding with a pole or leadline, a mate with a sextant taking the bearing to a landmark on shore or raising a flag to signal a shore crew with a theodolite. It was hard to imagine the focus, tenacity, and endurance that this all required.

Javier focused on steering the boat, responding to my occasional signals as I watched our route unfold on the GPS, achieving amazingly straight survey lines despite the waves and swell. Eduardo his brother seemed mostly bored, his thumbs hooked in the suspenders of his waterproof pants. The brothers would come alert when we passed flocks of birds, gulls bobbing in the waves, *faradelas* swooping above. Where there were birds, the brothers knew there were also fish. They were pros.

On the last leg of the survey, but still a mile or so from the harbor, the motor coughed and died. Javier pulled the motor's starter cord. He pulled it again. Nothing. He shook the gas tank. Plenty of gas. At least I could see the harbor of Puerto Sur in the distance. Being a bit of a worrywart, normally my anxiety would have soared. But this time, I felt calm. The brothers were tough Chilean fisherman. A little blip in technology was not going to be a big deal for them. Javier was the guy who'd nailed the board holding my echo sounder to the transom of his boat. Somehow I was sure that they would come up with something.

Eventually the brothers opened the motor's cover and fiddled with something. They'll figure it out, I told myself. After Javier had pulled the starter cord repeatedly with no result, he tried again while Eduardo leaned out over the transom, holding his finger on the critical part. The motor sputtered to life. The brothers carefully lowered the motor back into the running position while Eduardo maintained his contact with whatever his magic finger was holding. And in that mode, with Javier steering, and Eduardo draped over the motor, we limped the mile back to the harbor. Just another day at the office. Data in hand, I headed back to the house.

Analysis of the completed survey confirmed what Daniel and I saw on the first day. The island had subsided about five feet since the visit of the *Beagle* in 1835. The island was being dragged downward and to the east by the movement of the Nazca Plate in agreement with the theory of elastic rebound.

◆

While we waited for dinner the last evening on the island, the air was warm, the sea calm, the breeze mild, the afternoon light gentler than during the harsh sun of midday, and I realized that I didn't have a photo of Javier's boat at anchor. I walked to the beach, climbed onto the concrete boat ramp, and scanned the boats moored to buoys in the harbor. Spotting the distinctive diving compressor covered by an orange tarp, and the bright yellow air hoses hanging over the side, I snapped a quick photo.

It was only then that I saw the name painted just below the gunwale—*Jeremías*, Spanish for Jeremiah, the Old Testament prophet who foretells disaster.

CHAPTER SIXTEEN

The Chilean Earthquake of 2010

I feel it is quite impossible to convey the mingled feelings with which one beholds such a spectacle. . . . It is a bitter and humiliating thing to see works, which have cost men so much time and labour, overthrown in one minute; yet compassion for the inhabitants is almost instantly forgotten, from the interest excited in finding that state of things produced in a moment of time, which one is accustomed to attribute to a succession of ages.

—Charles Darwin

Six-year-old Constanza's eyes gleamed as she told her story. She lived just above the Tirúa River near its mouth, across the river from the town. Her grandmother, Doña Elena, the family matriarch who missed nothing—a stout, robust woman of eighty-six—lived across the road with her two dogs. But what happened at 3:34 A.M. that morning, February 27, 2010, caught even Doña Elena by surprise.

Ten miles beneath the coast, some 120 miles north of where Constanza and her family slept, the earth cracked. Straining under the forces accumulated since that day in 1835, 175 years ago, the rocks of the earth's crust gave way. The rupture ripped south beneath Concepción, Talcahuano, Santa María Island, and Tirúa, and north under Lipimávida and beyond, a total of three hundred miles. The South American Plate, previously dragged down by the Nazca Plate, lurched up and toward the sea.

But the geophysical details of the earthquake were far from the terrified mind of Constanza as she huddled in the dark with her family. Doña Elena, recalling that the shaking during the 1960 earthquake had seemed more intense, reasoned that the tsunami in the river shouldn't be too bad. Nonetheless, the family made its way up the hill, just to be sure. The first small wave came twenty minutes after the earthquake; the second after forty minutes.

Then . . . a black wall of water, sixty feet high.

◆

The earthquake rumbled up from beneath the hills of Valparaíso, headquarters of the Chilean navy, and up from beneath the central valley of Chile into the capital city of Santiago. The government had a system—and a plan—in place to issue a warning of a potentially dangerous tsunami. Sadly, the earthquake and tsunami did not cooperate with the plan. The plan involved seismologists looking at data from seismographs across the country fed into computers in Santiago. The plan incorporated examination of data by the navy in Valparaíso from tide gauges operated along the coast. The plan envisioned communication of a public alert by the national emergency management office in Santiago.

But the earthquake occurred in the middle of the night on a holiday weekend. The plan did not provide for 24/7 staffing of the seismograph network. The power went out in many parts of the country. Communications were a mess. Data were incomplete. The location of the epicenter of the earthquake, when it was determined, was east of the coastline, beneath the land. Nonetheless, around 3:50 or 3:55 A.M., an alert was transmitted to the emergency management office. But from there things went downhill fast.

Most landlines and cell phones were down. Communications with the most strongly affected areas were out. Officials in the government and navy were confused. If the earthquake was beneath the land, how could it create a tsunami? The officials were not seismologists or geologists, they were officials. The earthquake rupture had spread like a wildfire over the fault. The officials did not understand that the rupture from an earthquake this large would sweep over an area 300 miles long and nearly 100 miles wide, extending well out to sea. For a giant earthquake like this one, the epicenter, the point where the rupture starts, just is not that relevant. In addition, with communications out in the areas most strongly hit, the officials in Santiago and Valparaíso didn't know that as the minutes passed, tsunami waves were beginning to break on the coast. The tsunami alert was canceled. Around 5:00 A.M. the president of the country arrived at emergency operations center in Santiago. This situation was muddled. Conflicting information was flying around. But the president was not a seismologist. She did not understand that any giant earthquake along the coast should be considered as an extreme risk of a tsunami. Somehow, nobody told her. Various officials, including eventually the president, went on television to say that any tsunami wouldn't be too bad. Meanwhile, waves were striking the coast.

The whole warning system, from the scientists through the functionaries and bureaucrats to the politicians, simply fell apart.

Fortunately most people along the coast just ran for high ground when they felt the earthquake. Efforts at public education had had some effect, but it was also folk wisdom, as FitzRoy wrote about the citizens of Talcahuano in 1835, "Penco [obliterated by a tsunami in 1751 leading to the relocation of the city to Concepción] was not forgotten; apprehensive of an overwhelming wave, they hurried to the hills as fast as possible." But some, like Don José Luis in Lipimávida didn't get even that message. Others got the wrong message. For others, no message would have helped.

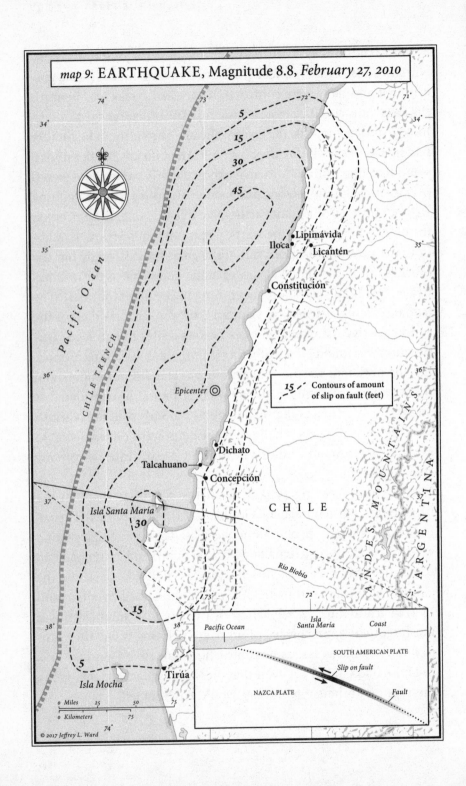

map 9: EARTHQUAKE, Magnitude 8.8, *February 27, 2010*

Constanza's father, Joel, was awakened by the earthquake. Five minutes after the earthquake, the house was still shaking. He woke Constanza and her sister. The power went out.

Their house was on the side of a hill sloping down to the Tirúa River, about fifty feet above the level of the water. Joel and his brother had built two cabins on the land below Joel's house. One was rented to Señor Montero, a school administrator and his son, Rodrigo.

After the ground stopped shaking, Joel, his wife, Marlene, Constanza, and her sister left their house and crossed the road to Doña Elena's. Their electric gate did not work because of the power outage, so they could not use their car. After about twenty minutes a low wave filled the marshy area around the river. Ten minutes later a young couple appeared on the road from the beach, shouting for help. They were completely naked. They had been camping on the beach and got caught in the first wave. Joel found them some clothes.

Señor Montero was also wakened by the earthquake in the cabin down the hill. He tried to stand but could not. The shaking was too strong. He was worried that the whole hill would slide into the river. His son, Rodrigo had been at a party the evening before, and the two had stayed up late talking afterward. Rodrigo slept through the earthquake. With difficulty Montero woke his son. Joel arrived, concerned about the water rising, and warned him that the electric gate was not working, but Montero continued talking with his son. Then, he heard a rumble from the town across the river. It was the sound of houses and tin roofs being smashed by the first wave. They too ran up toward Doña Elena's house.

Everyone went higher up the hill to a place where they could see the river and, across it, the town. Somehow Joel had opened the gate and Montero and Rodrigo went back to their cabin to get their car.

About twenty minutes after the first, a second low wave arrived, this time filling the river from side to side.

Joel remembered that he had a pair of borrowed oxen in the marshy area below his house. He went down the hill, and finding the oxen, brought them up the hill.

Rodrigo remembered his computer back in the cabin and insisted that he retrieve it. He and his father descended the hill, and in their car began driving down to the cabin.

Then the people on the hill saw the wall of water approaching. They began shouting for Montero and Rodrigo to come back.

Rodrigo too saw the wall. His father jammed the car in reverse, frantically backing up the hill. When the third wave struck, the water reached just up to the wheels of the car. They escaped.

◆

Across the river, the town of Tirúa extends from a low, flat area—behind the beach and along the river—up onto a hill. Doña Josefina, a retired school administrator, and her husband were asleep on the second floor of their house in the lower part of town. At first the shaking did not seem too bad. Then it grew stronger. She tried to stand, but could not keep her balance. She fainted. Her husband helped her downstairs. They did not have a flashlight, but climbed in their truck and drove to her daughter's house in the upper part of the town.

Meeting other members of her family, they realized that Doña Josefina's bedridden father had been left behind. Her brother drove to the lower part of the town to find their ninety-year-old father dressed and ready to leave. They too returned safely to the daughter's house before the waves struck.

◆

By sunrise in Tirúa, the sea was calm. In the early morning light, people began to descend from the high ground to check on their homes.

Joel found his house standing, but filled with nearly a foot of stinky black muddy sand, fish, crabs, and trash. The two cabins, including the one where Señor Montero and Rodrigo had stayed, were gone. Washed away. But Joel was lucky; his neighbors' houses were also gone.

Across the river in Tirúa, Doña Josefina's husband and son-in-law went down to check on their house. Watery sand and mud covered the streets. Cars had been tossed about. Some houses had been shifted off their foundations into the middle of the street. Some were simply washed away. Doña Josefina's house was destroyed. Her father's house was destroyed.

In the river—near the bank where Marco, Lisa, and I had traced sand layers on two earlier visits, hypothesizing that they were the result of earlier tsunamis—a house, swept clean from its foundation, floated, listing drunkenly to one side. Behind the bank, the field where we had encountered the dead penguin was covered with sand up to a foot thick. Any remaining penguin bones, lost pocketknives, or even trash carried in by the tsunami, were now a part of the geological and archeological record, perhaps to be exhumed some time in the distant future to the amazement of a curious scientist.

◆

Don Luis Bravo lived in Dichato, a town on the beach fifteen miles north of Talcahuano. He knew what to do. The earthquake woke him as he, his wife, and his two daughters were sleeping. Remembering the tsunami from the 1960 earthquake, he, his family and all his neighbors, ran for the hills. It was a virtual stampede, he said.

A former policeman, Don Luis was a sensible man. Seeing the destruction, he had instructed his family to grab their shoes. He knew that the streets would be covered by broken glass. As they dashed from their home—by some fluky reflex—he grabbed his wallet off the table.

When all the people had reached as high on the hills as they could, word spread that the authorities had announced that there would not be a tsunami alert. But Don Luis reasoned that with such a large earthquake there would have to be a tsunami. He wouldn't let his family descend. After about forty-five minutes, thanks to the bright light of the moon, he saw the sea rising. Then they heard a roar like the sound of breaking branches. To calm his family he suggested that it was only the sound of boats crashing on the coast. They heard this roar three times, each separated by about half an hour of silence.

About 7:30 in the morning he finally descended back to the town. He was horrified by the destruction. His home was totally destroyed. A barge rested on what had been the floor of his house.

Thanks to the little money in the wallet he'd grabbed during his escape, he was able to offer 150,000 pesos ($300) for a ride in a truck to take his family to his mother-in-law's in the south.

◆

Constitución is on the coast at the mouth of the Maule River. Throughout Chile the end of February marks the end of the summer vacation. Children, as well as students in colleges and universities, all return to school on the first of March. Consequently the last weekend in February is much like Labor Day in the United States, the last chance for a trip to the countryside, or to the beach. The last chance for a summer party.

Early that morning, February 27, on low, wooded Orrego Island in the Maule River, only two hundred yards off the waterfront of Constitución, a weeklong festival was almost over. Hundreds of people were camped on the island. Families with kids and grandparents. But there were not many boats. When the earthquake struck many realized that they should get off the island. Those who did have boats, open boats with outboard motors, many of them fishermen camped on the island with their families, began to ferry their relatives to shore. Two cousins, Osvaldo González and Osvaldo Gómez, each had boats and began to shuttle people to shore. Each round trip took about ten minutes. Arriving in Constitución after his the third trip from the island, Gonzalez had a problem with his motor. The water had receded. The boat rested on the muddy bottom. Then he saw the wave, more than thirty feet high. As Gonzalez ran for high ground, he caught sight of his cousin, in his boat on the crest of the wave.

"If I hadn't bogged down in the mud, I would have gone back into the water, like my cousin," he later said. "But I saw the wave and began to run." It was the last time he saw his cousin.

Mario Leal was on the island with his two children and pregnant wife, Mariela. He told his wife to hold the children while he went to find a boat. He dove into the dark water and swam to shore. As he searched for a boat in the darkness and confusion, he could hear shouting from the island. Then came the waves. He ran. The third and last was the largest.

"I lost my entire family," he later said. "Nobody warned us of anything."

A photo taken from space a few days after the tsunami shows Orrego Island swept bare, except for the largest trees.

◆

In April, 2009, when our little *Equipo* Darwin or Team Darwin, as we called ourselves, tried to brief Gastón Saavedra, the mayor of Talcahuano, about the 1835 earthquake and tsunami and the danger of future events, we met Doña Pilar Bermúdez. After the mayor excused himself from our meeting at the city hall, Doña Pilar was in the group that remained to listen. An attractive, stylishly dressed woman with short, reddish-brown hair, she sat in rapt attention, her notebook spread before her, as Lisa explained our efforts to reconstruct the history of similar events. Doña Pilar had worked in social services for the municipality of Talcahuano for twenty-seven years. Her special interest was the elderly. Her husband, Don Luis Fregonara, a math teacher and school administrator, was also the chief of the local fire brigade.

Pilar lived in an older, two-story brick building at Bulnes 178, just across the Plaza de Armas, or town square, from the city hall, and from where we parked our truck with the canoe on top.

February 26, 2010, was a special day for Pilar, her wedding anniversary. But it was complicated. That week her husband had gone on vacation with two of their three grown children to a place in the countryside, about thirty-five miles to the east. She had remained at home to care for her eighty-six-year-old mother, Doña María Cristina. Her husband drove back to the city to take her to lunch. After lunch he returned to the countryside, but he called a friend, asking him to deliver some flowers to Pilar that evening.

Pilar's mother was not well. Pilar had engaged the services of a pair of young nurses to help on alternate days. That evening, the nurse was twenty-four-year-old Karen Castillo.

Later that night, when Pilar's husband felt the earthquake, he tried to call her, but without success. Alarmed, he returned to the city.

The Talcahuano to which Don Luis returned in the dark that night was much like that described by Darwin and FitzRoy one hundred seventy-five years before. But instead of bales of yerba— the maté herb—forty-foot-long shipping containers were flipped around like piles of matchsticks. Streets in the lower parts of the city flooded. Tugboats washed into neighborhoods. The Plaza de Armas was piled with boats, shipping containers, cars, pieces of destroyed homes, and other debris washed in by the tsunami.

As Don Luis and his two sons, also both firemen, approached Bulnes 178 they could not but have feared the worst. The young men called to their mother. No reply. When they forced their way into the house, they saw that the second floor of the building had collapsed. They found the bodies of Doña Pilar and her mother in the rubble.

Other firemen who joined them found the young nurse, Karen, crushed but clinging to life. Amazingly, for some reason, water had not entered the house, or she surely would have drowned. Digging for forty minutes they were finally able to extricate her.

Meanwhile one of the firemen had spotted two empty coffins floating in the streets. He retrieved them for Doña Pilar and her mother.

A van was commandeered to take Karen to the hospital. Arriving there, she suffered cardiac arrest, but refused to die. She was transferred to the regional hospital, but was denied admittance. The doctors needed to concentrate their efforts on those for whom there was some hope of survival. She was taken back to the first hospital where, attempting to save her life, a surgeon amputated her leg.

Later she was transferred to the Central Hospital in Santiago, where she underwent more surgeries. After nearly a month, she was taken off a respirator. She survived.

Don Luis Fregonara buried his wife, Doña Pilar Bermúdez, on the Monday after the earthquake and tsunami. He then continued his work as part of the command responding to the emergency.

◆

When I learned about the earthquake in 2010, I was on a trip with my wife in Arizona. I was immediately conflicted about whether to drop everything and fly to Chile. From the outset it seemed as though this earthquake was similar to Darwin's earthquake in 1835. I spoke by telephone with Lisa, but it was a few days before we heard anything from Marco. He and his family were safe, but they had had a rough few days. He had used the battery from my echo sounder to provide some light in his house while the power was out.

After contact was reestablished, I learned that Marco and Daniel, who was still in Chile finishing his fieldwork, planned to survey the effects of the earthquake and tsunami. I was sorely tempted to join them. But a number of factors gave me pause. This was their earthquake, I told myself. Plus, I'd been involved in different ways in numerous investigations immediately following large earthquakes and volcanic eruptions. I don't really like the feeling. I am repulsed by the thought of seeming a ghoul, particularly a foreign ghoul, parachuting in to advance my own career in the face of those suffering loss and misery. Documenting ephemeral effects of a natural event is important, but much of the documentation can be done later, and in many cases more effectively, after the fog of disaster response has cleared.

Another factor was my pathetic Spanish. With my limited ability I feared I would be more of a burden than a help to my comrades.

I vacillated for days. I checked the Web for plane reservations. I imagined how I could rearrange my commitments for the next few weeks. But in the end I decided against going immediately.

I sent my GPS and echo sounder to Daniel via a USGS colleague who was headed down to help install instruments. Marco, Daniel, and another Chilean colleague took off on reconnaissance, living in Marco's camper truck. They documented the heights and deposits left by the waves of the tsunami. They measured the heights of the lines of newly stranded mussels that showed how the land along the coast had been uplifted.

Conditions were difficult. There actually was some civil unrest in Concepción in the days following the earthquake. When the group hitched a ride on a military plane to Santa María, the Concepción airport seemed more like a military base, lined with military planes and soldiers. Daniel knew that gasoline was in short supply on the island and that he would need some to repeat the echo sounder survey. He managed to smuggle a container of gas onboard the military plane.

I anxiously read Daniel's emailed report. "We just came back from Santa María—literally we saw what FitzRoy must have seen . . . ! on average 3 m [about ten feet] of uplift . . . !!! . . . The people are fine, they are thankful to the uplift, which reduced the tsunami wave from 8 to only 5 m [26 to 16 feet], and thus didn't destroy the houses at the shore or kill anyone. It only severely damaged all the boats."

Javier's boat, *Jeremías*, the prophet of doom, was among those lost. But Daniel was able to find a fisherman whose boat had not been lost to help with the survey.

What had happened to the island was obvious. The concrete boat ramp, though sagging from settlement and damage from the tsunami, was high and dry. It no longer reached the water. Much of the shallow bay that once formed the harbor of Puerto Sur was now a wet sandy flat. FitzRoy's rocky platform was again exposed above the tides. Mussels and other shellfish clinging to the rocks were dead or dying.

When Daniel was able to do the echo sounder survey, sure enough, the water in the bay was much shallower. There was no doubt. The island had been uplifted again.

◆

Marco was exhausted by the time Lisa and I finally reached Chile in May, not quite three months after the earthquake. He had been busy enough investigating the earthquake and tsunami, all the while fulfilling his teaching duties at his university, but he was one of the lead organizers of an international scientific conference to mark the fiftieth anniversary of the 1960 earthquake. He was scheduled to lead a three-day field trip to the area of the 1960 earthquake to show participants his evidence for earthquakes and tsunamis going back two thousand years. He'd been running on adrenaline for three months. Lisa and I wanted to visit Tirúa and other places, but Marco wouldn't be able to join us. He had had enough.

We teamed up with a colleague from the USGS, Alan Nelson. We didn't have much time, but we managed to make a long one-day trip from Concepción. The roads were twisty, with logging trucks. On the way to Tirúa, I drove too fast for Lisa's taste. She revoked my driving privileges.

Tirúa was still a mess. The lower part of the town had been inundated, destroying many houses. As we drove on the bridge across the Tirúa River toward the riverbank and field where we had worked, we could see that much of the field was still covered with fresh sand, trash, and debris.

Across the bridge, where Don Pedro's discotheque and restaurant had been, with the little cabin out back, where we had stayed the year before, there was nothing. Only the peripheral foundation, the concrete slab, and the tile floor of the bathroom remained. We had tried to stay here again only four months before when revisiting the riverbank. But as we were taking our things out of the truck, an outraged woman from an adjacent house began screaming at Don

Pedro—his aviator glasses nearly sliding off the end of his nose of his downcast face. Perhaps he had not paid the rent. We made other arrangements. But now, even the nearby house of Don Pedro's angry landlady had almost completely been washed away.

As we took a large sample of the sand in the field, two Chilean men with fishing poles approached. Lisa and Alan were busy wrapping the sample in plastic wrap, so I tried to explain what we were doing. The two men looked at me oddly. When Lisa's task was complete, she took over the explanation.

At a pause, one of the men pointed to me, and said "I know this guy." Remarkably, he was one of the two Carabineros, and fans of *The Simpsons*, who had given me a ride on Isla Santa María two years before. After the earthquake he had been deployed to Tirúa and was fishing with a friend on his day off.

After returning to Concepción, we drove through Talcahuano to see for ourselves the destruction from both the shaking of the earthquake and the tsunami. By then, much had been cleaned up, but damaged buildings and evidence of the destruction were widespread. I snapped a few photos of the worst examples.

Only much later did I discover that one of my photos showed the remains of the building at Bulnes 178 where Doña Pilar and her mother were killed.

◆

Lisa and I returned to Chile again in August. It was the rainy winter season in south central Chile, but at least Marco had had some time to catch his breath. We three were joined by three other tsunami researchers, two from Australia, and one from Puerto Rico.

We spent two weeks examining the coast where the earthquake and tsunami had struck. We planned to visit the sites where we thought we had found evidence of previous tsunamis, to see how those places had been affected by the 2010 event.

Our first stop was Licantén, where I had stayed with Brian Atwater on my first trip to Chile. It is a small town and there are not many options of where to stay. We pulled up to the small hostal where I had stayed before.

It was partially collapsed.

Through a hole in the wall I could see the room where I had eaten my first *pastel de choclo*, a traditional and favorite Chilean dish, a kind of corn casserole. The room was filled with adobe bricks that had fallen from walls above.

Nonetheless the hostal appeared to be functioning. Marco went inside and arranged for rooms for our group in the uncollapsed part of the building. Perhaps not the brightest idea, but we survived the night.

We drove along the coast through Iloca, which had been bustling with beach visitors in search of seafood on my earlier visit. Many if not most of the buildings had been washed away.

When we arrived in Lipimávida, the potato field where we thought we had found evidence of a prior tsunami was covered with sand. We dug new pits to measure the thickness of the sand and to try to find a way to date the older sand layer.

And this is when Marco and I met Don José Luis and heard his story of surviving the tsunami.

◆

Eventually we reached our southernmost stop, Tirúa. Lisa and I began to make a detailed map of the new sand in the field, recording the locations of its greatest thickness. Marco, Zamara Fuentes, our colleague from Puerto Rico, and I drove to the headland on the coast, on the side of the river opposite the town. There Daniel and Marco had measured a line of dying mussels on the rocks indicating a foot or more of uplift. Marco had also measured the elevation of some dying bushes on the headland above indicating that the crashing waters of the tsunami had reached a maximum height of about sixty feet. He wanted to see what changes might have occurred in the intervening six months. Indeed, the rocks showed a fresh line of young barnacles below the one above and, although the dead vegetation on the hillside was still apparent, the bushes were beginning to recover.

As we returned on the road from the headland toward Tirúa at dusk, we saw a light in a small house by the side of the road. Marco stopped the car to see if someone might be home. Stepping toward the fence surrounding the yard, he called. Immediately two unfriendly dogs sprang from house, barking ferociously. Country dogs in Chile are not something to be taken lightly. But before the dogs reached Marco, the door of the house opened and a short, stocky figure emerged, carrying a large stick, shouting at the dogs. This was Doña Elena. After the dogs calmed down, she consented to an interview, and told her story of the earthquake.

Doña Elena had lived in Tirúa a long time. She described the 1960 earthquake and tsunami fifty years before and how her recollections of the events of 1960 had guided her response to the 2010 earthquake. Doña Elena's house was on the uphill side of the road, but the house of her son, his wife, and their children was located downhill, just above the Tirúa River and less than half a mile from the beach. After she described the experiences of her family in the earthquake, Marco introduced Zamara and me. When she learned that I was from the U.S. she declared that I should take her home with me. When I protested that I was already married, she replied decisively, *"¡No importa!"* [No matter!]

◆

It was not as though the earthquake and tsunami were unexpected. Many Chileans have adopted a kind of stoic inevitability toward the challenges that the earth will throw their way: earthquakes, tsunamis, volcanic eruptions, floods . . . all just another challenge to be overcome. On the bare foundation of a former beach house at Iloca was a sign quoting Nietzsche, *"Lo que no te mata, te hace más fuerte"* [That which does not kill you, makes you stronger].

Since the 1970s, seismologists have argued about the notion of a "seismic gap," the idea that sections of earthquake zones that have not experienced a large earthquake recently are the most likely to experience one soon. The idea is so intuitively appealing and seemingly based on a simple view of the physics of elastic rebound that it strikes many as almost obvious. Nonetheless, the idea has engendered unending debate and serious statistical challenges.

Along the coast of Chile, some seismologists had identified such a gap, between the northern end of the rupture zone of the great Chilean earthquake of 1960 to the south, and the southern end of the rupture zone of a smaller, but very large earthquake that occurred in 1906. This gap roughly coincided with the zone thought to have produced the earthquake in 1835. Indeed, this gap had come to be called by some, the "Darwin gap."

The opening line of the abstract of one scientific paper published in 2009 read, "Concepción–Constitución area (35–37° S) in South Central Chile is very likely a mature seismic gap since no large subduction earthquake has occurred there since 1835." The paper's abstract closed with the line, "Therefore, in a worst case scenario, the area already has the potential for an earthquake of magnitude as large as 8–8.5, should it happen in the near future."

Indeed, this "gap" did have the potential for an even larger earthquake, but no one knew how soon it would come. The magnitude of the earthquake in 2010 was 8.8.

And even for those Chileans who might have read the article in *El Sur* with Marco's quote, and for the very few who were regular readers of *Physics of the Earth and Planetary Interiors*, the earthquake came as a very rude and unpleasant shock. Even for the scientific Jeremiahs who prophesized the event on a scale of years or decades, the instant of its occurrence was like a thunderbolt from a clear sky. And at that instant, there was no more time for planning, reinforcing, or rebuilding. No more time for mayors to prepare their cities. No more time to make up for opportunities lost. The individual and the society must respond based on the conditions that exist at that instant. For the individual, the primal impulse for survival kicks in. Woe be it to those near the beach who do not know to run for the hills.

◆

A year after the earthquake, in its final report, the Chilean government listed 524 people killed and a further 25 still missing. It was
believed that 156 of these deaths, plus all the still missing, were
as result of the tsunami. The greatest death toll was in the area
including Constitución with 91 fatalities. Talcahuano suffered 23,
Tirúa only two. About 370,000 homes were destroyed, with losses
estimated at $30 billion.

Modern buildings, on the whole, performed well during the
earthquake. The greatest loss of life resulted from older buildings
that did not meet modern codes and standards. Knowledge of the
importance of reaching high ground, and the ability to do so, were
the most important factors in reducing a potentially horrific loss
of life in the tsunami.

◆

Our work at Tirúa and Isla Santa María continued. At Tirúa, Brian
Atwater joined us for a visit in 2011. While Lisa and I dug and
logged a pit in the middle of the field, Brian worked with his WWII
trenching shovel to clear and scrape a vertical face along a section
of the riverbank about 450 yards up the river from the bridge,
extending about sixty feet along the bank. It became clear that
what was now the fourth sandy layer down had a wavy character
that did not look natural at all. It appeared that the sandy deposit
was draped over a sequence of regularly-spaced ridges and furrows
that clearly suggested cultivation. Had a tsunami flooded the field
of some Mapuche farmers long ago? It seemed very likely.

In 2013 Marco found a seed adjacent to the deposit. It turned
out to be the seed of *Calystegia* or false bindweed, a member of the
morning glory family, probably a weed that bedeviled the Mapuche
farmers. Regrettably, the age fell right on one of the troublesome
loops in the carbon-14 calibration curve. The lab concluded with
a 95% probability that the age of the seed was either 1450–1530

A.D., or 1570–1630 A.D. Based on its analysis—and the context that Lisa provided when she submitted the seed for identification and dating—the lab coldly concluded its report, "The Calystegia seed most likely reflects a weed growing with crops in the agricultural field immediately prior to the tsunami that deposited [the sand layer]." Sadly, we don't have a human face, the face of the human that excavated and tended the plants on the ridges between the furrows, battling the false bindweed, to go with this tsunami, or a story of how he or she, family, and friends might have run for the hills.

Given all the evidence, we concluded that at Tirúa we have very strong evidence for at least four tsunamis: 2010, 1960, 1751, and 1575, based on evidence associated with the historical earthquakes, and then preliminary evidence for another five earlier events. We believe that we see evidence for uplift in Tirúa in 2010 and 1751, but subsidence in 1960 and 1575, providing some constraint on the location and extent of rupture in the older earthquakes.

I returned to Isla Santa María with Daniel to repeat our survey of the bay in both 2011 and 2013. Javier had replaced *Jeremías*, the boat he lost in the tsunami, and he was again able to help us carry out the surveys. I was also able to find a very nice chart of the bay carried out by Spanish surveyors in 1804, and one by Chilean surveyors in 1868, to go with the *Beagle* chart from 1835. With some historical research to put the historical and modern pre- and post-earthquake surveys all on the same basis, we were able to develop a convincing picture of the vertical movements of bay adjacent to the island. It sat quite low in the water in 1804, rose seven feet in 1835, did not change much between 1835 and 1868, then sank about four and a half feet prior to uplifting about five feet in the 2010 earthquake.

A few years before the earthquake Daniel and Marcos Moreno, a Chilean colleague also working in Germany, began periodic, sophisticated GPS measurements on the island. Their measurements showed that before the earthquake, the island was sinking at a rate of almost half an inch per year, and moving toward the coast at a rate of more than an inch and a half per year. GPS measurements made on the island before and after the earthquake, and measurements of uplifted shellfish attached to rocks, gave estimates of the uplift of the island itself from about four and a half to about seven feet. Given that calculations based on elastic rebound theory predicted a gradual increase in the amount of uplift from east to west, the results from the bay (to the southeast of the island) were in excellent agreement with these other measurements. At the same time GPS measurements showed that the island lurched about sixteen feet toward the sea.

All these measurements were also in excellent agreement, not with Darwin's original idea of uplifts during earthquakes being monotonic steps in the uplift of the continent, but with the ups and downs corresponding to the release and accumulation of elastic strain as predicted by plate tectonics and elastic rebound theory.

But there was one wrinkle. It appeared that the uplift in 2010 was a bit larger than the subsidence in the period between 1835 and 2010. Daniel and colleagues had also measured the ages and elevations of a series of beach ridges formed on the eastern part of the island. These measurements suggested that over the last three thousand years the island had been gradually uplifting. Taken all together, these results suggested that Darwin had not been completely wrong. If the uplifts in past earthquakes were on average about the same amount larger than the preceding subsidence as in 2010, then the ups and downs of elastic rebound would result in a gradual uplift just as Darwin suggested. But the long-term uplift would occur in a saw-tooth pattern similar to two steps forward, one step back, rather than a monotonic, or ever-increasing, trajectory.

◆

In 2011 I was amazed to see with my own eyes the same uplifted rocky platforms, and even some remnants of dead shellfish, just as FitzRoy had seen here, and Darwin had pondered, one hundred seventy-six years before. Now it was clear that effects of the 2010 earthquake aligned with the predictions made by the hypotheses of plate tectonics and elastic rebound, with a nod to the gradual uplift that Darwin had predicted.

But now I could also see and feel for myself even more clearly than before, the human impact of these earthly paroxysms. Getting to know the people and learning the stories of those affected by the earthquake and tsunami did not reduce my scientific curiosity, but it dramatically increased my compassion for the victims. Perhaps if Darwin had had more than a few days in Talcahuano and Concepción, and not had to hurry off with the *Beagle*, he too might have learned more stories of the people affected, and not been quite so quick to allow his scientific excitement to replace his compassion.

These stories also led me to wonder about the role of the geologist, not only in thinking about the past, but also in imagining the future and what a constantly changing earth implies for the inhabitants of our planet.

Now You See Them, Now You Don't

> *All that at present can be said with certainty, is*
> *that, as with the individual, so with the species,*
> *the hour of life has run its course, and is spent.*
>
> —Charles Darwin

Juan Carlos Fernicola and I sat in his small office in the basement of the Museo Argentino de Ciencias Naturales in Buenos Aires. There was enough space to get out of the office in case of fire, but not much more. Bones and books filled the room. On a box behind his chair were three massive femurs, or thigh bones, of glyptodonts. On the table next to his desk was another thigh bone, and next to it an array of boxes containing remains of ground sloths, not giant ground sloths, but fossils of the older ground sloths before evolution turned them into monsters. On his desk sat two computers, a glass maté gourd with a metal straw, and a thermos of hot water.

Juan Carlos studies the creatures that intrigued Darwin. He is a jolly man, with a sharp sense of humor, but he is careful scientist, a man anxious to gather evidence and ready to be convinced by its preponderance. He writes serious scientific papers with titles like, "The evolution of armored xenarthrans and a phylogeny of the glyptodonts." He has carefully reviewed Darwin's fossils of the giant ground sloth, *Mylodon darwinii* and others from Punta Alta, and compared Darwin's and Owen's ideas with the adjustments and reinterpretations over the last 180 years.

He didn't want to answer my question. He didn't want to answer because he didn't know the answer, and he didn't think that anyone else did either.

It's not an easy question. Although some paleontologists think that they are ready to hazard a guess, there is not yet anything approaching general agreement. Darwin puzzled about it, but we still just don't know the answer. My question was, what happened to the ground sloths and their contemporaries? Why did they disappear?

Of course it was not just the giant ground sloths, it was almost all the really large mammals in North and South America. About 12,000–14,000 years ago, more or less, a bit later in the south than in the north, they all vanished. The relatively small tree sloths, armadillos, and anteaters are the only large survivors of the unique creatures that evolved in South America when it was isolated from the other continents.

"We need more data," Juan Carlos said. "We need to understand the paleoecology." So far no one has found a smoking gun that all can agree on. Not like the dinosaurs.

◆

Darwin would be enchanted by the story that has been pieced together over the last 180 years about the strange creatures he found at Punta Alta, and how their evolution and geographic distribution was related to plate tectonics. The correspondence among fossils of the same species, such as the fern *Glossopteris*—that lived during the same period of geologic time, but are now spread across the widely separated continents—proved to be one of the early and persistent arguments in support of continental drift. Now plate tectonics, its successor, possesses amazing explanatory power in understanding evolution and the distribution of species through time. Darwin foresaw this with the peculiar species, both fossil and living, both discovered by him and others in South America, writing, "This wonderful relationship in the same continent between the dead and the living, will, I do not doubt, hereafter throw more light on the appearance of organic beings on our earth, and their disappearance from it, than any other class of facts." George Gaylord Simpson, a preeminent American paleontologist and evolutionist, picked up the same theme a century later, but still before the general acceptance of plate tectonics. Writing about the oldest mammalian faunas of South America he said, "They also illustrate . . . on the most magnificent scale the deployment and evolution of a fauna developing in isolation but on a continental scale. They display an experiment in evolutionary principle conducted for us by nature in a space and over a period of time far greater than any laboratory research can encompass."

The story of the ground sloths and their ilk is a story of evolution with all its complexity, including elements of plate tectonics, migration, changing climate, competing species, and ultimately, of course, extinction. And a story for which the causes of the final act are still very much up for debate.

◆

Early in my explorations of Darwin's geologic explorations I learned the particulars of his discovery of *Mylodon darwinii*, the extinct giant ground sloth. Even earlier I had visited Cueva del Milodón in Chilean Patagonia where a life-size replica of this creature greets visitors, but I did not really bond with the species on that first meeting. Yet the more I learned about these peculiar creatures, the more intriguing their story became, their role in the history of paleontology—including the involvement of Thomas Jefferson—their formerly ubiquitous distribution in the Americas, and their relatively recent extinction. I confess to developing a fascination with them and their kin from the superorder Xenarthra.

Not to mention that the origin of Bruce Chatwin's obsession with Patagonia, as described in his book, *In Patagonia*, is a piece of a *Mylodon's* hide acquired by his grandfather and kept in a glass case in his grandmother's apartment in London.

Indeed, the discovery of a large piece of the very strange, but very fresh-appearing skin of a *Mylodon* at Cueva del Mylodón in the 1890s triggered an international wave of excitement that the creatures might still be hiding somewhere in Patagonia. The skin of the *Mylodon* is composed of an outer epidermis covered with long coarse hair, but beneath this is a layer containing small bony plates, or ossicles, an evolutionary nod to the scutes and platy exteriors of the *Mylodon's* relatives the Glyptodonts and modern armadillos. The search proved fruitless. The *Mylodon* was gone.

◆

One day I imposed myself on Greg McDonald, a paleontologist with the U.S. National Park Service, and one of the American experts on these creatures. In his office in Fort Collins, Colorado, Greg—a stocky man with white hair and a neatly trimmed white mustache—patiently explained to me some of the basics of the giant ground sloths. He squatted down, then put his hands on the floor in front of him with his fingers turned in, to illustrate the way that ground sloths walked on the knuckles of their front claws. They were knuckle walkers, not knuckle draggers. He explained that notwithstanding their fierce-looking claws, they were not carnivores, but vegetarians. Some species were grazers, other browsers. He explained that in North America at the end of the Pleistocene that there were four genera of ground sloths each represented by a single species. Two of these, the *Megalonyx*, or Jefferson ground sloth, and the *Paramylodon*, or Harlan ground sloth, were widely distributed across the continent. A third, the *Eremotherium*, was limited to the southeastern United States. And the fourth, the *Nothrotheriops*, or Shasta ground sloth, was widely distributed in the desert south-west of the United States. The Harlan ground sloth, like its close

South American relative *Mylodon darwinii*, was probably a grazer. In contrast, the Jefferson and Shasta ground sloths were browsers.

Almost all of these sloths were big. The *Eremotherium* weighed up to three tons or more and could reach up to 17 feet off the ground. The Harlan ground sloth was up to about ten feet high and weighed about a ton. The Shasta, the smallest, still reached about ten feet long from head to tail, and weighed five hundred pounds, the size of a large bear.

Riding back from lunch in the car, I asked Greg what he thought would have been the population of ground sloths in North America at their peak. Would they have been about as common as, say, bears are today. Yes, he replied, something like the population of bears.

And by about 12,000 to 13,000 years ago all the ground sloths on the mainland of the Americas were gone, although some varieties apparently persisted on islands in the Caribbean until only about 5,000 years ago.

◆

Darwin and Charles Lyell thought that the geology of continents derived from their repeated rising and sinking. They had no idea that the continents drifted around the globe, riding on their tectonic plates like so many kids in bumper cars at an amusement park, albeit a good deal more slowly. Like bumper cars, the continents were sometimes jammed into clusters, as Eurasia, India, and Africa are today. At other times one continent might be isolated, as Australia is now, moving along without another continent nearby. About 150 million years ago South America drifted away from the cluster of continents that had formed Gondwanaland. It was off drifting by itself, isolated, for 135 million years, until it reconnected with North America, first via a chain of islands beginning about 15 million years ago, and finally and firmly via the Isthmus of Panama about three million years ago. During this period of isolation, evolution kept humming along both in South America and elsewhere, species changing to improve their chances for survival. Connected continents could exchange species. But poor South America, off by itself in the corner like a lone bumper car, nurtured its own, turning out some very unique creatures that were unlike any on the other continents.

Two groups of these unique creatures from South America particularly fascinated Darwin. One, the *Xenarthra* (formerly called *Edentata*), includes creatures that still exist today—the armadillos, anteaters, and tree sloths—as well as the extinct giant ground sloths and glyptodonts. The second group, the *Notoungulates*, now all extinct, included Darwin's favorite the *Toxodon*, and *Macrauchenia*, the giant camel-like creature the bones of which Darwin discovered near Puerto San Julián in eastern Patagonia. All these animals were herbivores, munching plants, grass, and leaves. The only large carnivores in South America at the time were the "terror birds," or *Phorusrhacids*, a truly nasty bunch of ground fowl, some as much as ten feet tall with skulls as long as two feet. They were well adapted to running and whacking slower moving prey with their powerful beaks.

During this same time, Eurasia and North America, intermittently connected, were exchanging animals frequently. A different set of large animals (or megafauna) evolved there: the mammoths, mastodons, saber-toothed tigers, short-faced bears, camels, and *Equus*, the ancestor of the modern horse.

As North and South America began to approach one another, the continents began to swap their creatures. The first exchanges were of "island hoppers" and creatures that were adapted to swimming. (The ground sloths "were good swimmers," Daniel Perea, the Uruguayan paleontologist assured me with a smile.) Then, after the connection via the Isthmus of Panama three million years ago, the larger and less water-friendly animals joined in the trek. Eventually ground sloths, armadillos, *Glyptodons*, *Macrauchenia*, and a few terror birds streamed north. Even the *Toxodon* made it as far north as Central America. Similarly, the mammoths, mastodons, camels, horses, saber-toothed tigers, and bears headed south. This remarkable swap of animals, now called the Great American Biotic

Interchange, reached its peak shortly after the formation of the isthmus.

Many of the new arrivals in each continent did pretty well, continuing their evolution to meet the circumstances of their new environment. Others had problems. The terror birds ultimately lost out to other carnivorous mammals on both continents. Indeed the invasion of the carnivorous mammals from the north was a particularly violent shock to the animals of South America. But over a couple of million years things settled down. The ground sloths, for example, continued to do well in South America, and seemed to thrive in North America.

◆

As the continents of North and South America inched toward one another, eventually to be joined by the Isthmus of Panama, oceanic currents that flowed between the Atlantic and Pacific Oceans in the low latitudes were restricted and eventually blocked. The pattern of circulation within both oceans was altered. It seems likely that these changed currents had a profound effect on the planet's climate, some believe even triggering the onset of the Pleistocene and with it the ice ages.

In any case, about two and a half million years ago, not too long after the formation of the isthmus, the climate of the earth began to change dramatically. Studies of sediment cores taken in the deep ocean show that for the preceding few million years the planet's climate had been somewhat warmer than our present climate, and relatively stable, varying, but within a relatively narrow temperature range. However then, at the beginning of the Pleistocene, the climate began to grow colder on average, and also to oscillate between warmer and colder periods. For the first million and a half years, the oscillations were relatively regular. The warm and cold periods were about the same length, and one entire cycle of warm, cold, then back to warm, occurred about every 41,000 years. Then about one million years ago, the oscillations changed. The period of the oscillations changed from 41,000 to 100,000 years, the cold part of the cycle became longer, and cold periods became colder. These oscillations are recorded in ice cores from Greenland and Antarctica as well the deep sea cores. During the colder parts of these cycles the glaciers advanced, and during the warmer parts of the cycles they would retreat, with minor fluctuations along the way. The major cold periods are now termed glacial periods, or just glacials, the warm periods, interglacials. Depending on exactly how you count, the planet has been through about ten of these longer cycles over the last million years. For reasons that are simply historical and semantic, geologists declared the end of Pleistocene at the end of the last cold period, 11,700 years ago. This was the beginning of the current warm period, the Holocene. There is no reason whatever to believe that the current warm period is fundamentally any different (except for the possible effects of Homo sapiens), from the previous ten or so warm periods, and that this—our—warm period, will not eventually be followed by another cold period.

These warm and cold periods have been given different names in different parts of the world, and even in the United States. It can all get rather confusing. One ray of sunshine is that everyone calls the current warm period the Holocene. In North America, the most recent glacial period is called the Wisconsin in the Midwest and east, the Pinedale in the Rockies, the Tioga in the Sierra Nevada, and the Fraser in the Pacific Northwest. During this time, the glaciers reached their maximum extent, called the last glacial maximum, about 18,000 to 25,000 years ago. The previous warm period to our own is called the Sangamonian in North America and Eemian in Europe. During this last interglacial, the climate was similar to the Holocene, but a bit warmer.

Subtle changes in the earth's orbit called Milankovitch cycles seem to be responsible at least in part for this cyclic behavior. The most important of the Milankovitch cycles would seem to be a slight variation in the tilt of the earth's axis that correlates very well with the shorter 41,000 year cycles. Among the other Milankovitch cycles, there is a slight variation in the eccentricity of the earth's orbit, or the amount by which the orbit varies from a circle, with a period of 100,000 years, but current theory does not explain why this change in eccentricity should be strong enough to cause the cycles of warm and cold periods in the second half of the Pleistocene. One idea is that there is some as yet undiscovered resonance—like the ringing wine glass or the amplification of tsunami waves in the Bahía de Concepción—in the immensely complicated system of the global circulation of the oceans and atmospheres. Such a resonance could depend on the current configuration of the oceans, continents, and mountain ranges. The Milankovitch cycles are a major advance in understanding climatic variation as revealed in cores in the ice sheets and the deep ocean, but many questions remain.

One of the key puzzles about the extinction of the giant ground sloths and other megafauna is that these creatures successfully navigated through the many cycles of warm and cold during the Pleistocene, including about ten of the long 100,000-year cycles, only to exit the stage suddenly at the beginning of the most recent warm period, the Holocene.

◆

In my quest to learn about the ground sloths and their extinction I also pestered Bob Thompson, a paleoecologist with the U.S. Geological Survey who studies the changes in the ecology of the western United States since the last glacial maximum. Bob uses fossil pollen and plant macrofossils in his work to discover how the distribution of plants has changed, and from this to estimate the regional changes in climate. I was surprised to learn that Bob had a very intimate relation with the Shasta ground sloth.

Like some of the other ground sloths, the Shasta ground sloths liked to live in caves. Nice, cool, dry caves. Caves that preserved prodigious deposits of their desiccated poop in almost pristine condition.

"Hey, Rob wants to see the sloth poop!" Bob shouted to his assistant as we sat talking in his office. After a few minutes, a heavy-duty, ziplock bag appeared, filled with large fist-size balls of dry, grassy-looking sloth dung.

"Here take a look at this," Bob said, as he handed me a chunk. I confess that I accepted the dried, more than 13,000-year-old poop a bit reluctantly. As I turned it over in my hands, it vaguely resembled a large, extremely dry version of a road apple from a horse. But the sloth apparently had not masticated his food quite as well, because the poop seemed largely composed of long pieces of grass and stems, as if rolled in a ball. As I examined the sample, Bob explained what a boon this dung was for understanding the ecology and environment when the sloth lived, because the pollen, seeds, and remaining plant fragments provided such a rich record of what the hungry beast consumed.

Bob also told me the tragic story of Rampart Cave in Grand Canyon National Park. The cave was explored in the 1930s and 1940s, and discovered to contain layers of sloth poop that averaged several feet thick throughout the cave. Although the Park Service subsequently installed an iron gate, an apparently curious, or possibly ill-intentioned visitor, pried it open, and either accidentally or perhaps intentionally, started a fire that smoldered for several years, despite efforts to put it out, and ultimately consumed about 70% of the deposit.

Nonetheless, what is known from the deposit tells a fascinating story. The deposit consisted of three layers. The lowest and oldest layer was ground sloth poop deposited between about 26,000 to 42,000 years ago, that is, before the last glacial maximum. The second layer, about one foot thick, was composed of plant material left by ancient pack rats. The top layer was again made up of sloth dung, dated between 13,000 to 15,000 years ago. It would seem that the Shasta ground sloths decamped from the cave during the height of the last ice age. Perhaps they were denied access by high water levels on the Colorado River. Then they returned again before they finally vanished for good. Fortunately, despite the damage to Rampart Cave, other caves in the southwest have added to the story of the Shasta ground sloth and of the environment in which it lived.

◆

The extinction at the end of the Pleistocene that included the giant sloths is modest in comparison with other extinctions in geologic history. An even larger extinction led to the annihilation of the dinosaurs. They, along with half the genera and three-quarters of the species living on Earth at that time, died out 65 million years ago. This extinction affected plants and animals, large and small, on land and in the oceans, around the planet. Even the phytoplankton in the oceans, at the very bottom of the food chain, suffered. Individual species come and go all the time, but when so many species get whacked at once, paleontologists call the event a mass extinction. There have been only a handful of really big extinctions in the history of life on Earth, and many lesser events. Even this mass extinction that killed the dinosaurs, the K-T boundary event*, was not the largest, but it certainly ranks in the top five.

In contrast to most other extinction events, geologists have some good ideas about what caused the K-T boundary event. The whole planet was affected. It's pretty clear that plants and animals that had happily relied on sunlight and photosynthesis for their sustenance before the event, either directly or indirectly, suddenly found themselves in deep trouble. The sky went dark, most likely by an abrupt and tremendous increase in the amount of dust or ash in the atmosphere. The currently dominant view is that all these varieties of life were killed by the impact of an asteroid.

One of the best pieces of evidence for this theory is the fact that sediments deposited at the time of the K-T boundary event contain a spike of the metal iridium. Iridium is a silvery metal that is very, very rare on the surface of Earth, but is common in meteorites. If there is a smoking gun for the argument that the cause of this extinction came from space, it's the iridium. Nonetheless, some

---

* The K-T boundary is shorthand for the boundary between the Cretaceous (K) and Tertiary (T) geologic time periods. The term Cretaceous-Paleocene boundary is now increasingly preferred to Cretaceous-Tertiary.

geologists hold out that the extinction was caused by a period of intense volcanic activity.

The extinction that did in the ground sloths and others was not a mass extinction like the one at the K-T boundary. Almost all the damage done was to larger animals, especially those larger than about one hundred pounds. In South America alone, all the animals larger than twenty-two hundred pounds, including the ground sloths, the *Gyptodons*, the *Toxodons*, and *Macrauacchenia*, and all the very large invaders from the north, disappeared. The largest animals to avoid extinction in South America are the tapir, the jaguar, and the guanaco. The largest, the tapir, tips the scales at about seven hundred pounds. About 80% of the mammal species between one hundred to twenty-two hundred pounds also disappeared. Smaller animals seemed to make it through pretty well. The story is similar in North America, although the largest terrestrial animal in North America to survive is the bison, a little larger, at up to twenty-eight hundred pounds in the wild.

This extinction affected primarily the Americas, although the mammoths and mastodons in northern Eurasia died out at about the same time. This is very much unlike the mass extinction that included the dinosaurs, an extinction event that affected the whole world and nearly all living things, from the largest to the microscopic.

But what is perhaps most intriguing about the extinction in the Americas is that it just wasn't very long ago. We ought to be able to figure out why it happened.

Although Darwin didn't have a date for the time of this extinction, development of the absolute geologic time scale being in the future, he deduced that his "quadrupeds" had disappeared very recently. From the same gravel at Punta Alta where "the remains of several gigantic animals were extraordinarily numerous," he also collected shells. Upon returning to England he sent these shells to an expert, George Sowerby. Darwin concluded from Sowerby's identifications that "Of these shells it is almost certain that twelve species . . . are absolutely identical with existing species; and that four more are perhaps so; the doubt partly arising from the imperfect condition of the specimens. . . . Under these circumstances, I think, we are justified . . . in considering the shingle strata at Punta Alta, as belonging to an extremely modern epoch."

So what wiped them out?

◆

Something else happened about the time of the closing of the Isthmus of Panama, but in faraway Africa. The genus *Homo* turned up, walking around on two feet and banging stone tools. The picture sharpened between 100,000 and 200,000 years ago when the fossils began to look a lot, if not exactly, like us. Homo sapiens entered the scene. By about 50,000 years ago, these folks were making nice tools, hunting, fishing, conversing with one another, playing music and games, and trading over long distances. About 45,000 years ago they arrived in Europe, sharing the territory and likely interbreeding with *Homo neanderthalensis* until these evolutionary older cousins disappeared about 40,000 years ago. By about 15,000 years ago, Homo sapiens roamed Eurasia, Africa, and Australia. They hadn't started building the pyramids yet (the pharaohs of ancient Egypt wouldn't get going on the first pyramids until about 4,600 years ago), but they were wrapping up the Stone Age. Most importantly for our story, they were making their way to the Americas.

◆

As I drove through Uruguay with Daniel Perea and his student, Valeria Mesa, I asked them what they thought had happened to the ground sloths. They too, like Juan Carlos Fernicola, took a cautious view, allowing that the changing environment and other factors may have had a role, but declining to back a specific cause.

It is an understatement to say that the date of arrival of the first humans in the Americas—and the routes they followed—is an area of very active study and debate. The dominant view is that humans arrived in what is now Alaska from Siberia when the sea level was much lower. From about 30,000 to 15,000 years ago during the last glacial maximum, much of the Bering Sea was dry land. It is a misnomer to call this connection a bridge. At its greatest extent, this land, Beringia, was a thousand miles wide. But the connection narrowed as the glaciers melted and sea level rose, until the Bering Strait opened around 10,000 years ago. So far the oldest human tools found in Alaska date to 13,000 to 14,000 years ago, and the oldest human bones and teeth date to 11,500 years ago.

The oldest human remains found in North America so far include those of a male infant found in Montana, termed Anzick-1, dated at about 12,600 to 13,000 years ago, and the skeleton of a girl, termed Naia, found in submerged cave on the Yucatán Peninsula in Mexico dated to 12,000 to 13,000 years ago. DNA studies from both skeletons support the idea that they were part of populations derived from East Asia. The remains of the little boy were found with a distinctive assemblage of tools found widely in North America and referred to as Clovis culture. Sites have been found with tools that are pre-Clovis and well dated to about 14,000 years ago. Evidence for even older occupation, such as a Meadowcroft Rockshelter in Pennsylvania, remains debated.

Opinions about the route or routes humans followed south from Alaska vary. Perhaps they followed an ice-free corridor just east of the Rocky Mountains, or perhaps they came down the west coast. If they came down the west coast, much of the evidence is now likely submerged. Or perhaps they followed several routes. A key problem is evidence for human occupation at Monte Verde in Chile—only about forty miles from where Darwin and FitzRoy watched the eruption of Volcán Osorno—also dating to about 14,000 years ago. If these people came via Beringia, how did they get to southern Chile so quickly?

Valeria was inclined to think that the changing environment, subjecting the animals to stress, was the most important factor. From the backseat of the truck, I described a recent paper coauthored by Greg McDonald, who had directed me to Daniel. Greg and his coauthors made a very detailed and careful study of the thighbone of a *Megalonyx* that had turned up in the attic of small museum in Ohio. They concluded that a human had almost certainly hacked and scraped the flesh off the bone while it was still fresh. Of course this example only provides evidence that Homo sapiens and one ground sloth coexisted, and supported the notion that the paleo-indians hunted the ground sloths for meat. It is but one piece of permissive, but not persuasive, evidence about mankind's possible role in the extinction.

A group of paleontologists at the Universidad Nacional in La Plata, Argentina, have argued that it is time to get over the extinction debate as a dichotomy between humans and environment. They proposed what they call the Broken Zig-Zag hypothesis. The Zig-Zag comes from the idea that as the climate alternated between the colder, drier periods (the "glacials"), and the warmer, more humid periods ("interglacials"), the amount of open land would vary significantly, expanding during glacial periods and contracting during the interglacials. The populations of animals depending on these open areas would thus rise during the glacials and fall during interglacials. They go on to argue that the arrival of Homo sapiens as hunters at a time when the climate was warming and the range of these large animals was decreasing, thus coming at a time of great stress, was enough to push them over the brink.

◆

In summer, the rocky moraines and gravelly outwash plains left by the retreating glaciers of the Wrangell Mountains near my cabin in Alaska are covered with delicate white flowers, then later with the cottony white seed plumes, both from a plant named mountain avens, or more formally *Dryas octopetala*. This hardy dwarf evergreen shrub is one the first plants to colonize land newly exposed by the retreat of glaciers. It is very common in the northern regions and high mountains of the Northern Hemisphere. Indeed, it is the territorial flower of the Northwest Territories of Canada, and the national flower of Iceland.

Some 12,900 years ago, the glaciers in the northern hemisphere had retreated from their maximum extent almost to where they stand today. Then abruptly, and perhaps in a time period as short as a decade, the warming trend suddenly reversed, and the climate turned cold and remained so for twelve hundred years, before the warming resumed to begin the Holocene. This brief period, or stadial, of reversal in the general warming trend between the last glacial period and our current period has been named the Younger Dryas, in honor of *Dryas octopetala*. Leaves and pollen of the *Dryas* found in lake sediments and bogs were an early indication to researchers of this climatic reversal. (Yes, there are two earlier reversals in the warming trend from the last glacial period, called the Older and Oldest Dryas.)

There is ongoing debate about the cause of the Younger Dryas and about any relation that it may have with the extinction of the giant ground sloths and other large animals in the Americas at about the same time. The last decade saw the rise—and then fall—of enthusiasm among scientists for a hypothesis that the Younger Dryas was triggered by the impact or explosion of a comet on or just above the earth's surface that might also have led to the extinctions. At present the hypothesis with the most support is that the Younger Dryas was caused by a change in the circulation within the Atlantic Ocean, caused by the sudden release into the Atlantic of an enormous fresh water lake that covered much of the upper Midwest of the United States and central Canada. This lake, Lake Agassiz, was formed from the water of the melting glaciers and was impounded in part by the retreating glaciers themselves. The idea is that when the ice dams holding back Lake Agassiz failed, the resulting sudden influx of fresh, cold water into the Atlantic Ocean upset what is called the North Atlantic "Conveyor," a system of currents that brings warm water northward from the tropical regions.

◆

Leaving aside for a moment the current debate about the causes of global warming, there is little doubt that our species has had a profound effect on the ecology of our planet. Homo sapiens has mostly benefitted from the changes in the planet since the last time the glaciers were at their maximum. The current warm period, the Holocene, has truly been a Golden Age for our species. At the beginning of this period there were likely only a few million of us. Today, there are a thousand times more, over seven billion. Are there any other terrestrial mammals of our size or larger that have done as well over this period? Solid numbers are of course hard to come by, but it would seem pretty clearly that the answer is no.

We are now the most populous large mammal on earth. The only other large animals that come close, with numbers over a billion, are those that we raise to supply our own needs for food, such as cattle, sheep, and pigs. Animals with numbers between a half a billion and a billion are goats, cats, and dogs. Even the lowly domestic rabbit, butt of the pejorative phrase "breeding like rabbits," has a global population of under a billion, as do the large animals humans use for work such as water buffalos, horses, and donkeys. Maybe the phrase should be changed to "breeding like humans."

Archeologists and anthropologists who have studied the question carefully remain highly skeptical that humans were a major factor in the extinction of the late Pleistocene megafauna. They have found mass kill sites for bison, but not for the earlier animals. They just have not seen—so far at least—the evidence that they would expect to see that demonstrates a major role for *Homo*. But we are not off the hook. Homo sapiens—once it arrived at the scene—certainly did not help the ground sloths out. Whether hunting or other human activity made a major or only very minor contribution to their demise remains an open question.

Even though it is not proven that we hunted the Pleistocene megafauna to extinction, there is ample evidence that we hunted them. And the list of animals that have been hunted or otherwise pushed to or near extinction by humankind is a long one. The impact of humans on biodiversity as explored by David Quammen in *The Song of the Dodo* and Elizabeth Kolbert in *The Sixth Extinction* is immense. But what I find particularly scary is how much of what we—in the twenty-first century—have come to regard as the "natural environment" is actually the result of humankind's impact on what had come before. William Ruddiman has even argued that humankind's impact on the climate began with initiation of farming in the early Holocene. The recent surge of greenhouse gases is only the latest iteration of our unparalleled impact on the earth.

Attempts to analyze the human footprint on earth, while imprecise, suggest that much of the land surface of the planet, especially the whole of western Europe (excluding the northern regions of Scandinavia), the Middle East, eastern China, India, southeast Asia, central and east Africa, coastal regions of South America, Central America, and the eastern half of the United States is so trampled by humans that little "natural" area remains.

While in the Americas much of this land has been modified by farming, logging, and urbanization since the arrival of the first colonists beginning only five centuries ago, it is also argued that the native peoples may have had a significant impact before the arrival of the colonists, through the use of wildfire. In Europe and Great Britain there is little if any natural forest remaining. The moors of England and the Highlands of Scotland, for example, now protected for their natural qualities, were once forested, but since denuded of their natural forests by thousands of years of human activity.

There can be little doubt that Homo sapiens is special, in the sense that it has come, over the Holocene, to dominate the planet. Whether by luck or directed effort on our part, our numbers are no longer threatened by any natural predator in the normal sense. The threats we currently face from other species come from other branches of the tree of life, such as bacteria and viruses (if you are willing to consider a virus a living thing).

So I came to the conclusion that Juan Carlos was right. To understand why the giant ground sloths, the *Glyptodons*, the *Toxodons*, and the other megafauna vanished, we need more data. But as Javier—the estancia owner that I met in Uruguay who found a jaw of a *Toxodon*—showed by his collections, there is more evidence out there. We just need to find it. The mystery that puzzled Darwin, and many geologists and others since, hasn't yet been solved. But that's no reason to think that it can't be, or won't be. There may be several factors. Maybe it will be solved by a mass of slowly accumulated evidence. Or maybe by the discovery of some fossils in a unique circumstance. Or maybe by a flash of insight. But I am optimistic that, eventually, we'll learn the answer to the question of how and why they disappeared.

CHAPTER EIGHTEEN

Reflections:
What Does It All Mean?

Nothing can be more improving to a young naturalist, than a journey in distant countries. . . . The excitement from the novelty of objects, and the chance of success, stimulate him to increased activity. Moreover as a number of isolated facts soon become uninteresting, the habit of comparison leads to generalization.

—Charles Darwin

How are we to look upon Charles Darwin's geologic theorizing and adventures from our vantage point more than a century and a half on? What have I learned from my quest to follow his path? Certainly Darwin's observations were outstanding. He made the first geologic map of southern South America. His cross-sections of the Andes were pioneering. But while Darwin's geologic ideas were on the leading edge in the 1830s and 1840s, only a few have survived subsequent testing and challenge. Not including evolution by means of natural selection, the most notable survivor is his explanation for the formation of coral islands and atolls. His perception that the motions and deformation of the earth's crust could be related to movements in a more mobile substance is echoed in the modern view of plate tectonics, although the earth's mantle is hardly molten.

From a modern point of view many of Darwin's geologic ideas, such as the transportation of erratics by icebergs, may seem a bit naïve, if not even a little silly. But his ideas were no more naïve, and no more silly, than the ideas held by his nineteenth-century contemporaries in physics and chemistry. Then the physicists still believed that light propagated through a *luminiferous aether*. Eventually experiments five years after Darwin's death showed that the speed of light did not depend on its direction of travel as Planet Earth sped through this imaginary soup. Then two decades later Albert Einstein explained these results and others with special relativity, assuming that the speed of light is constant, and finally luminiferous aether fell by the wayside. The chemists of Darwin's time were still arguing about the atomic composition of matter, but then the physicists came to their aid. First they discovered radioactivity, then the atomic nucleus. Finally Einstein's explanation of Brownian motion—the random motion of particles suspended in a liquid or gas caused by their collisions with speeding atoms or molecules—proved atomic theory to be fundamentally correct. As Darwin said in the context of natural selection as a mechanism for evolution, atomic theory gave the chemists "a theory by which to work." Lord Kelvin is justly celebrated for contributions to thermodynamics and an explanation of absolute zero of temperature. But his ideas about the age and mechanical properties of the earth, in opposition to those of Darwin, were just plain wrong. In this light we should not judge Darwin harshly, but should understand those of his geologic ideas that have not survived as stepping-stones along the path of science.

His interpretations that the gravels high on the plains of Patagonia and in the rivers draining the Andes to the west were deposited on the shores of the sea, not to mention his "greatest blunder" at Glen Roy, were off track. Now it is clear that all of these were phenomena related to the ice ages. His long battle defending the notion that icebergs were responsible for the large-scale transportation of erratic boulders—and were the predominant cause of bedrock striations—seems quaint. These too were directly related to glacial activity. But the acceptance of innovative ideas in science frequently requires that they be defended at considerable length. This struggle requires a determined persistence, if sometimes not downright stubbornness.

Darwin's insights in the 1830s and 1840s were thrown off by the want of knowledge of the ice ages and their immense impact on the landscape and surface of the earth, not to mention altering sea level. The understanding that ice on the land—not floating as icebergs—was the agent responsible for erratics, striations, and the gravel deposits now known as glacial drift grew through the 1860s. By the late 1870s, near the end of Darwin's life, the idea of a former ice age was generally accepted by British geologists. A single idea will rarely explain all the facts. The difference between what may appear in hindsight to be genius, as opposed to wrong-headedness, may be as simple as the choice of which facts to cling to, versus those to ignore. Darwin's reluctance to accept the full impact of the impact of glacial theory—even after his field trip to Cwm Idwal—held him back, although the confounding evidence of the shell fragments high on Moel Tryfan may be blamed in part. He clung to the wrong fact. Indeed, Darwin's British geologic contemporaries took decades of further study and debate to finally accept that much of Great Britain had in fact been covered in ice. The shelly deposits on Moel Tryfan were one of the last "facts" to be explained by glacial theory.

Some of the shell deposits that Darwin relied on in Chile and Patagonia as evidence of recent uplift were not the result of uplift at all, but transported to the places where Darwin found them by indigenous people. In some cases, people perhaps transported the shells just to dine on shellfish in a location with a better view or to be closer to camp. In other cases, perhaps they transported the shells as trading material, or as special possessions. This is not always an easy call, and criteria for analyzing the origin of shell deposits were not as stringent as those commonly applied today. One need only look at the bitter controversy about Paleolithic humans in North America throughout the twentieth century to see how difficult the interpretation of similar deposits can sometimes be. His awareness of the potential for and the frequency of large horizontal movements within the earth's crust—along both what geologists now call thrust and strike-slip faults, was quite limited, leading to his fixation on the vertical movements. That understanding would only come later, as would the understanding of elastic rebound as the cause of earthquakes.

Where Darwin sought a single explanation for the apparent vertical motions of the earth's crust as measured by sea level, now geologists recognize at least five different processes at work. Over the long term, the motions of the plates and their interactions and deformation dominate. But the rise and fall of global sea level owing to climatic change and the amount of water stored as glacial ice on land is also key. The slower viscous response of the earth's mantle to the loading and unloading of glacial ice explains many raised beaches and the vertical movements of formerly glaciated regions. In subduction zones, as we have seen, long-term processes are modulated by earthquakes and elastic rebound. And finally, and most in line Darwin's original idea, the movements of molten magma within the earth's crust are indeed responsible for some vertical movements in volcanic regions. Movements of magma, for example, are now accepted as the explanation for the long-standing puzzle of the Temple of Serapis.

Fundamentally, to focus on the aspects of Darwin's geology at odds with modern thinking is to miss the point. Just as the physicists and chemists of today sit on pedestals of knowledge surrounded by the debris of discarded concepts, so do geologists. It would be foolish to fault Darwin for not discovering plate tectonics wholesale, which would require a whole additional century-plus of data gathering, insights, and argument by subsequent generations. Darwin simply did not have enough of the pieces to solve the puzzle.

At the same time Darwin's ingenious theory of the formation of coral atolls endures and fits elegantly into the framework of plate tectonics. This theory has survived largely intact, notwithstanding the sea level changes owing to the ice ages, and opposing concepts advanced and promoted by several others, including, prominently, Louis Agassiz's son, Alexander. The subsidence of the volcanic islands on which atolls grew—central to Darwin's theory—is now understood to be caused by the cooling, densification, and gradual subsidence of the oceanic plates as they are carried away by convection in the earth's mantle from their places of creation at the mid-ocean ridges. This phenomenon is apparent in even such a simple thing as a map of the depth of water throughout the Pacific Ocean, something that was not available until the 1950s and 1960s. Now as modern scientists strive to improve on Darwin's theory, they refer to his formulation as "Darwin's canonical model."

And while the notions of elastic rebound and plate tectonics seemed not to support Darwin's idea that earthquakes constituted steps in the uplift of South America, Daniel Melnick, my colleague from Islas Guafo and Santa María, has combined modern GPS measurements along the coast of Chile with computer modeling of the coastal terraces to conclude that, on average over the last three million years, contributions from deeper earthquakes lead to a rate of uplift of the coast of about half an inch per century. While this rate is slower than Darwin might have liked, Daniel's analysis supports Darwin's idea about earthquakes contributing a permanent component to the uplift.

Darwin's brilliance in making observations, in integrating these into hypotheses, and in testing the hypotheses, is on clear display in the development of his geologic theories. Following Darwin's effort to understand the processes of mountain building and continental formation is to witness a brilliant mind at work on a very difficult problem, one that wouldn't be solved for more than another century.

And, as we have seen, Darwin's insights into evolution by means of natural selection rested on this geological foundation as he combined his understanding of the geologic past and present with his observations of the distribution and character of living things.

If we have to work to stifle a knowing smugness as we think about Darwin's quainter geologic theories, it would be well to adopt some humility. How will some of our contemporary ideas be judged in the future? Which of the concepts on the frontier of science currently in favor—not to mention our cures and mitigation measures—will be overturned by a new discovery or insight in the next decade, not mention in the next century and a half? As we have seen in the development and acceptance of plate tectonics, science advances through observations and by posing and testing hypotheses. Debate is critical.

Beyond this, Darwin's geology provides the impetus for reflection on the relationship between the species that vexed Darwin the most, Homo sapiens, and the planet that it inhabits. Despite having evolved to be extremely successful occupants of our own ecological niche, Homo sapiens's relationship with the planet is far from always benign. And it brings us face-to-face with the troubling question of the ultimate fate of our species. Does the concept of extinction apply to us too?

◆

There are those who are not so willing to forgive the missteps in Darwin's geological explanations. While most modern geologists give Darwin his due, and recognize his ideas that did not work out as part of the inevitable trial and error of science, Steve Austin is not one of those. In preparing for my own trip to the Río Santa Cruz in Argentine Patagonia I spoke at length by telephone with Steve. Steve is a geologist with a PhD in sedimentology, but his point of view differs from that of most contemporary American geologists. The Institute for Creation Research—a creationist organization that purportedly "exists to conduct scientific research within the realms of origins and earth history" and to equip "believers with evidence of the Bible's accuracy and authority through scientific research"—publicizes his work.

In a poster Steve presented at a meeting of the Geological Society of America about his investigations along the Río Santa Cruz in Patagonia, he argued that what Darwin thought were beach deposits had in fact been carried down the valley by giant floods. Megafloods. This was, in fact, FitzRoy's preferred explanation after his turn away from science and toward fundamentalism. What Steve didn't include in his poster was his view that Darwin's interpretations on the Río Santa Cruz were based on "bogus methodology" that the valley had been formed during "the lapse of great geologic ages." Steve calls Darwin's views on the geology along the Río Santa Cruz, "Darwin's First Wrong Turn." He implies that Darwin's errors in Patagonia discredit his subsequent theories about evolution.

The conversation that I had with Steve was extremely polite and helpful. We discussed logistic options for visiting this remote area, and his evidence for flood deposits. Based on what I later saw, I agree with his interpretation about giant floods coming down the river. We did not discuss the timing or origin of those floods, or natural selection. My own view would be that the megafloods on the Río Santa Cruz likely occurred as the result of the failure of one or more moraine dams during the Pleistocene, that is more than 12,000 years ago. In his book, *Grand Canyon: Monument to Catastrophe*, Steve argues that the Grand Canyon was created by a literal interpretation of Noah's flood from the Bible's Book of Genesis. It would appear that he has similar views about the deposits in the valley of the Río Santa Cruz. Steve mentioned to me that Darwin would have been better served to have followed his first mentor, Reverend Adam Sedgwick, rather than the uniformitarian ideas of Charles Lyell. But as we know from Sedgwick's address to the Geological Society in 1831, even he had given up on Noah's flood as the cause for the deposits of gravel across Great Britain.

Nonetheless, despite the overwhelming preponderance of scientific evidence for a very old earth and the eons of geologic history—including the ice ages of the Pleistocene—as deciphered by multitudes of geologists, Steve Austin is not alone in his views.

◆

The scene resembled a war zone. Flattened trees, stripped of their leaves and branches, surrounded me. Big trees. Tree trunks tossed like so many matchsticks. Pieces of rock the size of grapefruit embedded in some. Spread everywhere was sandy gray volcanic ash. But the trunks of the smashed trees pointed away from the cause of their destruction. Turning to look, I could see the still-steaming hulk of Chaitén volcano.

I was visiting this devastated forest with Julia Jones, a geographer and ecologist from Oregon State University. She was hoping to learn how this temperate rain forest would recover from the destruction wrought by Chaitén's eruption.

Chaitén is not a large volcano. It escaped Darwin's notice when he sailed nearby with Lieutenant Sulivan's survey party, hidden in the lush green of the rainforest and tucked in the shadow of its much larger, snow-clad neighbor, Minchinmahuida. In this dense jungle, pink and red blossoms and sparkling raindrops accent the shiny green leaves of wild fuchsia. Enormous sword ferns block every path. Tightly packed stands of bamboo block the view of the sky. And the evil-looking nalca, with leaves three feet across and spine-covered stalks threaten any human who might dare to enter. Above this undergrowth rise magnificent vine-shrouded trees with magical sounding names: *ulmo, coihue, arrayán, tepa, tenío.*

Where we stood all but the odd snag had been toppled or obliterated by Chaitén's blast.

Prior to May 2, 2008, no one paid much notice to the unassuming crater of Chaitén volcano. The gaze of geologists since Darwin was also mostly attracted to the higher, larger, glacier-covered, more active peaks nearby. One paper by a Chilean geologist and an American colleague presented evidence of an eruption of Chaitén volcano about 10,500 years ago. But on May 2 this seemingly ninety-eight-pound weakling of volcanoes roared to life, almost without warning, as if to exact revenge on the sleepy little port at its foot, the town of Chaitén. Dark, boiling columns of ash jetted into the stratosphere, thirteen miles up, twice as high as jetliners fly. The most intense eruption lasted six hours.

The government evacuated the town. It was a good idea. Because things got worse.

Ash plumes continued for a week. Columns again rose high into the stratosphere on May 6 and 8. The southern Chilean rains poured, clouds obscuring views from even the snoopiest satellites. Then on May 12 all hell broke loose.

Of course from a geologic viewpoint it wasn't as bad as it could have been. Explosive blasts and pyroclastic flows—glowing avalanches of gas and molten rock, the most evil of volcanic phenomena—didn't reach the site of the town, even though they ripped through the temperate rain forest, and covered the foothills of the volcano.

But the little port of Chaitén suffered from the next worse volcanic plague, lahars—deluges of sandy mud. Torrential rains and newly fallen ash combined in a diabolic slurry that swept down the normally peaceful Río Chaitén, filling the river channel and overtopping its banks. Before the eruption, the Río Chaitén looped behind the town before it emptied into the Golfo de Corcovado. With the lahars, the river cut a new channel right through the middle of town, but not before dumping a three-foot layer of ashy ooze throughout the town, filling homes and buildings and burying streets, gardens, playgrounds.

After the eruption the government planned to relocate the town. The idea was not popular with those people from the town who wanted to return. On my way to Chaitén from Puerto Montt, while waiting for the overnight ferry, I encountered people displaced by the eruption. It was not difficult to distinguish these people with their packs and bags, suitcases and boxes tied with string. Many seem to know each other, greeting one another with hugs and quick kisses to the cheek. I overheard one man describing his situation. I could not understand much, but I did hear him use the word *volcán* many times. Later, after the night on the ferry, as I walked down the ramp at Chaitén I saw this same man, his parcels set beside him.

The Carabineros were questioning him. Was he trying to return without authorization?

Months later the government gave up. Those who wanted to return to Chaitén were allowed to. Just as the plants that Julia and her team documented reemerging, reclaiming the land devastated by the volcano, so too did the people.

◆

In 2004, Yuki Sawai—the careful, young Japanese geologist who I met on my first geologic trip to Chile with Brian Atwater—began studying sheets of sand on the Sendai Plain on the northeastern coast of Japan. As his studies with colleagues at the Geological Survey of Japan progressed over the next few years, it became increasingly clear that one of these sands had been deposited by an extremely large tsunami associated with an earthquake well-known in Japanese historical records as the Jogan earthquake of A.D. 869. Historical documents and the distribution of the sand showed that this tsunami ran more than a mile farther inland than any subsequent tsunami. Despite this history, the Jogan earthquake was not considered in the national seismic hazard map of Japan released in 2010, because—although the map included geologic data going back thousands of years for onshore faults—the map only considered subduction earthquakes on the coast of Japan for the previous 400 years.

In 2008, Yuki's colleagues used computer models to estimate the location and extent of the earthquake from the sand deposits, as well as the height of the tsunami along the coast. Yuki's boss, Yukinobu Okamura was alarmed by the implications of these results for coastal communities, which had been preparing for much smaller tsunamis—as was the Fukushima Daiichi Nuclear Power Plant, owned by the Tokyo Electric Power Company (Tepco), on the southern stretch of the affected coast. Okamura raised concern through a variety of official channels.

Independently, Kunihiko Shimazaki, professor at the University of Tokyo and one of Japan's leading seismologists, had been arguing since at least 2002 that the long-term evaluation of possible earthquakes—prepared by a government committee of which he was a leading member—should include the possibility of a large tsunami on this stretch of coast. His argument was based on the occurrence of other tsunami-generating earthquakes in other parts of the Japan Trench, in 1611, 1677, and 1896. Shimazaki's warning was not well received. At one meeting of a cabinet office committee in 2004, Shimazaki warned that the coast near Fukushima was vulnerable to tsunamis more than twice as high as the forecasts considered by Tepco and the regulators. But the minutes of the meeting showed that government bureaucrats moved to exclude Shimazaki's views from the report.

"They completely ignored me in order to save Tepco money," the *New York Times* quoted Shimazaki as saying, several years later.

Sadly, the warnings of Okamura, Shimazaki, and others about the possibility of a much larger tsunami along this coast were all ignored or pushed aside with the excuse that further study was needed.

When, in 2011, the giant Tohoku earthquake and tsunami struck this coast, the catastrophic consequences—including the release of radiation and near-meltdown of one the reactors at the Fukushima plant, as well as the horrific loss of life from flooding—were largely blamed on the unforeseen ferocity of the tsunami. The potential of a tsunami of the immensity of that in 2011 was not unforeseen. It was at best underappreciated, and at worst ignored.

◆

Although the eruption of Chaitén illustrates the vulnerability of mankind in the face of the earth's paroxysms, this eruption was a pipsqueak among eruptions, ten times smaller than Mount St. Helens in 1980, a hundred times smaller than Mount Pinatubo in 1991, and Krakatoa in 1883, a thousand times smaller than the eruption that created Crater Lake in Oregon about 5,700 B.C.E., and more than ten thousand times smaller than the eruption that created Yellowstone about 640,000 years ago.

The Tohoku earthquake in 2011 was the fourth largest earthquake since 1900[*]. The estimation of earthquake magnitudes from geologic data alone is not currently possible, and magnitudes estimated for earthquakes—from historical, rather than instrumental—data, even when possible, are highly uncertain. This leaves us with a sample that covers only an extremely short period of time.

Despite the horror of all these eruptions, earthquakes, and tsunamis, none alone seems likely to have presented an existential risk for the human species. We are a widely distributed and resilient lot. Even with the tragedy and pain, many of the consequences of these natural catastrophes can be mitigated and in some cases avoided. Our Stone Age ancestors survived the saber-toothed tiger and short-faced bear. Now the grizzly bear and other carnivores pose threats mostly to individuals, not to our species, notwithstanding our threat to them, which is far more existential. But just as predation by large carnivores no longer presents an evolutionary threat to Homo sapiens, neither do these geologic hazards. The potential for tragic loss abounds, but run-of-the-mill geologic hazards like earthquakes and volcanic eruptions will not lead to our extinction.

Although Darwin was fascinated by extinction, he believed that extinction generally occurred as a result of natural selection. As a strong follower of Lyell, he eschewed catastrophes in favor of gradual change to explain changes in the geologic past. Extinction too was a gradual, natural process, not a violent, catastrophic one. In contrast, the last century and a half of paleontology has shown that five episodes of mass extinction have jolted life on the planet, causing significant shifts in the character of its diversity in a relatively short time span. The current period of high extinction rates since the late Pleistocene does not yet rise to this level. While a consensus seems to be emerging about the cause of one of these five, the K-T boundary event—most likely caused by the impact of

---

[*] The seven largest earthquakes since 1900 with moment magnitudes are Chile, 1960, 9.5; Alaska, 1964, 9.2; Sumatra, 2004, 9.1; Japan, 2011, 9.0; Kamchatka, 1952, 9.0; Chile, 2010, 8.8; and Ecuador, 1906, 8.8.

an asteroid in the Yucatán Peninsula of Mexico—the causes of the other four are the subject of continuing debate, including the cause of the largest, the Permian extinction—in which an estimated 96% of the then-existing species on earth went extinct. Asteroid impacts, extreme volcanism, climate change, sea-level change, or a combination of these or other possibilities are all being considered. Thus while some extinction clearly occurs as a result of natural selection, as the American paleontologist, the late David Raup opined, "It may well be that most species have evolved ways of surviving anything that their environment can throw at them, as long as the stress occurs frequently enough for natural selection to operate."

So what could lead us the way of the ground sloths? For as macabre as it sounds, just as death is inevitable for the individual, so the geologic history of the planet suggests that for species, extinction is the rule. According to Raup, "Almost all species in the past failed. If they died out gradually and quietly and if they deserved to die because of some inferiority, then our good feelings about earth can remain intact. But if they died violently and without having done anything wrong, then our planet may not be such a safe place." The consequences of even those natural hazards that do not threaten our survival show the harsh, dangerous side of the planet.

Perhaps the most obvious natural candidate for our extinction would be the impact or atmospheric explosion of an asteroid or meteorite like the one that most likely led to the extinction of the dinosaurs. The earth is bombarded by small meteorites all the time, so key questions include how large it would have to be, could we tell if it was coming, and if so, what might we be able to do about it? Fortunately NASA and other national space agencies have been spending some of our tax dollars to answer these questions.

What about climate change? Humanity has clearly survived climate change in the past, including through the Pleistocene, and even the abrupt cold turn of the Younger Dryas just before the beginning of our current warm period. But what about the future? Could anthropogenic climate change lead to, or accelerate, our extinction? In a geologic context the earth's climate depends on the current configuration of the oceans and continents. The history of the climate through the ice ages of the Pleistocene shows that this system has two stable states, warm and cold. Suppose that, as the result of human actions, the climate continues to warm leading to catastrophic melting of the ice in Greenland and Antarctica. Then suppose that the flows of cold water into the oceans from the melting of ice in Greenland and the Antarctic were enough to trigger a change of state—from the warm climatic state to the cold one—similar to the likely triggering of the Younger Dryas cold period by the emptying of glacial Lake Agassiz.

Could our current global population survive? Are we as resilient as our Stone Age ancestors? Would the immensely more complicated social, political, and technological systems enjoyed by humanity today be as robust as the roaming bands of our hunter-gatherer ancestors? The global scramble for energy resources would surge just to keep people warm. How would our global political system adapt? One can imagine an abrupt climate change triggering a cascade of calamitous events, both geologic and man-made.

Similarly one can imagine that damage to the oceanic food chain by continuation of the current acidification of the oceans could trigger a similar cascade. Other existential threats might come not from fellow mammals, but from other parts of the tree of life. Could a super pandemic lead to our extinction? Then there is the ultimate "human error," a global nuclear war in which we obliterate populations in parts of planet, but contaminate much of the rest of it with radioactivity inhospitable to our continued existence.

The failure of the flood control system in New Orleans in Hurricane Katrina, and the overtopping of the tsunami wall and crisis at the Fukushima nuclear plant during and following the Japanese earthquake of 2011 do not provide ringing endorsements of the robustness of our systems. And we have yet to demonstrate that we can take sufficient collective action with regard to the anthropogenic component of climate change and the acidification of the oceans.

Philosopher Nick Bostrom of the University of Oxford argues that improving humanity's ability to deal with existential risk will require "collective wisdom, technology foresight, and the ability when necessary to mobilize a strong global coordinated response." A tall order indeed. I would argue—and I would like to imagine that Darwin would agree—that these efforts must include a realistic and cold-eyed assessment of humanity's place in the geologic history of a changing planet and the recognition of both our ability for harm, and of the limitations in our ability to influence the change in ways that we desire.

◆

United States Supreme Court Justice Anthony Kennedy wrote in 1992—reaffirming the central holding of the famous Roe v. Wade decision—"At the heart of liberty is the right to define one's own concept of existence, of meaning, of the universe, and of the mystery of human life." Insofar as Planet Earth, its origin, its history, and the history of the creatures that live upon it fall within this realm, Justice Kennedy's opinion seems to give license to those who would challenge the findings of geologists and other natural scientists—and even medicine.

Certainly science and scientists have much more confidence in some results and theories than others. But it seems bizarre to view science as an elaborate tapas menu from which one can choose the results and theories that are consistent with a person's taste. To rationalize the radiometric estimates of the 4.6-billion-year-old age of the earth, young earth creationists—arguing from a supposedly scientific point of view—appeal to "a burst of nuclear decay" during the "week of creation," and a second at the time of Noah's flood to make up for the billions of years implied by radioactive decay at the measured present rates. Believing this may be contained within a right "at the heart of liberty," but it can hardly be called science.

Newton's laws seem not to raise much quibble, but more complex or less familiar theories certainly do. A theory is little use if it has no predictive power. And the predictive power of scientific theory is the foundation of technology. How can it be consistent for those who doubt the application of physics and chemistry to understanding the history of the earth, to then enjoy the technological fruits of these same sciences every day of their lives?

Every tap on a smartphone affirms quantum mechanics, because quantum mechanics are the basis for the millions of transistors in these devices. Users of GPS affirm both Einstein's theories of special and general relativity as positions from GPS would be worthless without the corrections made for relativistic effects. People who receive annual flu shots tacitly accept the basic idea of natural selection, the process by which the mutation of flu viruses leads to new strains capable of overcoming last year's antibodies. Even much of the oil—the raw material of the gas we pump into our cars—is found based on an understanding of geologic history.

The same physics of radioactivity used in treating cancer forms the basis for estimating the age of the earth. Justice Kennedy supports a person's right to believe that the earth is only six thousand years old. But how can such people visit a radiologist with a clear conscience when they or a loved one becomes ill?

Gallup reported in 2014 that 42% of Americans preferred the statement that "God created human beings pretty much in their present form at one time within the last 10,000 years or so," over alternatives favoring evolution, either guided (31%) or unguided (19%) by God. The proportion holding this view has held nearly constant since 1982 when Gallup began asking the question. If this attitude is any reflection on the success of geologists and scientists of all stripes to explain their understanding of the history of the planet and its inhabitants, it is little wonder that the current debate about climate change is so shallow.

◆

In my quest to follow Darwin's trail, I rekindled a passion for deciphering the history of the earth and the processes that shape it from unvarnished observations in the field, some made by the human eye alone, others assisted by the wonders of modern technology. I learned to care even more deeply about the impacts of the earth's seeming hostile behavior on my fellow humans. And I learned about the humanity, brilliance, and stubbornness of my geologic forebears. No one was right about everything, nor can one advance the frontier without entering the fray. While it may be fair to fault Darwin for his reluctance to accept ideas about the ice ages, it is also fair to fault Agassiz for overselling them—not to mention his later refusal to accept Darwin's ideas about natural selection.

Notwithstanding the difficulties that some have found in reconciling the results of geology with aspects of their faith, I conclude that geologists, broadly speaking, have a special role in society. And it is not necessarily an antireligious role, although it is sometimes perceived that way. One only needs to recall that one of Darwin's first mentors was a minister and he himself had once been on the path to the clergy himself.

The geologist's role begins with the responsibility to lay out the "facts," as Darwin would have said, about the history of earth and of the creatures upon it, to explain the processes that have shaped the planet—and its inhabitants—in the past, continue to shape them today, and will continue to shape them in the future. But beyond that, geologists need to point out to society how the history of the earth—and the understanding of the processes that shape it—can inform society's choices for a sustainable future.

It's not an easy row to hoe, trying to convince this assembly of human beings—individuals with average life spans of only several decades, bills due at the end of the month, and a need to put something on the table for dinner tonight—that events of a few hundred, a few thousand, or even millions of years ago have some relevance to the choices that each of us and society make each day. But they do. Perhaps we do sound a bit like Jeremiah, with our lamentations and warnings about earthquakes, volcanoes, floods, the conservation of topsoil, the fragility of wetlands, the impermanence of barrier islands, the limitations of resources, and the changing—and fragile—climate.

One of the most difficult tasks is not to explain what scientists generally believe to be true, but to find useful and helpful ways to explain what we don't yet fully understand. And why even partial knowledge can be extremely important. No, geologists don't yet understand every cause of the climatic variations through the Pleistocene, but clearly the system is a complex and likely delicate one. Similarly, while the theories of plate tectonics and elastic rebound go a very long way toward understanding large earthquakes along the boundaries between plates, large earthquakes in the interiors of plates also present a significant hazard, even if their causes remain the subject of debate.

But this is the geologist's job. We need to give it our best shot. Even if it sometimes makes us feel like Dr. Stockmann in Henrik Ibsen's *An Enemy of the People*, a man who was ostracized for being the bearer of bad news.

◆

While we ponder the fate of our species and the role of science, we should not forget particular individuals who have played important parts in this story. FitzRoy's obsession for service continued his entire life. He served as a Member of Parliament, then as governor of New Zealand. Although his interest in science persisted, his attention to geology, geologists, and their ideas largely withered as they appeared to contradict his faith. Yet in 1854 he combined his instinct for service and passion for weather in founding the UK Meteorological Office. He pioneered the rapid compilation of the reports of weather conditions from an array of outlying stations to formulate forecasts, taking advantage of the new technology of the telegraph. Initially FitzRoy provided only warnings of gale conditions for shipping.

FitzRoy was embarrassed by Darwin's publication of the *The Origin of Species* in 1859, believing that Darwin's conclusions contradicted Biblical "truth." FitzRoy attended the famous debate at Oxford between Bishop Wilberforce and Thomas Henry Huxley. There he spoke against Darwin. In 1861 he began to provide weather forecasts to newspapers. This proved problematic. Criticism about the inaccuracy of the forecasts was devastating to his increasingly fragile psyche.

Several times during his career in service, dating back to the refitting of the *Beagle* and the purchase of the *Adventure*, he had dipped into his own pocket—without subsequent repayment—to supplement funds available from the government. By the early 1860s, he was significantly in debt.

I pondered this troubled history as I sat in the last pew of Fitz-Roy's church. Because religion was so important to FitzRoy, it seemed only proper and respectful to attend a service at the church where he lies buried. The neighborhood of All Saints Parish in Upper Norwood, southeast London, is not now what it was when FitzRoy lived nearby. In contrast to the blue bloods of FitzRoy's time, the parish is now quite diverse, and includes many immigrants to the UK from Africa and the Caribbean.

In preparation for the Sunday service, the choir director, a tall slender young man with black, flowing wavy hair, goatee, and serious black-framed glasses, rehearsed the congregation in a choral response. He found the congregation's first attempt wanting, but our second attempt he found satisfactory. "Brilliant, thank you very much," he said. Likely FitzRoy would have approved of this earnest young man, Thomas Hewitt Jones, a graduate of Cambridge and a serious composer.

As the procession of priests and a visiting bishop filed down the aisle toward the altar, the church filled with the rich smell of incense. The smoke rose through the sunbeams entering the south windows, and upward to the peak of the sanctuary. I wondered whether incense was a part of the services of the Reverend J. Watson, MA Cantab [Master of Arts, Cambridge], who according to a plaque on the wall was the vicar during FitzRoy's last years. Was his ministry a help to FitzRoy during the last decade of his life as struggled with his troubles in the Meteorological Office, anguished over his role on the *Beagle* in facilitating Darwin's studies, and grappled with his personal financial problems?

Outside after the service, as the parishioners chatted in the fall sunshine, I studied FitzRoy's neatly fenced grave. The morning sun gave a feeling of warmth to the headstone with its carved anchor entwined on a cross. "Sacred to the memory of Robert FitzRoy, Vice Admiral. Born July 5th, 1805, Died at Norwood, April 30th, 1865." The footstone inscription read, "Vice Admiral Robert FitzRoy, First Head of the Meteorological Office, 1854." Nothing about the *Beagle*, nothing about his chain of longitudes around the planet, nothing about New Zealand, nothing about his erstwhile colleague whose theories had so mortified him.

Sadly, FitzRoy's faith and stoicism were not enough to carry him through his anxieties and depression. On that Sunday morning in 1865, at the age of fifty-nine, FitzRoy used his razor to cut his own throat. Darwin's shipmate, now Captain, Bartholomew Sulivan, was one of only a handful of mourners outside the family at his interment in this churchyard. He sent a report on to Darwin. Darwin's wife, Emma, said that Charles "was very sorry about FitzRoy—but not much surprised." Hearing about the sorry financial plight of FitzRoy's widow, Darwin sent a generous check, perhaps to salve his own mixed feelings about his former captain.

After FitzRoy's death, the Meteorological Office ceased issuing weather forecasts, only to resume them more than a decade later.

◆

The most remarkable thing about Charles Darwin's tombstone in Westminster Abbey is its simplicity. A plain slab of light gray marble, inscribed:

Charles Robert Darwin
Born 12 February 1809
Died 19 April 1882

No interpretation. Just the facts.

As I gazed at these simple words on Darwin's tomb, Pat, a female docent, dressed in an official-looking green smock, her white hair neatly trimmed in a pageboy haircut, stood nearby gently urging the lines of tourists to keep moving. "I'm just here to tell people who's here and where to go," she explained when I asked her about Darwin. "He didn't really want to be buried here, you know," she said. "He's here . . . for what he did, not for what he believed." In a whisper Pat recalled the story that Darwin had professed his faith on this deathbed. But Pat wasn't having any of it, concluding her story, "His daughter said 'No, he didn't.'"

In fact Darwin had hoped to be buried in the churchyard at St. Mary's Church, Downe, with his wife and two of the children that they had lost. But there really can't be too much confusion about Darwin's religious views. In late November, 1880, only fifteen months before his death, Darwin received a rather strange letter from one Frederick McDermott. McDermott politely explained that he would like to read Darwin's books, but did not wish to do so if reading them would challenge his faith. "Do you believe in the New Testament?" he wrote, appealing to Darwin, "So [if] you will write on the back of this page Yes or No you will be doing a real kindness which I will certainly not abuse by sending a paragraph to the theological papers."

In a monument to Victorian politeness, and no less to directness, Darwin responded immediately in a one-sentence note bearing the notation "Private": "Dear Sir, I am sorry to inform you that I do not believe in the Bible as a divine revelation, & therefore not in Jesus Christ as the son of God. Yours faithfully, Ch. Darwin."

It had been quite a long journey from the days almost fifty years before when Darwin's father had him on track to become a country parson, tending the souls of his flock as a vocation, with a plan to pursue his interests in natural science as a hobby.

Darwin wrote in his notebook as far back as 1838, "Man in his arrogance thinks himself a great work, worthy the interposition of a deity. More humble and I believe true to consider him created from animals." *The arrogance of man.* Today we use a fancy word, anthropocentrism. Self-centeredness at the species level. Notwithstanding the body of Darwin's work, which struck a blow at the self-importance of Homo sapiens, this characteristic continues undiminished, and may continue as long as the species exists.

The only tomb that rivals Darwin's for simplicity is Newton's, only a few yards away, which simply reads *"Hic depositum est, quod mortale fuit Isaaci Newtoni."* [Here lies that which was mortal of Isaac Newton]. Lyell's tombstone, ten yards down the walk, stands in stark contrast. Almost as if prescient that Lyell's contributions to the science of geology—and indirectly to evolution through Darwin—might be forgotten, the inscription with brass letters on a beautiful fossiliferous piece of marble rivals an encyclopedia entry in length.

On the Sunday following Darwin's funeral the Bishop of Carlisle spoke at Westminster Abbey. Reflecting on Darwin's passing, he said, "It would have been unfortunate if anything had occurred to give weight and currency to the foolish notion . . . that there is a necessary conflict between a knowledge of Nature and a belief in God." Nearly a century and a half later, some are still having difficulty working that out.

◆

In early 2015, nearly five years after the earthquake and tsunami in Chile, Marco, our team, and I once again visited Lipimávida. Marco and I found Don José Luis with his shovel in one of the potato fields across the road from his house. His life was better than when we met him just after the tsunami. His face was animated, with the hint of a smile when he recognized us. Sure it would be okay if we dug some more holes in the unused potato field, as long as we filled them well afterward so a cow wouldn't step in and break a leg.

He had rebuilt his house in its proper place, no thanks to the government, which had given him no help. Ordinary people came from the cities, Santiago and others, he said—his voice turning bitter—came to help people rebuild, but he got nothing from the government. Then the local mayor had even had the nerve come to ask for his vote before the next election. No, he'd told him, his eyes flashing as he told us the story.

As we talked he used his shovel to open and close ditches that led water into the rows of potato plants. His neighbor who had left to check on his house in the city the night of the earthquake had abandoned the lot where his summer cabin once stood. But now other people were coming to build more houses. Most people had forgotten about the tsunami, he said.

Later Marco and I walked by his house. It was neatly repaired and painted. He had raised the house up about two feet. Pink and blue hydrangeas and bright orange flowers bloomed gaily around the wooden posts of the foundation. The road in front of his house was repaved.

We had hoped to learn how the sand dumped by the tsunami into the potato field had been assimilated into the natural environment over the intervening five years, to learn how the record of the 2010 tsunami might be preserved in the profile of the soil in the field together with the record of earlier tsunamis.

But Marco had asked Don José Luis about the sand—sand that in places on his field and near the road had been more than a foot thick after the tsunami. Don José Luis's matter-of-fact answer struck a serious blow to our plan. The sand had been scraped and hauled away.

NOTES

Basic references for Darwin's life are the two-volume biography by Janet Browne, *Charles Darwin, Voyaging* and *Charles Darwin, The Power of Place*, as well as David Quammen, *The Reluctant Mr. Darwin*. References for FitzRoy's life include John and Mary Gribbin, *FitzRoy: The Remarkable Story of Darwin's Captain and the Invention of the Weather Forecast*; Peter Nichols, *Evolution's Captain: The Story of the Kidnapping That Led to Charles Darwin's Voyage Aboard the* Beagle; and H.E.L. Mellersh, *FitzRoy of the* Beagle. The basic reference for Darwin's geology is Sandra Herbert, *Charles Darwin, Geologist*.

For the voyages of the *Beagle*, I relied most heavily on the original 1839 editions, King and FitzRoy, *Narrative*, Vol. 1, for the first expedition, 1826–1830, then FitzRoy, *Narrative*, Vol. 2 and Appendix, and Darwin, *Narrative*, Vol. 3, for the second expedition, 1831–1836. Derivatives of Darwin's, *Narrative*, Vol. 3, are now available in a variety of modern editions as the *Voyage of the* Beagle. I have also relied on *Charles Darwin's* Beagle Diary, edited by R. D. Keynes. On a few interesting points, Darwin's thinking changed or developed in the few years subsequent to 1839, and I have quoted from Darwin, *Journal of Researches*, the 1845 edition.

For Darwin's correspondence I have relied most heavily on the wonderful Darwin Correspondence Project at Cambridge University. All of the letters cited here include the code for the letter at the Project's website in the form, DCP-LETT-xxx, so that each letter can be accessed there at http://www.darwinproject.ac.uk/DCP-LETT-xxx. All of Darwin's letters cited here were accessed between October 17 and November 18, 2016. The correspondence has also been published in Darwin et al., *The Correspondence of Charles Darwin*, and the portion of the correspondence while on the *Beagle* is in Darwin, *The Beagle Letters*.

Darwin Online, http://darwin-online.org.uk/, is another marvelous resource, also at Cambridge University. While many of Darwin's notebooks have been published, his *Geological Diary* (http://darwin-online.org.uk/ EditorialIntroductions/Chancellor_GeologicalDiary.html) from the *Beagle* has not as of yet. This website provides access to virtually all the early editions of the published books; Darwin's notes, notebooks (with both original images and transcriptions), and personal "Journal"; and many secondary books and articles. This website is an excellent place to look for anything related to Darwin other than the letters. Citations from this website include the Cambridge University Library reference number. Darwin's shorter papers are collected in Darwin and Van Wyhe ed., *Darwin's Shorter Publications, 1829–1883*. Another great source for books and journals related to Darwin from this period is the Biodiversity Heritage Library, http://www.biodiversitylibrary.org.

As with Darwin's letters, all website addresses cited below were accessed between October 17 and November 18, 2016.

During the period of Darwin's contributions to the Geological Society, summaries of the papers presented at the meetings, typically prepared by the Secretary, were reported in the *Proceedings of the Geological Society of London*, but the written papers were later published, if at all, in the *Transactions of the Geological Society of London*, or elsewhere. Exceptions were the Presidential Addresses, which appeared in full in the *Proceedings*. In some cases I have quoted from the summaries in the *Proceedings* as if they were the words of the author. Points made in the papers presented in the meetings were not always included in the published papers, and vice versa. Discussion at the meetings was not formally recorded, but is reconstructed to some extent in Thackray, *To See the Fellows Fight: Eye Witness Accounts of Meetings of the Geological Society of London and Its Club, 1822–1868*.

Four highly readable books that trace the development of the society in which Darwin did science and several of the key personalities that preceded and surrounded him, are Holmes, *The Age of Wonder: The Romantic Generation and the Discovery of the Beauty and Terror of Science* (about Joseph Banks, William Herschel, and Humphrey Davy); Winchester, *The Map That Changed the World: William Smith and the Birth of Modern Geology*; Snyder, *The Philosophical Breakfast Club: Four Remarkable Friends Who Transformed Science and Changed the World* (about John Herschel, Whewell, Babbage, and Richard Jones); and finally, McCalman, *Darwin's Armada: Four Voyages and the Battle for the Theory of Evolution* (about Thomas Huxley, Hooker, Wallace, and Darwin himself).

The complete references may be found in the Bibliography.

"Abstract of Meteorological Journal." The entry for this day also records the time of the earthquake. FitzRoy, *Narrative*, Vol. 2 Appendix, p. 38.

xvi   The events of this day are described in the narratives of both FitzRoy, *Narrative*, Vol. 2, p. 402ff., (the sketches of the town follow p. 398) and Darwin, *Narrative*, Vol. 3, p. 368ff.

xvii   "It came on suddenly," Darwin, *Narrative*, Vol. 3, p. 368.

xvii   "At Valdivia the shock began gently," FitzRoy, *Narrative*, Vol. 2, p. 415.

PART ONE: DARWIN IN THE FIELD

CHAPTER ONE: THE LIEUTENANT AND THE BEETLE COLLECTOR

3   "If ever I left England again," King and FitzRoy, *Narrative*, Vol. 1, p. 385.

3   "Geology is a capital science to begin," Darwin to W. D. Fox [August 9–12]. 1835, DCP-LETT-282.

5   "Rethinking Seismicity Declustering," AGU Fall Meeting, 2007, Seismology Program, http://abstractsearch.agu.org/meetings/2007/FM/sections/S.

6   For *Time* magazine's list of the 100 for 2005, see http://content.time.com/time/specials/packages/article/0,28804,1972656_1972712_1974230,00.html.

7   "Imperial Century," see Hyam, *Britain's Imperial Century: 1815–1914: A Study of Empire and Expansion*.

7   Rudyard Kipling's "The White Man's Burden" may be found, for example, at http://sourcebooks.fordham.edu/halsall/mod/kipling.asp.

7   Hutton's ideas were summarized in his three-volume *Theory of the Earth*, published in 1795. They were then explained to some extent by his friend, Playfair, *Illustrations of the Huttonian Theory of the Earth*, in 1802. Lyell would take up the cause with his own spin beginning in the 1820s.

8   The development of geology and the central issues at play into and through the nineteenth century are also discussed in Rudwick, *Bursting the Limits of Time* and *Worlds Before Adam*; Greene, *Geology in the Nineteenth Century*; and Davies, *The Earth in Decay*.

9   For Sir Francis Drake's handling of pilots, see Bawlf, *Secret Voyage*, pp. 147–8, 157–9.

9   For the superiority of Dutch charts of English coasts, see Reidy, *Tides of History*, pp. 34–35, and Blewitt, *Surveys of the Seas*, p. 20.

9   The story of the development of the chronometer as a tool for measuring longitude is wonderfully told in Dava Sobel, *Longitude*.

10   Both navigational and surveying instruments of the period are discussed in Turner, *Scientific Instruments, 1500–1900*, including the sextant (pp. 32, 47), theodolite (p. 42), and station pointer (p. 83). For a contemporary discussion, see Mackenzie and Horsburgh, *A Treatise on Marine Surveying*. Although this book is not listed as being in the "library" on the *Beagle*, it was almost certainly on board as a reference for the surveyors. Mackenzie invented the station pointer. For more history of British hydrography, see, Blewitt, *Surveys of the Seas*, and Ritchie, *The Admiralty Chart*.

10   British commercial interests in Latin America in the 1800s are discussed in Bethell, "Britain and Latin America in historical perspective," and Hyam, *Britain's Imperial Century*.

10   The sad story of Pringle Stokes is told in King and FitzRoy, *Narrative*, Vol.1, pp. 150–182, and Stokes and Campbell, *Journal*, pp. 141–251.

11   "Around us," and "The weather," quoted in King and FitzRoy, *Narrative*, Vol. 1, p. 179.

11   "The Seasons," Thomson, *The Seasons*, https://books.google.com/books/about/The_Seasons_A_poem.html?id=GWcaeJnb5MkC.

12   This discussion of FitzRoy's appointment and family connections relies on Gribbin and Gribbin, *FitzRoy*, p. 47ff.

12   Both Robert FitzRoy and Princess Di were descended from Henry FitzRoy, the 1st Duke of Grafton, illegitimate son of Charles II, see, for example, http://www.englishmonarchs.co.uk/stuart_33.html.

12   "Restless he rolls from whore to whore," John Wilmot, Earl of Rochester, "A Satyr on Charles II," https://andromeda.rutgers.edu/~jlynch/Texts/charles2.html.

13   "We had another Examination," quoted in Gribbin and Gribbin, *FitzRoy*, p. 15.

13   See Browne, *Voyaging*, for more on Darwin's boyhood.

14   "An inordinate fondness for beetles," quoted in Evans and Bellamy, *An Inordinate Fondness for Beetles*, p. 9.

14   "Fanny, as all the world knows," Darwin to W. D. Fox [December 24, 1828], DCP-LETT-54.

14–15   "Why did you not come home this Xmas?" Fanny Owen to Darwin [January 27, 1830], DCP-LETT-77.

15   "lecture on some plant or other object," quoted in Browne, *Voyaging*, p. 122.

15   "the man who walks with Henslow," Darwin, *Autobiography*, p. 55.

15   "Alexander Humboldt's description of his travels," Humboldt, *Personal Narrative*.

16   "I am at present mad about Geology," Darwin to C. T. Whitley [July 12, 1831], DCP-LETT-102A.

16   "My trip with Sedgwick," Darwin to J. S. Henslow [August 30, 1831], DCP-LETT-107.

16   The tale of the stolen whaleboat and subsequent capture of the Fuegians is told in King and FitzRoy, *Narrative*, Vol. 1, p. 391ff.

19   "That it would be a useless undertaking," Darwin to R. W. Darwin [August 31, 1831], DCP-LETT-110.

19   "The undertaking would be useless as regards his profession," Josiah Wedgwood II to R. W. Darwin [August, 31 1831], DCP-LETT-109.

19   For the schedule of coaches to Cambridge, see *The Cambridge Guide*, pp. 266–71.

20   For weather in London, summer, 1831, see http://booty.org.uk/booty.weather/climate/1800_1849.htm.

20   "full of zeal and enterprize," F. Beaufort to R. FitzRoy [September 1, 1831], DCP-LETT-113.

21   For a modern treatment of Erasmus Darwin, see Fara, *Erasmus Darwin*.

21   "nauseated," Coleridge wrote, "I absolutely nauseate Darwin's poem," quoted in Fara, *Erasmus Darwin*, p. 44.

21   "How can anyone," Darwin described FitzRoy's concerns about the shape of Darwin's nose in *Autobiography*, p. 61.

21   "Mr. Chester," Darwin to Henslow [September 5, 1831], DCP-LETT-118.

21   "many books, all instrument(s), guns," "Shall you bear being told," "thought it his duty," and "I scarcely thought of going to Town," Darwin to S. Darwin [September 5, 1831], DCP-LETT-117.

22   "Gloria in excelsis," Darwin to J. S. Henslow [September 5, 1831], DCP-LETT-118.

23   "Whilst I, luckless wretch," F. Watkins to Darwin [September 18, 1831], DCP-LETT-130.

CHAPTER TWO: FIELD TRIP WITH A MASTER: THE STATE OF GEOLOGY

25   "The tour was of decided use," Darwin, *Autobiography*, p. 59.

25   This account of Darwin's geologic tour in North Wales with Adam Sedgwick is largely based on Barrett, "The Sedgwick-Darwin Geologic Tour," Roberts, "Just before the *Beagle*," and Browne, *Voyaging*. For more on Sedgwick, see Roberts, "Adam Sedgwick."

25   For the night before, see Browne, *Voyaging*, p. 140.

26   For a description of the Holyhead Road that Darwin and Sedgwick followed, see Harper, *The Holyhead Road*, Vol. 2, p. 142ff.

27   "put all the tables in my bedroom," Darwin to J. S. Henslow [July 11, 1831], DCP-LETT-102.

27   For Darwin's geological explorations near Shrewsbury, see Roberts, "Darwin at Llanymynech," Herbert and Roberts, "Charles Darwin's Notes on His 1831 Geological Map," and Roberts, "I Coloured a Map."

27   "so easy as I expected," Darwin to W. D. Fox [July 9, 1831], DCP-LETT-101.

30   Steno's story is told in Cutler, *The Seashell on the Mountaintop*.

30   For Arduino's names, see Rudwick, *Bursting the Limits of Time*, pp. 90–94.

30   For Cuvier and Brongniard's terminology, see, Rudwick, *Worlds Before Adam*, pp. 11–23.

31   Sedgwick's correspondence with Murchison about their summer plans is in Sedgwick, *Life and Letters*, Vol. 1, p. 376. The full story of the sorting out of the rocks beneath the Old Red Sand Stone and through the Transition Rocks is told in Rudwick, *The Great Devonian Controversy*, and Secord, *Controversy in Victorian Geology*.

32   "The bank facing" and "The contrast between this," quoted in Barrett, "The Sedgwick-Darwin Geologic Tour," pp. 155–6.

32   "nearly drowned" and "the greywacke hills," Sedgwick, *Life and Letters*, Vol. 1, p. 378.

33	A digital copy of Greenough's map is included in the DVD by Wigley et al., *Strata Smith*.
33	"From several observations," quoted in Barrett, "The Sedgwick-Darwin Geologic Tour," p. 157.
33	"may pass for Old Red," Sedgwick, *Life and Letters*, Vol. 1, p. 378.
33	"after a day or two" and "In the evening," Sedgwick, *Life and Letters*, Vol. 1, p. 380–381.
34	"The Old Red all round Orm Head," Sedgwick, *Life and Letters*, Vol. 1, p. 378.
34	"I spent some days," Sedgwick, *Life and Letters*, Vol. 1, p. 378.
35	"generally consist of an altered slate" and "There is a very large mass of Basalt," quoted in Barrett, "The Sedgwick-Darwin Geologic Tour," p. 157.
35	"from basalt," Sedgwick to Darwin [September 4, 1831], DCP-LETT-116.
35	This part of Darwin's trip is covered in Roberts, "Darwin's Dog-Leg."
36	"[Darwin] is doing admirable work," Sedgwick, *Life and Letters*, Vol. 1, p. 380.
37	The history of the Geological Society of London may be found in Herries Davies, *Whatever is Under the Earth*, and Lewis and Knell, *The Making of the Geological Society of London*, with special highlights covered in Thackray, *To See the Fellows Fight*.
37	"stupid red nosed waiter," Sedgwick to Darwin [September 4, 1831], DCP-LETT-116.
37	"We left Conway early," Sedgwick, *Life and Letters*, Vol. 1, p. 381.
38	Sedgwick's presentation is Sedgwick, "Address on Announcing." The history of the somewhat tortured relationship between William Smith, George Bellas Greenough, and the Geological Society is told in Winchester, *The Map That Changed the World*.
38	"Father of English Geology," Sedgwick, "Address on Announcing," p. 279.
38	"90 merry Philosophical faces," quoted in Herries Davies, *Whatever is Under the Earth*, p. 79.
38	Sedgwick's presidential address is Sedgwick, "Address to the Geological Society," pp. 281–316.
38	"nineteen twentieths," Ibid., p. 311.
38	"instruction I received in every chapter," Ibid., p. 302.
38	"I cannot but regret," Ibid., p. 303.
39	"Bearing on this difficult question," Ibid., p. 313.
41	For trilobites, see for example, Walker and Ward, *Fossils*, pp. 56–65.

CHAPTER THREE: SETTING OUT: THE ADVENTURES BEGIN

43	"It then first dawned on me," Darwin, *Autobiography*, p. 68.
43	"being in a ship is being in a jail," quoted in Bartlett, *Familiar Quotations*, p. 354.
44	"dear bought experience," Beagle *Diary*, p. 17.
44	"the heart of a sea-sick man" and "great & unceasing suffering," Darwin and Keynes, Beagle *Diary*, p. 18.
44	"We have left perhaps," Beagle *Diary*, p. 19.

44   "This was a great disappointment," FitzRoy, *Narrative*, Vol. 2, p. 49.

45   "one of the best tho' full of Wernerian nonsense," Sedgwick to Darwin [September 18, 1831], DCP-LETT-129.

45   "The structure of the Cape de Verde Islands," Daubeny, *A Description of Active and Extinct Volcanos*, p. 264.

45   "geologizing," This term became one of Darwin's favorites to describe his geologic work in the field. His first use of it that I have found is in Beagle *Diary*, p. 36.

46   "The vicinity of Port Praya," FitzRoy, *Narrative*, Vol. 2, p. 52.

46   "first saw the glory" and "It has been for me," Beagle *Diary*, p. 23.

46   "The task before him," Pearson and Nicholas, "'Marks of extreme violence,'" give a particularly interesting and insightful reprise of Darwin's fieldwork on St. Jago.

46   "it has the exact appearance," Darwin, *Geological Diary*, CUL-DAR32.15-20, http://darwin-online.org.uk/content/frameset?viewtype=side&itemID =CUL-DAR32.15-20&pageseq=1.

46   "as far as my knowledge goes," Ibid.

47   "The geology of St. Jago," "But the line of white rock," and "It then first dawned on me," Darwin, *Autobiography*, p. 68.

48   "A considerable difference," quoted in FitzRoy, *Narrative*, Vol. 2, p. 24.

48–49   Good resources on the methods and challenges of navigation in the early to mid-1800s may be found in Bowditch, *The New American Practical Navigator*, and Williams, *From Sails to Satellites*. See Calder, *How to Read a Nautical Chart* for a quick overview. Also see http://penobscotmarinemuseum.org/pbho-1/ history-of-navigation/history-navigation-introduction.

49–50   For a discussion of the analemma and an Excel workbook to calculate your own, https://web.archive.org/web/20060323145857/http://www.wsanford. com/~wsanford/exo/sundials/analemma_calc.html.

51   *"Nautical Almanac,"* Elliot, *Nautical Almanac and Astronomical Ephemeris for the Year 1832*.

51   For Darwin removing a drawer to hang his hammock, see Francis Darwin, The *Life and Letters* of Charles Darwin, p. 219.

CHAPTER FOUR: THE FIRST YEARS OF THE VOYAGE

63   "To the southward," quoted in FitzRoy, *Narrative*, Vol. 2, p. 26.

63   For a summary of Bento Sanchez Dorta, see Meireles Gesteira, "Observações astronômicas e físicas no Rio de Janeiro setecentista (1781–1787)," http:// www.sbhc.org.br/conteudo/view?ID_CONTEUDO=838.

63   "Baron Roussin," Roussin, *Navigation aux Côtes du Brésil*.

63–64   As was the custom at the time, each of FitzRoy's predecessors penned an account of his voyage, Owen, *Narrative of Voyages to Explore the Shores of Africa, Arabia and Madagascar*; Beechey, *Narrative of a Voyage to the Pacific and Beering's Strait*; King and FitzRoy, *Narrative*, Vol. 1; and, Webster and Foster, *Narrative of a Voyage to the Southern Atlantic Ocean*.

64   "reclining beneath the awning," Webster and Foster, *Narrative*, Vol. 2, pp. 190–191.

65   "suspended in gimbals," FitzRoy, *Narrative*, Vol. 2 Appendix, p. 325ff.

66   "At Rio de Janeiro," quoted in FitzRoy, *Narrative*, Vol. 2, p. 26.

67   "his calculation of 43°8′45″W," FitzRoy, *Narrative*, Vol. 2 Appendix, p. 65.

67   "a calculation of 43°9′W," FitzRoy, *Narrative*, Vol. 2 Appendix, p. 323.

67   "The whole neighboring country," and "The whole country moreover," Darwin, *Geological Diary*, CUL-DAR32.3-8, http://darwin-online.org.uk/content/frameset?viewtype=side&itemID=CUL-DAR32.3-8&pageseq=1.

67   "The structure of the country," Ibid.

68   "I was the Undertaker," Fanny Owen to Darwin [December 2, 1831], DCP-LETT-151.

68   "Darwin received the news," Catherine Darwin to Darwin [January 8–February 4, 1832], DCP-LETT-154.

69   "I find that my thought," Caroline Darwin to Darwin [April 1–6, 1832], DCP-LETT-164.

69   "Remember you will always," Fanny Owen to Darwin [March 1, 1832], DCP-LETT-162.

69   "the real work of the expedition," quoted in FitzRoy, *Narrative*, Vol. 2, p. 26.

69   For a modern look at what Darwin saw north of Buenos Aires, see Iriondo and Kröhling, "From Buenos Aires to Santa Fe."

70   "its odd little grunt beneath my head," Darwin, *Narrative*, Vol. 2, p. 90.

70   "The bird was cooked and eaten," Darwin, *Narrative*, Vol. 2, p. 108.

70   "those immense plains of Buenos Ayres," Darwin, *Narrative*, Vol. 2, p. 52. For a modern discussion, see Zárate and Folguera "On the Formations of the Pampas."

72   "pleasant cruize" and "The smooth water," Darwin, Beagle *Diary*, p. 106.

72   "I never knew before" and "enough to make," Darwin to J. S. Henslow [c. October 24–26, 1832], DCP-LETT-192.

72   "some rocks," Darwin, Beagle *Diary*, p. 106.

72   "There is nothing like geology," Darwin to Catherine Darwin [April 6, 1834], DCP-LETT-242.

72   "To my great joy," Darwin, Beagle *Diary*, p. 107.

73   "My friend's attention," FitzRoy, *Narrative*, Vol. 2, p. 106.

73   "dinosaur," Owen, *Report on British Fossil Reptiles. Part II*, p. 103.

73   "great quadruped," Darwin, *Narrative*, Vol. 3, p. 103.

73   "four kinds of giant ground sloth," Fernicola et al., "The Fossil Mammals."

73   This account on the discovery and naming of the *Megatherium* is based on Rudwick, *Georges Cuvier*, pp. 25–32.

74   "huge beast," Ibid., p. 26.

74   "This animal differs" and "indicative characters," Ibid., p. 28.

74   "The great thickness," Ibid., p. 30.

74   The story of the *Megalonyx* is from Jefferson, "A Memoir on the Discovery."

74   "at a depth of two or three feet," Ibid., pp. 246–247.

74   "giant claw," Ibid., p. 248.

75   The following lines were included in Thomas Jefferson's written instructions to Meriwether Lewis: "Other objects worthy of notice will be . . . the animals of the country generally, & especially those not known in the US.; the remains & accounts of any which may be deemed rare or extinct." https://www.loc.gov/exhibits/lewisandclark/transcript57.html. In addition, Thomas Jefferson's collaborator in studying the bones of the *Megalonyx*, Caspar Wistar, tutored Lewis in paleontology prior to the expedition. Jefferson had written in *A Memoir on the Discovery*, "In the present interior of our continent there is surely space and range enough for elephants and lions, if in that climate they could subsist; and the mammoths [mastodons] and megalonyxes who may subsist there. Our entire ignorance of the immense country to the West and North-West, and of its contents, does not authorise us to say what it does not contain." (p. 252) The details of this story may be found in E. E. Spamer and R. M. McCourt, *Lewis and Clark's Lost World: Paleontology and the Expedition*, http://www.lewis-clark.org/channel/372. https://www.loc.gov/exhibits/lewisandclark/transcript57.html. This episode is also mentioned briefly in Ambrose, *Undaunted Courage*, p. 91.

76   See Owen, *Fossil Mammalia*. For a fascinating and thorough modern view of the history, paleontology, and paleoecology of the fossil animals that Darwin found including the giant ground sloths, the *Glyptodonts*, and the *Toxodon*, with abundant illustrations, see Fariña et al., *Megafauna*.

76   Darwin's contact with d'Orbigny is described in Darwin, *Geological Observations on South America*, p. iv; the shells that he identified for Darwin from Punta Alta are listed on p. 83.

76   "lived whilst the sea was peopled," Darwin, *Journal of Researches*, p. 83. This was an opinion that Darwin actually came to some years later after returning to England as described by Keynes in Beagle *Diary*, pp. 176–177.

77   The best references in English about Teresa Manera's work are Bayón et al., "Following the Tracks," and Fariña, et al., *Megafauna*, pp.165–166. Original papers in Spanish include Aramayo and Manera de Bianco, "Edad y Nuevos Hallazgos;" Manera de Bianco et al., "Trazas de Pelaje;" and "Yacimiento Paleoicnológico de Pehuen Co;" Vizcaíno et al., "Viaje al Sepulcro de Los Gigantes, and Farinati et al., "La Bahía que Iluminó a Darwin."

79   For Teresa's Rolex Award, see http://www.rolexawards.com/profiles/laureates/teresa_manera_de_bianco.

81   "having heard of some giant's bones," Darwin, *Narrative*, Vol. 3, pp. 180–181.

83   "Sierra del Pedro Flaco" and "The view," Darwin, *Narrative*, Vol. 3, p. 173.

CHAPTER FIVE: PATAGONIA: THE GREAT WORKSHOP OF NATURE

85   "Such is the history," Darwin, *Narrative*, Vol. 3, pp. 207–208.

86   David's paper on the density of the ancient atmosphere is Som et al., "Air Density 2.7 Billion Years Ago."

86   "'whole' gale," FitzRoy, *Narrative*, Vol. 2 Appendix, p. 12. For details on the HMS *Beagle* itself, see Marquardt, HMS *Beagle*.

87   "The sea filled our decks," Beagle *Diary*, p. 132.

87   "much injured," FitzRoy, *Narrative*, Vol. 2, p. 126.

87   For Magellan, see Bergreen, *Over the Edge of the World*, pp. 133–71.

87   For Sir Francis Drake, see Bawalf, *Secret Voyage*, pp. 100–6.

88   "a stick of wood," Anson, *A Voyage Round the World*, p. 64.

88   For Lieutenant Sholl's death, see King and FitzRoy, *Narrative*, Vol.1, p. 121.

88   "an earthy mass," and "several of the vertebrae," and "the skeleton," Darwin, *Geological Observations on South America*, p. 95.

88   For guanaco, see Chester, *A Wildlife Guide to Chile*, pp. 303–305.

91   A reproduction of Conrad Martens's watercolor *Christmas Day, 1833*, showing slinging the monkey may be found in Keynes, *Beagle Record*, p. 173, and at https://www.repository.cam.ac.uk/handle/1810/194378. Other of Martens's watercolors showing Port Desire (now Puerto Deseado) may be found in the same book on p. 124 and p. 172, and his sketches of Port Desire in FitzRoy, *Narrative*, Vol. 2, following p. 316, and at https://www.repository.cam.ac.uk/handle/1810/194347.

92   The discovery of the Río Santa Cruz is described in Bergreen, *Over the Edge of the World*, p. 156.

92   "a few sheets of copper" and "Keel Point," FitzRoy, *Narrative*, Vol. 2, p. 336. Conrad Martens's sketches of the *Beagle* on its side and of the party on the Río Santa Cruz follow p. 336. Here and also at Port Desire, Darwin also examined what he called the Great Patagonian Tertiary Formation. He found this formation in cliffs along five hundred miles of the coast and up the Río Santa Cruz, and considered its breadth to be two hundred miles. Abundant shells, the most common of which was a "gigantic oyster" showed that it was of marine origin (Darwin, *Journal of Researches*, pp. 170–171). The widespread existence of this clearly marine formation likely contributed to his incorrect conclusion that the gravels above it, the Patagonian Shingle Formation, were also of marine origin. See also Parras and Griffin, "Darwin's Great Patagonian Tertiary Formation," and Casadío and Griffin," Sedimentology and Paleontology."

93   "Nothing could be more favorable" and "dry & sterile," *Beagle Diary*, p. 232.

93   "During the former voyage," FitzRoy, *Narrative*, Vol. 2, p. 336.

93   "Perhaps its most remarkable feature," Darwin, *Narrative*, Vol. 3, p. 213.

94   "one-half of our party" and "Many were the thorny bushes," FitzRoy, *Narrative*, Vol. 2, p. 342.

94   "brownish yellow," FitzRoy, *Narrative*, Vol. 2, pp. 337–338.

94   "Scattered herds" and "Is it not remarkable," FitzRoy, *Narrative*, Vol. 2, p. 338.

95   "some of the party felt the cold," *Beagle Diary*, p. 233.

95   "7 feet in circumference," Chancellor and van Whye, *Notebooks*, p. 271.

95   "An old Mr. Cotton," Darwin, *Autobiography*, p. 45.

95   "I felt the keenest delight," Darwin, *Autobiography*, p. 46.

96   Darwin's note and paper on icebergs are, respectively, "Note on a Rock Seen on an Iceberg," and "On the Power of Icebergs."

96   "Later he even suspected," Darwin to Hooker [August 7, 1856], DCP-LETT-1940.

96   "few doubtful looks," Beagle *Diary*, p. 235.

96   "two spells," Chancellor and van Whye, *Notebooks*, p. 272.

96   "My great puzzle," Chancellor and van Whye, *Notebooks*, p. 273.

96–97   The procedure for using the barometers is described in FitzRoy, *Narrative*, Vol. 2, *Appendix*, pp. 308–309, and Baily, *Astronomical Tables*. Some of Darwin's measurements are in Chancellor and van Whye, *Notebooks*, pp. 271–285, and some of his actual calculations using logarithms are available at http://darwin-online.org.uk/content/frameset?pageseq=1&itemID=CUL-DAR34.101-102&viewtype=image.

97   "This day I found," Beagle *Diary*, p. 236.

98   "This plain," Chancellor and van Whye, *Notebooks*, p. 277.

98   "Like St[rait] of Magellan?" Chancellor and van Whye, *Notebooks*, p. 279.

98   "generally of an angular form," Darwin, "On the Distribution of the Erratic Boulders," p. 415.

98   "yet partially retaining their color," Chancellor and van Whye, *Notebooks*, p. 280. It was later argued by Egidio Feruglio, *Descripción Geológica de la Patagonia*, vol. 3, that these shells were likely carried there by indigenous people. See also Martínez et al., "Charles Darwin and the first scientific observations on the Patagonian Shingle Formation," p. 91.

98   "from the high land" and "The river here is very tortuous," Beagle *Diary*, p. 237.

99   "was very unpleasant," "The Captain decided," "We crossed a desert plain," and "We took a farewell look," Beagle *Diary*, p. 238.

101   For discussion of sheep ranching in Patagonia and its impact see "In Patagonia, sheep ranches get another chance," *New York Times*, July 23, 2003, "Grasslands of Patagonia," in *Grasslands of the World*, http://www.fao.org/docrep/008/y8344e/y8344e09.htm; and "Overgrazing and Desertification," http://www.conservacionpatagonica.org/whypatagonia_mtp_overgrazing.htm.

102   For rainbow trout in the Río Santa Cruz, see Pascual et al., "First Documented Case of Anadromy."

102   For the dams, see "Companies defend Santa Cruz dams project," *Buenos Aires Herald.com*, http://www.buenosairesherald.com/article/205937/companies-defend-santa-cruz-dams-project.

103   The fossils found at this site are described in Fernicola et al., "Fossil Localities."

104   Conrad Martens's drawing is reproduced in FitzRoy, *Narrative*, Vol. 2, following p. 348; in Keynes, *The Beagle Record*, p. 205; and at http://www.lib.cam.ac.uk/exhibitions/Darwin/conradmartens.html.

106   "Every one excepting myself," Darwin, *Narrative*, Vol. 3, p. 226.

106   "I came to another" and "series of lesser," Darwin, *Narrative*, Vol. 3, p. 204.

107   "there would be formed" and "Let the elevations," Darwin, *Narrative*, Vol. 3, p. 205.

108   "There cannot be any doubt," and "the land was depressed," Darwin, *Geological Diary*, CUL-DAR34.104-111, http://darwin-online.org.uk/content/record?itemID=CUL-DAR34.104-111. Darwin's interpretation

that the terraces along the coast as being of marine origin has largely been substantiated, see Pedoja et al., "Uplift of Quaternary Shorelines in Eastern Patagonia." In contrast, the terraces up the Río Santa Cruz were created by the river and glacial processes. For the modern interpretation of the deposits around the Río Santa Cruz, quite at odds with Darwin's, see Strelin and Malagnino, "Charles Darwin and the Oldest Glacial Events in Patagonia," Martínez et al., "Charles Darwin and the First Scientific Observations on the Patagonian Shingle Formation," and Dott and Dalziel, "Darwin the Geologist in Southern South America."

108   "the great workshop of nature," Darwin, *Narrative*, Vol. 3, p. 204.

108   Charpentier, "Notice Sur La Cause."

108   For Agassiz and Charpentier, see Lurie, *Agassiz*, p. 94, Imbrie and Imbrie, *Ice Ages*, pp. 22–28, or Bolles, *The Ice Finders*, pp. 41–44, 51–60, 83–90.

CHAPTER SIX: MARIA GRAHAM AND THE DEBATE ON THE CAUSES OF ELEVATION

111   Kölbel-Ebert's "Observing Orogeny" is a modern account of the story of Maria Graham, later Maria Graham Callcott. The story is a rich one and deserves wider attention.

111   "Among the subjects," Greenough, "Address Delivered," p. 54.

111   Greenough, "Address Delivered," pp. 42–70.

112   For a perspective on historical theories of mountain building see, Oldroyd, *Thinking about the Earth*, pp. 167–91.

112   "If by some fiat," McPhee, *Basin and Range*, p. 194.

113   Darwin read Herschel's *A Preliminary Discourse on the Study of Natural Philosophy*, and was so affected by it that he recalled five decades later, "During my last year at Cambridge I read with care and profound interest Humboldt's *Personal Narrative*. This work and Sir J. Herschel's *Introduction to the Study of Natural Philosophy* [sic] stirred up in me a burning zeal to add even the most humble contribution to the noble structure of Natural Science. No one or a dozen other books influenced me nearly so much as these two." Darwin, *Autobiography*, p. 57.

113   "scientists," Whewell proposed the use of this term seriously in 1840, although he apparently had used it anonymously (and satirically) earlier. Whewell, *The Philosophy*, Vol. 1., p. cxii.

113   "*vera causa*," Herschel assigns the definition of this term to Newton. "Experience having shown us the manner in which one phenomenon depends on another in a great variety of cases, we find ourselves provided, as science extends, with a continually increasing stock of such antecedent phenomena, or causes (meaning at present merely proximate causes), competent, under different modifications, to the production of a great multitude of effects, besides those which originally led to a knowledge of them. To such causes Newton has applied the term *veræ causæ*; that is, causes recognized as having a real existence in nature, and not being mere hypotheses or figments of the mind," and then Herschel examines the case of "shells found in rocks, at a great height above the sea." Herschel, *Preliminary Discourse*, pp. 144–145.

113   "the transition," Whewell, *Philosophy*, Vol. 2, p. 318.

113–114 "The notices of the banditti," Callcott, *Three Months Passed*, p. v.

114   The swashbuckling story of Lord Cochrane, including his interactions with Maria Graham, is told in Cordingly, *Cochrane*.

114   The fascinating stories of Bernardo O'Higgins and his father Ambrosio are told in Clissold, *Bernardo O'Higgins*.

114   Maria Graham Callcott, *Journal of Residence*.

116   "Though not handsome" and "His conversation," Callcott, *Journal of Residence*, p. 188.

116   "very still and clear," and "three minutes," Graham, "Account," p. 413.

116   "that of the earth," "the alluvial valley," and "It appeared on," Ibid., p. 414.

118   "although deeply sensible," "raised above its former level," "But by what standard," and "By what means," Greenough, "Address," p. 56.

118   Women were not allowed to attend meetings of the Geological Society until 1901, but it was not until 1919 that the first female Fellow was elected, see Herries Davies, *Whatever is Under the Earth*, and Burek, "The First Female Fellows."

118   Maria Graham, by now Maria Graham Callcott, wrote her scalding letter addressed to the President and Members of the Geological Society. Because there was to be a delay before the letter could be read before the Society, she had the letter printed privately (Callcott, *A Letter to the President*), so it does not formally appear in the *Proceedings of the Society*. Interestingly, however, Maria's privately printed letter is bound into the copy of the *Proceedings of the Geological Society*, Vol. 2, held by the University of California Library, and scanned for Google Books (https://play.google.com/books/reader?id=bl68AAAAIAAJ&printsec=frontcover&output=reader&hl=en&pg=GBS.PA707), where it can be seen, steaming with Maria's outrage. Maria's original letter, the relevant section of Greenough's Presidential Address, and Maria's privately printed response to Greenough were all also reprinted in Silliman, "On the reality."

119   "This attack implies," Callcott, *A Letter to the President*, p. 3.

119   "such an absurdity" and "a regular geological survey," Ibid., p. 4.

119   "a naval officer or naturalist," Greenough, "Address," p. 57.

119   Lyell cited Maria Graham's report in Lyell, *Principles*, Vol. 1, pp. 402–403.

119   "If I am to pronounce," Greenough, "Address," p. 58.

120   Cuvier's view on catastrophes are discussed in Rudwick, *Worlds Before Adam*, pp. 89–91.

120   "In the course of the last century," Lyell, *Principles*, Vol. 2, p. 161.

121   "the difficulty of proving," and "the scientific investigator," Lyell, *Principles*, Vol. 1, p. 416.

121   For the engraving of the Temple of Serapis, see Ibid., p. iii. Dvorak and Mastrolorenzo In "Mechanisms of Recent Vertical Crustal Movements" give a detailed history of the observations of the Temple of Serapis and their interpretation including the the modern understanding.

122   "At the bottom of the cavities," Ibid., p. 453.

122   "We must, consequently," Ibid., p. 454.

123   "The pavement of the Temple of Serapis," Playfair, *The Works*, Vol. 1, p. 441.

123   "This celebrated monument of antiquity," Lyell, *Principles*, Vol. 1, p. 449.

123   The three theories are described in Rudwick, *Worlds Before Adam*, p. 113.

124   "I at one time felt," "Dr. Daubeny," and "the different height," Darwin, *Geological diary*, CUL-DAR32.21-36, http://darwin-online.org.uk/content/frameset?viewtype=side&itemID=CUL-DAR32.21-36&pageseq=1.

124   "the people of Östhammar," Ekman, *The Changing Level of the Baltic Sea*, pp. 22–25, and Ekman (1991).

124   "During recent years," Ekman, *The Changing Level of the Baltic Sea*, p. 22.

125   For Celsius's paper, see Ekman, "A Concise History."

126   Lyell's trip to Sweden convinced him. See Lyell, "The Bakerian Lecture."

CHAPTER SEVEN: DARWIN'S EARTHQUAKE

129   "To my mind since leaving England," Beagle *Diary*, p. 302.

129   "enough to make," "furious gales," and "great sea," Beagle *Diary*, p. 244.

130   "who have read works" and "it was as surprising," Beagle *Diary*, p. 250.

130   "going up," Darwin to Caroline Darwin [August 9–12, 1834], DCP-LETT-253.

130   "there is nothing," Darwin to FitzRoy [August 28, 1834], DCP-LETT-254.

130   "a very nice hacienda," "several very pretty Signoritas," "turned up their charming eyes," and "The absurdity of a Bishop," Beagle *Diary*, p. 257.

130   "fringes of gravel," "to suppose," and "I cannot doubt," Darwin, *Geological Observations on South America*, p. 65. These features were actually of fluvial origin, i.e., owing to a river. See for example, Dott and Dalziel, "Darwin the Geologist in South America," p. 332.

133   "I am so surrounded by troubles of every kind," FitzRoy to his sister Fanny [November 6, 1834], CUL MS Add. 8853/46, f. 125, http://www.lib.cam.ac.uk/exhibitions/Darwin/bigpics/FitzRoy_depression_letter.jpg.

133   "I confess that my own feelings," FitzRoy, *Narrative*, Vol. 2, p. 362.

134   "Although the houses," and "A bad earthquake," Darwin, *Narrative*, Vol. 3, p. 369.

134   "that not a house in Concepción," "I soon saw abundant proof," and "Besides chairs," Darwin, *Narrative*, Vol. 3, p. 370.

135   "Both towns" and "In Concepción," Darwin, *Narrative*, Vol. 3, p. 371.

135   "It is generally thought," Darwin, *Narrative*, Vol. 3, p. 371.

135   Earthquakes that have struck Concepción are listed in Lomnitz, "Major Earthquakes of Chile."

135   "At Talcahuano the great earthquake," "Nearly all the inhabitants," "About half an hour after the shock," and "This terrific swell," FitzRoy, *Narrative*, Vol. 2, p. 406.

136   "It broke over" and "In a few minutes," FitzRoy, *Narrative*, Vol. 2, pp. 406–407.

136–137   "After some minutes" and "Quickly retiring," FitzRoy, *Narrative*, Vol. 2, p. 407.

137   The harbor at Isla Santa María was the site of the dramatic recapture of the slave ship by Amasa Delano (see his *Narrative*) that provided the basis for Melville's novella *Benito Cereno*, in Melville, *Billy Budd and Other Stories*.

138   "beds of dead muscles," "proofs of the upheaval," and "It was concluded," FitzRoy, *Narrative*, Vol. 2, p. 413.

138   "to settle the matter" "When we landed," FitzRoy, *Narrative*, Vol. 2, p. 413.

138   "I took many measures," FitzRoy, *Narrative*, Vol. 3, p. 414.

139   "extensive rocky flat" and "covered with dead," FitzRoy, *Narrative*, Vol. 3, p. 414.

145   Daniel's paper is Melnick et al., "Coastal Deformation."

CHAPTER EIGHT: THE ANDES ARISING

147   "It is an old story," Darwin, *Narrative*, Vol. 3, p. 390.

147   Caldcleugh, *Travels in South America*.

148   "The snow was so deep," Caldcleugh, *Travels in South America*, Vol. 1, p. 315.

148   "The madrina is a mare," Beagle *Diary*, p. 304.

148   "Since leaving," Darwin to Susan Darwin [April 23,1835], DCP-LETT-275.

148   For the wonderful story of the French expedition, see Ferriero, *Measure of the Earth*.

149   See Humboldt, *Personal Narrative*.

151   "irregularly-stratified mass" and "The rivers," Darwin, *Narrative*, Vol. 3, p. 385.

151   "should rather be called," "Amidst the din," and "The sound spoke," Darwin, *Narrative*, Vol. 2, p. 385.

152   "The valley takes its name," Darwin, *Narrative*, Vol. 3, p. 389.

152   These layers of gypsum, also referred to as evaporites, are now recognized as providing the failure surfaces for the thrust, or detachment faults, in this region, Giambiagi et al., "Cenozoic deformation."

152   "zigzag track" and "The cordillera is this part," Darwin, *Narrative*, Vol. 3, p. 389.

152–153   "The inhabitants" and "Upon finding," Darwin, *Narrative*, Vol. 3, p. 393.

153   "porphyritic conglomerate," Darwin, *Geological Observations on South America*, p. 177.

153   "Gypseous Formation," Darwin, Ibid., p. 178.

153   "which alternates," Darwin, *Narrative*, Vol. 3, p. 390.

153   "These great piles of strata," Darwin, *Narrative*, Vol. 3, p. 390.

153   "Neither plant nor bird," Darwin, *Narrative*, Vol. 3, p. 394.

153   "Even at the very crest," Darwin, *Narrative*, Vol. 3, p. 390.

153   "*Gryphaea*, or 'devils toenails,'" see Walker and Ward, *Fossils*, p. 101; http://www.kgs.ku.edu/Publications/ancient/f18_clams.html. For similar fossils in Kansas, see Everhart, *Oceans of Kansas*.

153   "undoubtedly the most famous," Gould, *The Evolution of Gryphaea*. For Gould's proposed variation on evolution, see Gould, *Punctuated Equilibrium*.

154   "without thunder" and "The peril is imminent," Darwin, *Narrative*, Vol. 3, p. 396.

155   "coarsely-crystallized," "immense granitic dikes," and "central mass," Darwin, *Narrative*, Vol. 3, p. 391.

155   "perfectly rounded" and "immense quantities," Darwin, *Narrative*, Vol. 3, p. 392.

155   "the circumstance that rivers," Darwin, *Narrative*, Vol. 3, p. 393.

CHAPTER NINE: CORAL REEFS AND THE SINKING BOTTOM OF THE SEA

"One Hundred and Thirty Years." For historical views, see McCalman, *Reef*, and Dobbs, *Reef Madness*.

169   "I am glad we have visited these Islands," Beagle *Diary*, p. 418.

170   "Is there a large proportion," Chancellor and van Whye, Notebooks, p. 398.

171   The *Endeavour*'s encounter with reefs is described in McCalman, *Reef*.

171   See La Pérouse, *Journal*.

171   For the *Pandora*, see Wahlroos, *Mutiny and Romance in the South Seas*.

172   "Naval men," Woodward, *The History of the GSL*, p. 201.

172   "a very interesting inquiry," quoted in FitzRoy, *Narrative*, Vol. 2, p. 38.

173   The original work is in Quoy and Gaimard, *Mémoire sur l'accroissement*.

173   The sketch of the island is in Beechey, *Narrative*, following p. 188.

173   "In regard to the thickness of the masses of coral," Lyell, *Principles*, Vol. 2, p. 286.

174   "nothing more," Ibid., p. 290.

175   "gloomy region that extends far," Beagle *Diary*, p. 364.

175   "daily the sun shines brightly," Beagle *Diary*, p. 364.

175   "during the two previous years," Darwin, *Autobiography*, p. 82.

176   "one of the low coral" and "Our observations," FitzRoy, *Narrative*, Vol. 2, p. 506.

176   "Honden Island," Burney, *Discoveries in the South Sea, Part II*, p. 377.

176   "peculiar character" and "understood by [the] drawing," Beagle *Diary*, p. 364.

176   "These [islands]" and "The width of dry land," Beagle *Diary*, p. 365.

177   "Tahiti, an island" and "Crowds of men," Beagle *Diary*, p. 365.

177   "between two and three thousand" and "interior mountains," Beagle *Diary*, p. 368.

177   "The island is completely encircled," Beagle *Diary*, pp. 368–369.

177   "The effect was very pleasing," Beagle *Diary*, p. 369.

178   "paddled for some time" and "It is my opinion," Beagle *Diary*, p. 378.

178   "a union of the two prevailing kinds," Beagle *Diary*, p. 380.

179   For Darwin's essay, see Stoddard, "Coral Islands," available at http://www.sil.si.edu/DigitalCollections/atollresearchbulletin/issues/00088.pdf. Also see Darwin, C. R. "Coral Islands." (1835) CUL-DAR41.1-12. Transcribed and edited by D. R. Stoddart (*Darwin Online*, http://darwin-online.org.uk/).

179   "High Islands encircled," quoted in Stoddard, "Coral Islands," p. 6.

179   "looked in vain," Ibid., p. 10.

179   "If then the two following postulates," Ibid., p. 13.

180   "if circumstances are favourable," quoted in FitzRoy, *Narrative*, Vol. 2, p. 33.

180   "before our little ship," FitzRoy, *Narrative*, Vol. 2, p. 628.

180   "When a number of gannets," FitzRoy, *Narrative*, Vol. 2, p. 628.

180   "a long but broken line of cocoa-palm trees" and "within five miles," FitzRoy, *Narrative*, Vol. 2, p. 629.

180   "Little or no notice was taken," FitzRoy, *Narrative*, Vol. 2, p. 630.

180   "till 1823," FitzRoy, *Narrative*, Vol. 2, p. 631.

181   "about a year afterward," FitzRoy, *Narrative*, Vol. 2, p. 632.

181   "very worthless character," Beagle *Diary*, p. 413.

181   "every one was actively occupied," "but two moderate days," and "Only a mile from the southern," FitzRoy, *Narrative*, Vol. 2, p. 630.

181   "The [southern] cluster of islets," FitzRoy, *Narrative*, Vol. 2, p. 630.

181   "The outer edges," FitzRoy, *Narrative*, Vol. 2, p. 630.

182   "Among the great variety," FitzRoy, *Narrative*, Vol. 2, p. 634.

182   "to ascertain if possible" and "Judging however," FitzRoy, *Narrative*, Vol. 2, p. 634.

182   "crabs eat cocoa-nuts," FitzRoy, *Narrative*, Vol. 2, p. 635.

182   "strip of dry land," Beagle *Diary*, p. 414.

183   "eye of the body," "eye of reason," and "We feel surprised," Beagle *Diary*, p. 418.

183   "Hence we must consider" and "If the opinion," Beagle *Diary*, p. 418.

184   "We see certain Isds in the Pacifick" and "In time the central," Beagle *Diary*, p. 418.

185   For the drilling on Bikini Atoll, see Rosen, "Darwin, Coral Reefs."

186   "We may finally conclude," Darwin, *Structure*, p. 146.

PART TWO: DARWIN THEORIZING
CHAPTER TEN: FAITH COMFORTS, FACTS PERSUADE

189   "But the busiest time," Darwin to W. D. Fox [November 6, 1836], DCP-LETT-319.

189   "This zig-zag," Darwin to Susan Darwin [August 4, 1836], DCP-LETT-306.

190   "I feel inclined," Darwin to Caroline Darwin [July 18, 1836], DCP-LETT-305.

190   "My dear Henslow," Darwin to J. S. Henslow [July 9, 1836], DCP-LETT-304.

191   "dreadfully stormy," Beagle *Diary*, p. 447.

191   "He came to see," "fly-catcher," "stone-pounder," "the coral insects," and "a most agreeable," Fox, *Memories of Old Friends*, p. 9.

191   "Good bye," Darwin to FitzRoy [October 6, 1836], DCP-LETT-310.

191   "Dearest Philos," FitzRoy to Darwin [October 19–20, 1836], DCP-LETT-312.

192   FitzRoy assessed his results in FitzRoy, *Narrative*, Vol. 2 Appendix, p. 352.

192   "the great men" and "anxious to dissect," Darwin to J. S. Henslow [September 30–October 1,1836], DCP-LETT-317.

193   "Mr. Lyell" and "I am out of patience," Darwin to J. S. Henslow [September 30–October 1, 1836], DCP-LETT-317.

193   "My London visit has been quite," Darwin to W. D. Fox [November 6, 1836], DCP-LETT-319.

194   "settle my jolted brains," and "It will be a most," Darwin to W. D Fox [November 6, 1836], DCP-LETT-319.

194   "All the outlyers of the family," Litchfield, *Emma Darwin*, p. 385.

194–195 "distribute proper shares," "plied him with questions," "several little geological lectures," and "was rather anxious," S. E. Wedgewood to Hensleigh Wedgewood [November 16,1836], DCP-LETT-322.

195   "Charles Darwin, Esq.," *Proceedings GSL*, Vol. 2, p. 435.

195   "Such raised beaches!" Sedgwick to Whewell [October 7, 1836], in Sedgwick, *Life and Letters*, Vol. 1, p. 462.

196   "'Mr. President," "evidently in a great rage," "Then, rose immediately," and "was foaming," Murchison to Sedgwick [December 14, 1836], quoted in Thackray, *To See the Fellows Fight*, pp. 68–67.

196   "Will you come up," "with greatest pleasure," and "The idea of the Pampas," Lyell to Darwin [December 26, 1836], DCP-LETT-335.

196–197 The attendance and alcohol consumption at this dinner are from the record book of the Geological Survey Club held in the archive of the Geological Society of London.

197   "that the coast around the Bay of Dublin," *Proceedings GSL*, Vol. 2, p. 437.

197   "you stand the first," Lyell to Darwin [December 26, 1836], DCP-LETT-335.

198   "I was in the meeting," quoted in Thackray, *To See the Fellows Fight*, pp. 70–71.

199   Caldcleugh, "Some Observations."

199   Darwin, "Observations of proofs."

199   "previously to his return," Caldcleugh, "Some Observations," p. 444.

199   "his full conviction," Ibid., p. 444.

199   "but of which seamen," Ibid., p. 445.

199   "gave an account," Ibid., p. 446.

199–200 "El Ara[u]ncano" and "strongly support[ing]," Ibid., p. 446.

200   "Close to the mouth," Darwin, "Observations of Proofs," p. 446.

200   "the great number of shells," Ibid., p. 447.

200   "large proportional number," Ibid., p. 447.

200   "he met no intelligent person," Ibid., p. 447.

200   "under three feet," Ibid., p. 448.

200   "the coast of Chili has risen" and "a change effected imperceptibly," Ibid., p. 448.

200   "earthquakes are never experienced," Ibid., pp. 448–449.

201   "Proofs of Modern Elevation," Ibid., p. 501.

201   "the elevatory movement," Ibid., p. 507.

201   "argued that their position," Ibid., p. 535.

201–202 "It was not until after" and "Greenough still sticks," quoted in Thackray, *To See the Fellows Fight*, p. 74.

202   "Sedgwick made" and "a kind of tiger-cat," quoted in Ibid., p. 74.

202   "Dr. Buckland will never come to the scratch," quoted in Ibid., p. 74.

202   "We have had the subject," quoted in Ibid., p. 75.

202   "of great extent" and "all fall on the areas of elevation," Darwin, "On Certain Areas of Elevation and Subsidence," p. 554.

202–203 "I am very full" and "Let any mountain," Lyell, *Life, Letters*, Vol. 2, p. 12.

203   "mixed up [with] Capt. FitzRoy's," Hensleigh Wedgwood to Darwin [December 20, 1836], DCP-LETT-332.

203   "snowed up" and "the profits if any," FitzRoy to Darwin [December 30,1836], DCP-LETT-337.

204   "arranging [his] general collection," John van Wyhe ed., Darwin's "Journal" (1809–1881). http://darwin-online.org.uk/content/frameset?itemID=CUL-DAR158.1-76&viewtype=text&pageseq=1. CUL-DAR158.1-76 (*Darwin Online*, http://darwin-online.org.uk/).

205   "such assistance," FitzRoy, *Narrative*, Vol. 2, p. 430.

206   "Captain FitzRoy," FitzRoy to Darwin [November 15, 1837] DCP-LETT-386.

206   "My dear Darwin," FitzRoy to Darwin [November 16, 1837], DCP-LETT-387.

206–207 "Most people" and "[Lyell] does not seem," FitzRoy to Darwin [November 16, 1837], DCP-LETT-387.

207   "I am happy to say," FitzRoy to Darwin [February 26, 1838], DCP-LETT-403.

207   "slow coach," FitzRoy to Darwin [February 26, 1838], DCP-LETT-403.

208   "A very few remarks" and "I suffered," FitzRoy, *Narrative*, Vol. 2, p. 657.

208   "One of my remarks," FitzRoy, *Narrative*, Vol. 2, p. 658.

208   "It beats" and "Although I owe much," Darwin to Caroline Wedgwood [October 27, 1837], DCP-LETT-542.

208   "Faith comforts," Quammen, *The Song of the Dodo*, p. 305.

CHAPTER ELEVEN: THE THEORY COMES TOGETHER

209   Darwin's theories about elevation and subsidence are discussed in Rhodes, "Darwin's Search for a Theory of the Earth," and Herbert, *Charles Darwin, Geologist*.

209   "Geology of the whole world," Darwin and Herbert, *The Red Notebook*, p. 50.

209   "by reference to causes," Lyell, *Principles*, Vol. 1, subtitle.

211   "one does not exactly," De la Beche, *Researches in Theoretical Geology*, p. 122.

211   De la Beche's cartoon is reproduced in Rudwick, *Worlds Before Adam*, p. 326.

212   "The Earthquake and Volcano," Beagle *Diary*, p. 302.

212–214 "the best singers," Sulivan, *Life and Letters*, p. 43.

214   "We shall have a large bag," Sulivan, *Life and Letters*, pp. 43–44.

214   "The day rose," "another great Volcano," and "the lofty peaked," Beagle *Diary*, p. 265.

215   "During this night" and "It was a very magnificent sight," Beagle *Diary*, p. 280.

215   "Fitzroya patagonica," Hooker, "Fitz-roya Patagonica."

215   "over the greatest part," Douglas to Darwin [January 5, 1836], DCP-LETT-292.

216   "The Volcanos," "burning stones," "the thundering Corcovado," Ibid.

216   "the grandest Volcanic," "like an imense," and "I[t] appeared to be," Ibid.

216   The attendance at this dinner on Wednesday, March 7, 1838 are from the record book of the Geological Survey Club held in the archive of the Geological Society of London.

217   "After the dinner." This account of the meeting relies on Lyell's letter quoted in Thackray, *To See the Fellows Fight*, pp. 81–83, and Darwin's papers "On the Connexion" in the *Proceedings of the GSL*, then published two years later in the *Transactions of the GSL*.

218   "opened upon," Thackray, *To See the Fellows Fight*, p. 81.

218–219 "seems to have been" and "this fact," Darwin, "On the Connexion," p. 602.

219   "evident that the volcanic chain," Ibid., p. 604.

219   "We see, therefore, that, in 1835," Ibid., p. 605.

219   "force which elevates Continents," Ibid., p. 606.

219   "the idea of water splashing," Ibid., p. 607.

219   "earth's crust" and "The facts appear," Ibid., p. 608.

219   "periods of increased Volcanic," Ibid., p. 610.

220   the "subterranean forces" and "then bursting," Ibid., p. 615.

220   "It cannot be otherwise" and "to one conclusion alone," Ibid., p. 615 .

220   "we may," "a violent rending," and "comparatively rare," Ibid., p. 619.

220   "the earthquake . . . relieves," Ibid., p. 619.

220   "Therefore we must conclude," Ibid., p. 624.

220   "paroxysmal violence" and "the important fact," Ibid., p. 625.

221   "mountain-chains" and "theoretical reasoning," Ibid., p. 625.

221   "without the very bowels," Ibid., p. 626.

221   "volcanic forces either now," Ibid., p. 629.

221   "there is little hazard," "When we reflect," and "and do not forget," Ibid., p. 630.

221   "The furthest generalization," Ibid.,p. 631.

222   "done his description" and "a long oration," Thackray, *To See the Fellows Fight*, p. 82.

222–223 "pronounced a panegyric," "pristine forces," "with the same intensity," "warmth of his eulogy," "in the habit of admiring," "coming up occasionally," "finished his drollery," "most of the critics," and "we ought not to try," Thackray, *To See the Fellows Fight*, p. 82.

223   "a vigorous defiance," "able to measure," "a diminishing fire," and "I restored him," Thackray, *To See the Fellows Fight*, p. 83.

223   "I found Sedgwick," Darwin to Lonsdale [May 15, 1838], DCP-LETT-412.

223–224 "The concluding," Sedgwick to GSL [after May 1838], DCP-LETT-414.

224   "The main facts," Sedgwick to GSL [after May 1838], DCP-LETT-414.

224   "same power." The title of the paper as published was "On the Connexion of certain Volcanic Phenomena in South America; and on the Formation of Mountain Chains and Volcanos, as the Effect of the same Power by which Continents are elevated."

225   "I cannot doubt," Darwin to Lyell [September 14, 1838], DCP-LETT-428.

225   "a bit of theory," "to float," and "a mixture of fixed rock," Herschel to Lyell, quoted in Babbage, *The Ninth Bridgewater Treatise*, p. 227.

225   The meeting in March 1834 is in *Proceedings of the GSL*, Vol. 2, pp. 72–76.

225   Babbage, *Observations on the Temple of Serapis*, p. 29.

CHAPTER TWELVE: EXTENDING THE THEORY: THE PARALLEL ROADS OF GLEN ROY

227   This account benefitted greatly from Rudwick, "Darwin and Glen Roy," and Rudwick "The Parallel Roads . . . A Field Guide" and "The Origin of the Parallel Roads," as well as the original sources, Darwin's Glen Roy Notebook (Barrett et al., *Charles Darwin's Notebooks, 1836–1844*), and Darwin, "Observations on the Parallel Roads of Glen Roy."

227   "The results to which," Darwin, "Observations on the Parallel Roads," p. 39.

228   "As everything that is wonderful" and "On the sides," Greenough quoted in Rudwick, "Hutton and Werner Compared," p. 131.

228   "They are carried along," Pennant, *A Tour of Scotland*, pp. 266–267.

228 "I cannot learn" and "traces of buildings," Pennant, *A Tour of Scotland*, p. 268.

229 "The country people," Pennant, *A Tour of Scotland*, p. 269.

229 "as false and absurd" and "It is really wonderful," Greenough quoted in Rudwick "Hutton and Werner Compared," p. 131.

229 "immense lake," Ibid., p. 132.

230 See MacCulloch, "On the Parallel Roads of Glen Roy" and Lauder "On the Parallel Roads of Lochaber."

230 "all these different shelves," Lauder "On the Parallel Roads of Lochaber," p. 5.

231 "Descending a few," Robins, *Coastal Passenger Liners of the British Isles*, p. 16.

231 "My trip in the steam packet," Darwin to Lyell [August 9, 1838], DCP-LETT-424.

232 "The point of the Pencil," Barrett et al., *Notebooks*, p. 141.

232 "in gigs & carts," Darwin to Lyell [August 9, 1838], DCP-LETT-424.

232 "beach & channel precisely as with Isld," Barrett et al., *Notebooks*, p. 149.

234 "Chief Points to be Attended to," quoted in Rudwick, "Darwin and Glen Roy," pp. 179–181.

234 "2. Organic remains," Ibid., p. 179.

234 "12th. The great problem," Ibid., p. 179.

235 "Even on Lauder Dicks," Barrett et al., *Notebooks*, p. 151.

235 "the gneiss is worn," Darwin, "Observations on the Parallel Roads," p. 41.

235 "the water had remained" and "Standing on," Ibid., p. 42.

236 "so complicated," Barrett et al., *Notebooks*, p. 162.

236 "alternating layers," Barrett et al., *Notebooks*, p. 153.

236 "some full grown men," Darwin to Lyell [August 9, 1838], DCP-LETT-424.

236–237 "enjoyed five days," "I think without exception," "far the most remarkable," and "fully convinced," Darwin to Lyell [August 9, 1838], DCP-LETT-424.

237 "Glen Roy has astonished," Darwin to Lyell [August 9, 1838], DCP-LETT-424.

237 "forty thieves," *Autobiography*, p. 32.

237 "Golden Square" and "It is one of the squares," Dickens, *Nicholas Nickleby*, p. 6.

238 "On a summer's night," *Nicholas Nickleby*, p. 6.

238 "got together quite" and "For I am sure the first," Darwin to Lyell [August 9, 1838], DCP-LETT-424.

238 "the mere explaining," Darwin, "Observations on the Parallel Roads," p. 39.

238 "if proved," Ibid., p. 39.

238 "It is admitted," Ibid., p. 39.

238–239 "matter in the form" and "to the ancient beaches," Ibid., p. 40.

239 "These four cases," Ibid., p. 43.

239–240 "it is far easier to assert," Ibid., p. 48.

240 "The conclusion is inevitable," Ibid., p. 48.

240 "an expanse of slowly," Ibid., p. 56.

240 "is due to the rising," Ibid., p. 58.

240 "It can scarcely be doubted," Ibid., p. 56.

240 For many examples of "thought experiment," see Isaacson, *Einstein*.

240 "would be produced," Darwin, "Observations on the Parallel Roads," p. 58.

240   "There is clear evidence," Ibid., p. 59.

240   "I believe, then," Ibid., p. 60.

240   "the non-extension," Ibid., p. 60.

241   "attentively examined" and "may be considered," Ibid., p. 63.

241   "the fragments of shells" and "the marine origin," Ibid., p. 64.

241   "alone worthy of consideration" and "one would have anticipated," Ibid., p. 70.

241   "cannot be attributed" and "it may be granted," Ibid., p. 79.

241–242   "We may almost venture," Ibid., p. 80.

242   "left . . . by the slowly retiring," Ibid., p. 81.

242   "at least 1278 feet," "so equably," and "was effected by," Ibid., p. 81.

242   "*Ki te kahore hoki,*" Misses Horner to Darwin [March 17, 1837–December 8, 1838], DCP-LETT-350.

242–243   "As for a wife," Darwin to Whitley, [May 8, 1838], DCP-LETT-411A.

243   "I have so much more pleasure," Darwin, C. R. "Work finished If not marry" [Memorandum on marriage]. (1838) CUL-DAR210.8.1. (*Darwin Online,* http://darwin-online.org.uk/).

243   "Marry," Darwin, C. R. "This is the Question Marry Not Marry" [Memorandum on marriage]. (7.1838) CUL-DAR210.8.2. (*Darwin Online,* http://darwin-online.org.uk/).

243   "the day of days!" John van Wyhe ed., Darwin's "Journal" (1809–1881). http://darwin-online.org.uk/content/frameset?itemID=CUL-DAR158&viewtype=text&pageseq=1. CUL-DAR158.1-76. (*Darwin Online,* http://darwin-online.org.uk/).

244   "I cannot get people," Darwin to Whitley [November 23, 1838], DCP-LETT-443.

244   "some few dozen drawers" and "I trust to be able," Darwin to Emma Wedgwood [December 31, 1838–January 1,1839], DCP-LETT-466.

244   "I hope you will manage," Emma Wedgwood to Darwin [January 3, 1839], DCP-LETT-482.

244   "First week January," John van Wyhe ed., Darwin's "Journal" (1809–1881). http://darwin-online.org.uk/content/frameset?itemID=CUL-DAR158&viewtype=text&pageseq=1. CUL-DAR158.1-76 (*Darwin Online,* http://darwin-online.org.uk/).

244–245   "contains much original research," "I do not think," "The discussion of the erratic," "theory of expansion," "at least far too long," and "There are some short," Sedgwick to Royal Society, 26 March 1839, quoted in Rudwick, "Darwin and Glen Roy," pp. 181–183.

245–246   "an article in the Edinburgh," Agassiz, "Upon Glaciers."

246   "the presence" and "during the gradual," Darwin, *Narrative,* Vol. 3, p. 617.

246   "never follow" and "What explanation," Darwin, *Narrative,* Vol. 3, p. 618.

246   "these very curious facts," Darwin, *Narrative,* Vol. 3, p. 619.

246–247   "planted vertically" and "at least, as simple," Darwin, *Narrative,* Vol. 3, p. 620.

247   "only veræ causæ" and "the hypothesis," Darwin, *Narrative,* Vol. 3, p. 625.

248   "principle of exclusion," Autobiography, p. 70.

PART THREE: BACK ON DARWIN'S TRAIL
CHAPTER THIRTEEN: FROM UPLIFT TO EVOLUTION

251   "I wish with all my heart," Darwin to Lyell [September 14, 1838], DCP-LETT-428.

252   "Present to Covington on leaving me £2," quoted in Freeman, *Charles Darwin, A Companion*, p. 61.

253   "is as full of good original," Owen to Darwin [June 11, 1839], DCP-LETT-519.

253   "What I like best" and "who has often wished," Fitton to Darwin [June 13,1839], DCP-LETT-520.

253   "strength of talent" and "You told me," Humboldt to Darwin [September 18, 1839], DCP-LETT-534.

253–254 "earthquake paper," Darwin, "On the Connexion."

254   A beautiful translation and reproduction of Agassiz's book is Agassiz and Carozzi, *Studies in Glaciers*.

254   "I have read his," Sedgwick, *Life and Letters*, Vol. 2, p. 18.

254   Agassiz, "On Glaciers, and the Evidence," pp. 327–32.

255   "I should much like to hear," "I cannot give up," and "Though I am very sure," Darwin to Buckland [November 1840–February 17, 1841], DCP-LETT-641A.

255   "I have lately enjoyed," Darwin to Agassiz [March 1, 1841], DCP-LETT-593.

255–256 "The difficulty about ice-barrier," Darwin to Lyell [March 12, 1841], DCP-LETT-595.

256   The presentation of Darwin's paper on erratics and icebergs is recorded in *Proceedings of the GSL*, Vol. 3, pp. 425–430. The full paper is Darwin, "On the Distribution of Erratic Boulders." Evenson et al., in "Enigmatic Boulder Trains," have done a fascinating study of the boulders Darwin found at Bahía San Sebastian, Tierra del Fuego. Evenson et al. present a persuasive case that these boulders fell in a rock avalanche onto a glacier in the Cordillera Darwin and were then transported on the surface of the glacier, first northwest, then northeast to the location where Darwin found them, a distance of about 150 miles.

256   Buckland, "On the Glacia-diluvial Phaenomena."

256   "Yesterday (and the previous day)" and "it convinces me that my views," Darwin to Fitton [c.28 June 1842], DCP-LETT-632.

258   The shells that Joshua Trimmer found are described in Darwin, "Notes on the Effects," p. 185.

258   The pencil sketch is described in Browne, *Voyaging*, pp. 436–9; see also http://darwin-online.org.uk/content/frameset?itemID=CUL-DAR6&viewtype=image&pageseq=1 and http://darwin-online.org.uk/content/frameset?itemID=CUL-DAR217.2&viewtype=image&pageseq=1.

259   "in case of my," Darwin to Emma Darwin [July 5, 1844], DCP-LETT-761.

259   "The Editor must be," Darwin to Emma Darwin [July 5, 1844], DCP-LETT-761.

259   "dabbled," Darwin to Hooker [September 10, 1845], DCP-LETT-915. In this letter, Darwin also subtly announces his intention to work on species, modestly suggesting that from this effort he will receive "more kicks than half-pennies."

259   "I am not inclined," Hooker to Darwin [September 4–9, 1845], DCP-LETT-914

259–260 "slaving to finish" and "the collection of facts," Darwin to Lyell [August 8, 1846], DCP-LETT-990.

260 "I am going to begin," Darwin to Hooker [2 October 1846], DCP-LETT-1003.

260 "I have lately been," Darwin to Lindley [c. October 10, 1846], DCP-LETT-999.

262 "there never was a more futile theory," Darwin to Horner [August 17– September 7, 1846], DCP-LETT-993.

262 See Milne, "On the Parallel Roads of Lochaber."

262 "seismometer," Davison, Founders of Seismology, p. 43.

263 "You will, I fear, think me very obstinate," "staggered [him] in favor," and "I am not, however," Darwin to Milne [September 20, 1847], DCP-LETT-1120.

263 For an engaging account of this period of Darwin's life, see Stott, Darwin and the Barnacle. For a discussion of some common barnacles, see Ricketts and Calvin, Between Pacific Tides.

263 "Where does your father," quoted in Browne, Darwin's Origin of Species, p. 55.

264 Darwin's 1863 paper is "On the Thickness of the Pampean Formation."

264 "Mr. Arthobalanus," Darwin to Hooker [October 18, 1846], DCP-LETT-1015.

264 Darwin's books on the living barnacles are Darwin, A Monograph on the Sub-class Cirripedia, Vol. 1, The Lepadidae; or, pedunculated cirripedes, and Vol. 2, The Balanidae, (or sessile cirripedes); the Verrucidae; and on the fossil barnacles, Darwin, A monograph of the fossil Lepadidae, and Darwin, A Monograph on the Fossil Balanidae and Verrucidae.

265 "wretched digestion," Darwin to Fox [March 27, 1851], DCP-LETT-1396.

265 "Finished packing up," John van Wyhe ed., Darwin's "Journal" (1809– 1881). http://darwin-online.org.uk/content/frameset?itemID=CUL-DAR158&viewtype=text&pageseq=1. CUL-DAR158.1-76 (Darwin Online, http://darwin-online.org.uk/).

265 Darwin's paper is "On the power of icebergs."

265–266 For Darwin's experience with a wire rope and inquiring if anyone has experience with a "gutta percha," or rubber, bucket, see Darwin to Gardeners' Chronicle [before January 10, 1852], DCP-LETT-2127.

266 Darwin's receipt of the letter and manuscript from Wallace is covered in detail by Quammen, The Reluctant Mr. Darwin, p. 153ff., and Browne, Charles Darwin, The Power of Place, p. 14ff.

267 "The most remarkable thing," Quammen, The Reluctant Mr. Darwin, p. 163.

267 "abstract of [his] notions," Darwin to Eyton [August 4, 1858], DCP-LETT-2319.

267 "13 months & 10 days," John van Wyhe ed., Darwin's "Journal" (1809–1881). http://darwin-online.org.uk/content/frameset?itemID=CUL-DAR158&viewtype=text&pageseq=1. CUL-DAR158.1-76 (Darwin Online, http://darwin-online.org.uk/).

267–268 The troubling letter was Jamieson to Darwin [September 3, 1861], DCP-LETT-3242A.

268 "that these parallel roads," Ibid.

268 "I thank you sincerely," Ibid.

268 "I am smashed to atoms," Darwin to Lyell [September 6, 1861], DCP-LETT-3246.

271 Jamieson's paper is Jamieson, "Parallel Roads of Glen Roy." For a modern view, see Sissons, *The Evolution of Scotland's Scenary*, and Palmer et al., *Quaternary of Glen Roy*.

271 Darwin's distinctive boulders in Cwm Idwal are discussed in Roberts, "Buckland, Darwin," and Herbert, *Charles, Darwin, Geologist*, pp. 279–284. See also Addison, *Ice Age in Cwm Idwal*.

271 For the history of how the glaciation of the British Isles was finally accepted by British geologists, see the very readable accounts in Davies, *The Earth in Decay*.

CHAPTER FOURTEEN: FROM NATURAL SELECTION TO PLATE TECTONICS

273 "I found, on landing," FitzRoy, *Narrative*, Vol. 2, p. 365.

274 The account of the 1960 tsunami arriving at Guafo is from Sievers et al., "The Seismic Sea Wave of 22 May 1960 along the Chilean Coast."

275 "five or six times greater," Newton, *Principia*, p. 336.

275 "One can say," Smallwood, "Bouguer Redeemed," p. 14.

276 A rigorous, but quite readable account of the development of the concept of isostasy maybe found in Chapter 1 of Watts, *Isostasy and Flexure of the Lithosphere*.

276 Jamieson, "On the Cause of the Depression and Re-Elevation."

276–277 The original story is in Gilbert, *Lake Bonneville*. For more on this exceptional geologist, see Pyne, *Grove Karl Gilbert*. An important second look at the deformation of the shorelines, reinforcing Gilbert's view, is Crittenden (1963). Oviatt et al., (1992) treat the history of the lake and its relation to the changing climate.

277 Darwin's original estimate of three hundred million years included in the original 1859 edition of *Origin of Species*, p. 287, was little more than a back-of-the-envelope calculation based on the denudation of a valley, the Weald, in the south of England. He quickly came to regret this hasty estimate because it became an easy target for his critics. Darwin struggled with this issue in later editions of *Origin*. See Burchfield, *Lord Kelvin and the Age of the Earth*, p. 70ff., for this story. Burchfield also explores Kelvin's attempts to estimate the age of the earth. For an overview of the whole question of the age of the earth, including radioactivity and the developments leading to the modern estimate, see Lewis, *The Dating Game*, or Dalrymple, *The Age of the Earth*.

277 Lord Kelvin's work is Thomson, "On the Rigidity of the Earth."

278 "throughout a solid," G. H. Darwin, "On the Stresses," p. 219.

278 The development of seismology is discussed by Shearer, *Introduction to Seismology*, pp. 2–15, and Agnew, "History of Seismology." For the early development of instruments see Dewey and Byerly, "The Early History of Seismometry."

279 A brief but wonderful biography of Mohorovičić, including images of the seismograms and travel-time curves that he used to discover the Moho, his famous discontinuity, may be found at http://www.gfz.hr/sobe-en/andrija.htm.

279 "a line or narrow zone," California. State Earthquake Investigation Commission, *The California Earthquake of April 18, 1906*, Vol. 1, p. 25.

280 The elastic rebound theory for earthquakes is in Reid, "Mechanics of the Earthquake," Ibid., Vol. 2, pp. 16–28.

280 Suess, *The Face of the Earth*.

281 "continental drift," Wegener, *The Origin of Continents and Oceans*.

281–282 Du Toit's now famous work is *Our Wandering Continents*. Arthur Holmes wrote a textbook used in various editions by generations of students, including the author, Holmes, *Principles of Physical Geology*. The last chapter (which in the author's recollection was skipped over very lightly in his class during the era before plate tectonics) described Holmes's views about continental drift and convection currents in the earth's mantle. Notwithstanding his heretical views about continental drift, the standing joke for decades about competing textbooks—generally judged to be inferior—was "There is no place like Holmes." For much more about Holmes, see Lewis, *The Dating Game*. Reginald Daly was a long-time professor of geology at Harvard. In addition to supporting Wegener's ideas about continental drift, he also worked on the interrelated problems of coral reefs, Pleistocene glaciation, changing relative sea level, isostasy, and the rheological properties of the earth's mantle. See, for example, Watts, *Isostasy and Flexure of the Lithosphere*, pp. 34–36.

282 For Harold Jeffreys's views about continental drift, see *The Earth*. As a graduate student I spent many hours reading the fourth edition of this book. In retrospect I might have been considerably more imaginative in my selection of material for inspiration.

282 For a brief history of the Seismo Lab see http://www.seismolab.caltech.edu/ history.html.

282–283 The multiple editions of Gutenberg and Richter, *The Seismicity of the Earth*, from 1941 through 1965, provide a record of the progress in understanding the spatial distribution of earthquakes on the planet.

283 For Wadati-Benioff zones, see Shearer, *Introduction to Seismology*, pp. 6–8.

283 See http://earthquake.usgs.gov/data/crust/ for a very clear and modern view of how the thickness of the earth's crust varies around the planet.

283 The geologic origin of the 1960 earthquake was a huge puzzle at the time of the event. For a fascinating snapshot of the state of thinking about great earthquakes just before the dam broke leading to plate tectonics, see Allen, "Earthquakes and Mountains Around the Pacific," available at http:// calteches.library.caltech.edu/2197/1/allen.pdf. Also see Dobrovolny and Lemke, "Engineering geology and the Chilean earthquakes of 1960," Duke, "The Chilean Earthquakes of May 1960," and St. Amand, "South Chile Earthquake Swarm of 1960."

284 For a biography of George Plafker, see Fuis et al., "A Tribute."

284–285 Press and Jackson, "Alaskan earthquake [March 27, 1964]: Vertical extent of faulting and elastic strain energy release," and Plafker, "Tectonic deformation associated with the 1964 Alaska earthquake."

285–286   The efforts of seismologists to relate their observations of the waves from earthquakes to the actual motions on a fault are related in Agnew, "The History of Seismology," pp. 8–9. For the application to the 1964 earthquake, see Stauder and Bollinger, "The focal mechanism of the Alaska earthquake."

286   The meeting was the Annual Meeting of the AGU in Washington, D.C. The paper was Plafker, "Possible evidence for downward-directed mantle convection."

286   For the closing of the case, see Plafker, "Tectonic," Plafker and Savage, "Mechanism of the Chilean earthquakes," and "Alaskan Earthquake of 1964 and Chilean Earthquake of 1960."

287   F. J. Vine and D. H. Matthews, "Magnetic anomalies over oceanic ridges."

288   For a brief introduction to the modern theory, see Molnar, Plate Tectonics.

290   D'Orbigny bestowed the name Petricola patagonica on this little mollusk in the encylopedic ten-volume description of his voyage to South America and the flora and fauna he found, d'Orbigny, Voyage dans l'Amérique Méridionale, vol. 5, pp. 547–548. (Available at http://www.biodiversitylibrary.org/ bibliography/85973.) D'Orbigny's collections and descriptions of them were extraordinary. Darwin was justifiably concerned, when he wrote Henslow upon learning that d'Orbigny had preceded him in Patagonia, "I am very selfishly afraid he will get the cream of all the good things, before me," Darwin to Henslow [c. October 26–November 24, 1832], DCP-LETT-192.

290   Fortesan and the Chilean nutritional program are discussed in Shurtleff and Aoyagi, History of Soy Flour, pp. 1064, 1178, and Monckeberg and Chichester, "Chilean Experience with Fortified Children's Formulas."

294   The chart is Officers of the HMS Beagle, Santa Maria Island.

295   A very good collection of "facts" was beginning to emerge, indeed. See Melnick et al., "Present and past land-level changes" and "Transient evolution of plate coupling."

CHAPTER FIFTEEN: IN THE BEAGLE'S WAKE

297   "The most remarkable effect," Darwin, Narrative, Vol. 3, p. 379.

297   "It will require no great expenditure of time," FitzRoy, Narrative, Vol. 2, p. 31.

297   "A dangerous coast," Beagle Diary, p. 293.

297   "that the anchor had been accidentally let go," FitzRoy, Narrative, Vol. 2, p. 402.

298   The difficult story of the relations between the Mapuche and the Spanish, then the colonists, and finally the Republic of Chile is told in Clissold, Bernardo O'Higgins, and, Rector, The History of Chile. This sad story continues to the present day.

298   For Sir Francis Drake, see Bawalf, Secret Voyage, pp. 124–125.

298   "I reflected on the multiplied sufferings," FitzRoy, Narrative, Vol. 2, p. 399.

298   "Probably the finest district," FitzRoy, Narrative, Vol. 2, p. 400.

299   Previously, Marco found evidence of great earthquakes and tsunamis preceding 1960, Cisternas, "Predecessors of the Giant 1960 Chile Earthquake."

303 See Burbank and Anderson, *Tectonic Geomorphology*, pp. 33–52, for a discussion of carbon-14 dating and other methods currently in use to date relatively recent geologic features.

305 *"El Salto Gigante"* [The Great Leap], see http://www.talcahuano.cl/alcalde/, for background on the mayor.

305 Marco Cisternas was quoted in the paper, *El Sur*, April 19, 2009, p. 5.

310 Marco Cisternas interviewed Gabriel Gutiérrez Vargas on Isla Santa María, January 12, 2010.

315 For a discussion of seismogenic zones, see Scholz, *The Mechanics of Earthquakes and Faulting*, p. 152. For other discussion of giant subduction earthquakes, see Satake and Atwater, "Long-Term Perspectives on Giant Earthquakes."

CHAPTER SIXTEEN: THE CHILEAN EARTHQUAKE OF 2010

319 "I feel it is quite impossible," Beagle *Diary*, p. 302.

319–320 The story of Constanza, her grandmother, Elena Rosa Riquelme Carrillo, her father, Joel Castro Riquelme, and Joel's tenants, José Luis Montero Bernales, and his son, Rodrigo, is based on interviews by Marco Cisternas and Zamara Fuentes in Tirúa on August 26 and 27, 2010.

320 Geologic and geophysical details of the earthquake may be found in Madariaga et al., "Central Chile Finally Breaks," Vigny et al., "The 2010 M-W 8.8 Maule Megathrust Earthquake," and Moreno et al., "Toward Understanding Tectonic Control."

320 The on-again, off-again local tsunami warning is a sad and embarrassing episode for the proud country of Chile, not unlike the *Challenger* disaster or the Galveston hurricane and flood of 1900 for the U.S. Nonetheless the story remains an instructive example of a social and political system overwhelmed by a natural reality. The facts remain confused and in dispute. They have been the subject of investigation and litigation. The account here relies on Dengler et al., "Factors That Exacerbated or Reduced Impacts," and Soulé, "Post-Crisis Analysis of an Ineffective Tsunami Alert." The legal consequences are discussed in Pallardy and Rafferty, "Chile earthquake of 2010," *Encyclopedia Britannica*, https://www.britannica.com/event/Chile-earthquake-of-2010.

324 Marco Cisternas interviewed Josefina Rosario Astorga Lagos in Tirúa on August 26, 2010.

325 Marco Cisternas interviewed Luis Bravo Torres in Dichato on August 23, 2010.

326–327 The stories of Osvaldo González, Osvaldo Gómez, and Mario Leal, including quotes, are from Juan Andres Guzman, "I saw wave, began to run," *Miami Herald*, March 2, 2010, p. 1A. See also *McClatchy DC*, March 2, 2010, http://www.mcclatchydc.com/news/nation-world/world/article24575215.html.

327 Doña Pilar Bermúdez. This very sad account is assembled from "Karen jamás hubiera dejado sola a la abuelita que cuidaba," *Diario El Sur*, March 27, 2010, http://diarioelsur.cl/base_elsur/site/artic/20100327/pags/20100327004500.html; Sonnia Mendoza, "Historias para no olvidar: 27/F, tres minutos en el infierno," *Revista nos*, February, 2011, http://www.revistanos.cl/2011/02/

historias-para-no-olvidar-27f-tres-minutos-en-el-infierno/; and "Funcionarios rinden emotivo homenaje a destacada y querida funcionaria Pilar Bermúdez," https://www.facebook.com/notes/i-municipalidad-de-talcahuano/ funcionarios-rinden-emotivo-homenaje-a-destacada-y-querida-funcionaria-pilar-ber/10150112580527692/.

333–334 Daniel and Marco also measured the uplift along the coast from the elevated positions of intertidal mussels, just as FitzRoy had done, Melnick et al., "Estimating Coseismic Coastal Uplift."

334–335 For a discussion of the concept of seismic gap, see Scholz, *The Mechanics of Earthquakes and Faulting*, pp. 244, 284–7.

335 "Concepción–Constitución area (35–37° S)" and "Therefore, in a worst case scenario," Ruegg et al., "Interseismic strain accumulation." See also Moreno et al., "2010 Maule earthquake."

336 This list, "Informe Final de Fallecidos y Desaparecidos Por Comunas" dated January 31, 2011, was posted on the website of the Subsecretaria del Interior de la Republica de Chile, but has since been removed.

337 "The Calystegia seed," Puseman et al., "Identification and AMS."

337 The results of our work are in Ely et al., "Five centuries of tsunamis."

337 The results of our surveys at Santa María are Wesson et al., "Vertical deformation."

CHAPTER SEVENTEEN: NOW YOU SEE THEM, NOW YOU DON'T

341 "All that at present can be said with certainty," Darwin, *Narrative*, Vol. 3, p. 212.

341 Fernicola et al., "The Fossil Mammals."

342 Vizcaíno et al., "Young Darwin."

343 "This wonderful relationship," Darwin, *Journal of Researches*, p. 173.

343 "They also illustrate," Simpson, "The Beginning of the Age of Mammals in South America," p. 11. See also Simpson, "Splendid Isolation."

343–344 For more information on the Cueva de Milodón, see Fariña, *Megafauna*, pp. 166–167, and Brandoni et al. (2010). See Martin et al., "Land of the Ground Sloths," for an up to date review and recent research near the site. The excitement after the initial discovery, urged on by the British press baron, Cyril Arthur Pearson, was the impetus for the expedition in 1900 described in Prichard (later Hesketh-Prichard), *Through the Heart of Patagonia*, and reprised by his great-grandson, Charlie Jacoby, a century later (see Daily Mail, February 8, 2001, http://www.charliejacoby.com/dailyexpress080201.htm).

344 The skin of the *Mylodon* was described by Woodward—complete with photographs—in Pritchard, *Through the Heart of Patagonia*, pp. 305–314.

344–345 For more on the North American ground sloths, see Lange, *Ice Age Mammals of North America* and Kurtén and Anderson, *Pleistocene Mammals of North America*.

345 For the persistence of the small ground sloths on Caribbean islands, see Steadman et al., "Asynchronous Extinction," and MacPhee et al., "Prehistoric Sloth Extinction."

346–347   For the Great American Biotic Interchange, see Stehli and Webb, *Great American Biotic Interchange* and Cione et al., *Great American Biotic Interchange: A South American perspective*. Recently arguments have been advanced that the Isthmus of Panama closed ten million years ago or earlier and that the triggering of Interchange was related to climatic changes associated with the glaciation, see Bacon et al., "Quaternary Glaciation." Proponents of the classical view are fighting back hard, concluding that the current evidence continues to support the formation around 2.8 million years ago leading to the Interchange, see O'Dea et al., "Formation of the Isthmus of Panama."

347   The triggering of the Pleistocene is discussed in Haug and Tiedmann, "Effect of the Formation of the Isthmus of Panama" and Bartoli et al., "Final Closure of Panama."

347   The story of unraveling of the ice ages is told in Imbrie and Imbrie, *Ice Ages*.

347   The change in the period of the oscillations is discussed in Head and Gibbard, "Early-Middle Pleistocene transitions."

348–349   For the Milankovitch cycles, see Imbrie and Imbrie, *Ice Ages*.

349   Remaining questions are discussed in Muller and MacDonald, *Ice Ages and Astronomical Causes*.

350   For Rampart Cave, see Justin Tweet, "Rampart Cave," http://nature.nps.gov/geology/nationalfossilday/2016_rampart.cfm, 2016. See also Long et al., "Extinction of the Shasta Ground Sloth."

351   For the K-T boundary event, see Frankel, *The End of the Dinosaurs*, and Alvarez, *T. Rex and the Crater of Doom*.

351–352   For the extinction involving the ground sloths, see Haynes, *American Megafaunal Extinctions*.

352   "remains of several gigantic animals," Darwin, *The Zoology of the Voyage*, p. 7.

352   "Of these shells it is almost certain," Ibid., p. 9. For a modern view, see Quattrocchio et al., "Geology of the Area of Bahía Blanca," and Vizcaíno et al., "Young Darwin."

353   Current understanding of the evolution of the genus *Homo* is discussed in Grine et al., *The First Humans*, and McHenry, "Human Evolution." The exact dates for the events listed here before about 15,000 years ago are the subject of very active research and vary among authors and seem to be subject to rapid change. The dates for *Homo sapiens* acquiring culture is from Klein and Edgar, *Dawn of Human Culture*; for the arrival in Europe from Bosch et al., "New Chronology;" and that for the extinction of the Neanderthals, from Higham, et al, "The Timing and Spatiotemporal Patterning."

353   The longstanding controversies and recent developments through about 2008 are reviewed by Meltzer, *First Peoples in a New World*.

354   Beringia and its role in the peopling of the Americas, including discussion of the oldest tools found in Alaska, are reviewed in Hoffecker et al. "Beringia and the Global Dispersal of Humans." The skeletal remains of a child dated to 11,500 years ago are described in Potter, "A Terminal Pleistocene Child Cremation." A

fascinating animated map of Beringia as the glaciers melted and sea level rose is at https://www.ncdc.noaa.gov/paleo/parcs/atlas/beringia/lbridge.html.

354   For Anzick-1, see Rasmussen et al., "The genome of a Late Pleistocene human." For Naia, see Chatters et al., "Late Pleistocene human skeleton."

354   Monte Verde in Chile. The detailed story of Monte Verde is spelled out in Dillehay, *Monte Verde*. Dillehay et al., "Monte Verde: Seaweed," provide a recent update.

355   For the thighbone of the *Megalonyx*, see Redmond et al. (2012).

355   See Tonni et al., "The Broken Zig-Zag." The late Paul Martin of the University of Arizona was the leading advocate for the notion of "overkill" by the first humans in the Americas, Martin, *The Twilight of the Mammoths*. See also Koch and Barnosky, "Late Quaternary Extinctions," and Barnosky and Lindsey, "Timing of Quaternary Megafaunal Extinction." See Meltzer, "Pleistocene Overkill," for a technical review of the whole controversy. For a highly readable account, although perhaps not reflecting the full extent of the ongoing debate, see Kolbert, *The Sixth Extinction*.

355   *Dryas octopetala*. See for example, http://www.fs.fed.us/wildflowers/plant-of-the-week/dryas_octopetala.shtml

356   For Younger Dryas, see Carlson, "The Younger Dryas Climate Event."

356   For the cause, see Carlson, "What Caused the Younger Dryas Cold Event?"

357   For population at the beginning of the Holocene, see for example Goldewijk et al., "Long-Term Dynamic Modeling of Global Population."

357   See "Most populous large mammal on earth," http://www.worldatlas.com/articles/most-populous-mammals-on-earth.html; FAO, *FAO Statistical Pocketbook*; and Lukefahr and Cheeke, "Rabbit Project Development Strategies."

358   For examples of "natural environment" that has actually been significantly modified by humans, see Roberts, *The Holocene*, and Rackham, *Woodlands*.

358   William Ruddiman, *Plows, Plagues and Petroleum*.

358   For our human footprint see Venter et al., "Sixteen years of change."

CHAPTER EIGHTEEN: REFLECTIONS: WHAT DOES IT ALL MEAN?

361   "Nothing can be more improving," Darwin, *Narrative*, Vol. 3, p. 607.

361   Darwin's geologic map of southern South America is discussed in Zappettini and Mendía, "The First Geological Map of Patagonia," and Dott and Dalziel, "Darwin the Geologist in South America."

362   For the history of thought about the luminiferous aether, see Swenson, *The Ethereal Aether*; also Isaacson, *Einstein*.

362   For Brownian motion, see Isaacson, *Einstein*.

363   For a more detailed modern assessment of Darwin's geological work in southern South America, including his map of the bedrock geology not treated here, see Dott and Dalziel, "Darwin the Geologist in South America." Also see Mills, "Darwin and the Iceberg Theory."

363   The battle for acceptance of the former ice ages is described in Davies, *The Earth in Decay*; Imbrie and Imbrie, *Ice Ages*; and Bolles, *The Ice Finders*.

364 The bitter controversy about Paleolithic humans in the Americas is described by Meltzer, *The Great Paleolithic War.*

365 "Darwin's canonical model," Toomey et al., "Profiles of Ocean Island Coral Reefs." Also see Terry and Goff, "One Hundred and Thirty Years" for another modern view.

365 A relation between the uplift of terraces along the Chilean coast from earthquakes has been demonstrated by Melnick, "Rise of the Central Andean Coast."

365 The debates leading to plate tectonics are discussed in Oreskes, *Rejection of Continental Drift*, Oreskes and LeGrand, *Plate Tectonics*, and Menard, *Ocean of Truth.*

366 Steve Austin's poster is Austin and Strelin, "Megafloods," https://gsa.confex.com/gsa/2011AM/webprogram/Paper191286.html.

366 "exists to conduct scientific research," http://www.icr.org/who-we-are.

366 "believers with evidence," http://www.icr.org/zlp-gaw/bible-creation-science-003/.

367 "bogus methodology" and "the lapse of great geologic ages," Austin, "Darwin's First Wrong Turn," http://www.icr.org/articles/view/4346.

368 For Julia Jones's results, see Swanson et al., "Effects of Volcanic and Hydrologic Processes."

368 The previous eruption of Chaitén was noted by Naranjo and Stern, "Holocene Tephrochronology."

369–370 The eruption sequence of Chaitén and its impacts are reported in Lara, "The 2008 eruption of the Chaitén Volcano," and Major and Lara, "Overview of Chaitén Volcano."

370 See Sawai et al., "Challenges of anticipating the 2011 Tohoku earthquake."

371 For the voicing of concerns about earthquake and tsunami hazards at the site of the power plant by Okamura and Shimazaki, see Willacy, *Fukushima*; Bricker, *The Fukushima Daiichi Nuclear Power Station Disaster*, pp. 104–106; and "State ignored predictions 10 years before 3/11 tsunami, says seismologist," *Japan Times*, March 23, 2016.

371 "They completely ignored me," Fackler, "Nuclear Disaster in Japan Was Avoidable, Critics Contend," *New York Times*, March 9, 2012.

371–372 The size of these eruptions is rated by the Volcanic Explosivity Index (VEI) as listed by Global Volcanism Program, 2013. Volcanoes of the World, v. 4.5.1. Venzke, E (ed.). Smithsonian Institution. http://dx.doi.org/10.5479/si.GVP.VOTW4-2013.

371–372 See "Largest earthquakes in the world since 1900," https://earthquake.usgs.gov/earthquakes/world/10_largest_world.php.

373 "It may well be that most species," Raup, "The role of extinction in evolution."

373 "Almost all species," Raup, *Extinction*, pp. 4–5.

373 The NASA and European Space Agency's programs to track "near earth objects" are described at http://neo.jpl.nasa.gov/ and http://www.esa.int/Our_Activities/Operations/Space_Situational_Awareness/Near-Earth_Objects_-_NEO_Segment.

373–374   Abrupt climate changes and research into their causes is described in National Research Council, *Abrupt Climate Change*, and Rashid et al., *Abrupt Climate Change*. O'Leary et al., "Ice Sheet Collapse," offer a scenario for the events leading to the collapse of the ice sheets during the last interglacial.

375   "collective wisdom," Bostrom, "Existential Risk Prevention." Also see Rees, *Our Final Hour*.

375   "At the heart of liberty," U.S. Supreme Court, Planned Parenthood v. Casey.

375   "a burst of nuclear decay" and "week of creation," DeYoung, *Thousands . . . Not Billions*," p.137.

376   For a discussion of how the Anglican Church has dealt with geologic time, see Roberts, "Genesis Chapter 1 and Geologic Time."

376   See Newport, "In U.S., 42% believe creationist view of human origins," Gallup, June 2, 2014, www.gallup.com/poll/170822/believe-creationist -view-human-origins.aspx.

378   FitzRoy's role in founding the Meteorological Office and the subsequent controversies is described in Walker, *History of the Meteorological Office*, p. 23ff.

381   "Do you believe," McDermott to Darwin [November 23, 1880], DCP-LETT-12845.

381   "Dear Sir, I am sorry to inform you," Darwin to McDermott [November 24, 1880], DCP-LETT-12851. The original of this letter was auctioned by Bonhams in New York on September 21, 2015, for $197,000, reportedly a world record price for a Darwin letter, https://www.bonhams.com/ press_release/20025/.

381–382   "Man in his arrogance," Barrett et al., *Charles Darwin's Notebooks*, p. 300.

382   "It would have been unfortunate," quoted in Nelson, *Darwin, Then and Now*, p. 64.

382–383   Interview with Don José Luis by Marco Cisternas and the author in Lipimávida, Chile, January 13, 2015.

Bibliography

The Cambridge Guide . . . A New Edition, Etc. Cambridge, England: J. & J. Deighton, 1837.

Addison, Kenneth. *Ice Age in Cwm Idwal.* Broseley, Shropshire, England: K & MK Addison, 1988.

Agassiz, Louis. "Upon Glaciers, Moraines, and Erratic Blocks." *Edinburgh New Philosophical Journal* 24 (1838): 364–83.

———. "On Glaciers, and the Evidence of Their Having Once Existed in Scotland, Ireland, and England." *Proceedings of the Geological Society of London,* no. 72 (1840): 327–32.

Agassiz, Louis, and Albert V. Carozzi. *Studies on Glaciers: Preceded by the Discourse of Neuchâtel.* New York: Hafner, 1967.

Agnew, Duncan Carr. "History of Seismology." *International Handbook of Earthquake and Engineering Seismology* 81 (2002): 3–11.

Allen, Clarence R. "Earthquakes and Mountains around the Pacific." *Engineering and Science* 26, no. 4 (1963): 24–8.

Alvarez, Walter. *T. Rex and the Crater of Doom.* Princeton, N.J.: Princeton University Press, 1997.

Ambrose, Stephen E. *Undaunted Courage: Meriwether Lewis, Thomas Jefferson, and the Opening of the American West.* New York: Simon & Schuster, 1997.

Anson, George, and Richard Walter. *A Voyage Round the World in the Years 1740–44.* London: John and Paul Knapton, 1769.

Aramayo, Silvia A, and Teresa Manera de Bianco. "Edad y Nuevos Hallazgos de Icnitas de Mamíferos y Aves en el Yacimiento Paleoicnológico de Pehuen Có (Pleistoceno Tardío), Provincia de Buenos Aires, Argentina." [In Spanish]. *Asociación Paleontológica Argentina, Publicación Especial* 4, no. 1 (1996): 47–57.

Armstrong, Patrick. *Under the Blue Vault of Heaven: A Study of Charles Darwin's Sojourn in the Cocos (Keeling) Islands.* Perth: Indian Ocean Centre for Peace Studies, 1991.

———. *Darwin's Other Islands.* London; New York: Continuum, 2004.

Austin, Steven A. *Grand Canyon: Monument to Catastrophe.* Dallas: Institute for Creation Research, 1995.

Austin, Steven A, and Jorge A Strelin. "Megafloods on the Santa Cruz River, Southern Argentina." Paper presented at the 2011 GSA Annual Meeting in Minneapolis, 2011.

Babbage, Charles. *The Ninth Bridgewater Treatise. A Fragment.* second ed. London: J. Murray, 1838.

———. *Observations on the Temple of Serapis, at Pozzuoli, near Naples, with an Attempt to Explain the Causes of the Frequent Elevation and Depression of Large Portions of the Earth's Surface in Remote Periods; and to Prove That Those Causes Continue in Action. With a Supplement. Conjectures on the Physical Condition of the Surface of the Moon. [with Two Plates].* London: Privately printed, 1847.

Bacon, Christine D., Peter Molnar, Alexandre Antonelli, Andrew J. Crawford, Camilo Montes, and Maria Camila Vallejo-Pareja. "Quaternary Glaciation and the Great American Biotic Interchange." *Geology* 44, no. 5 (2016): 375–78.

Baily, Francis. *Astronomical Tables and Formulæ Together with a Variety of Problems Explanatory of Their Use and Application.* London: Printed by R. Taylor, 1827.

Barnosky, Anthony D., and Emily L. Lindsey. "Timing of Quaternary Megafaunal Extinction in South America in Relation to Human Arrival and Climate Change." *Quaternary International* 217, no. 1 (2010): 10–29.

Barrett, Paul H. "The Sedgwick-Darwin Geologic Tour of North Wales." *Proceedings of the American Philosophical Society* 118, no. 2 (1974): 146–64.

Bartlett, John, and Emily Morison Beck. *Familiar Quotations: A Collection of Passages, Phrases, and Proverbs Traced to Their Sources in Ancient and Modern Literature.* 15th and 125th anniversary ed. Boston: Little Brown, 1980.

Bartoli, G., M. Sarnthein, M. Weinelt, H. Erlenkeuser, D. Garbe-Schönberg, and D. W. Lea. "Final Closure of Panama and the Onset of Northern Hemisphere Glaciation." *Earth and Planetary Science Letters* 237, no. 1 (2005): 33–44.

Bawlf, R. Samuel. *The Secret Voyage of Sir Francis Drake, 1577–1580.* Vancouver: Douglas & McIntyre, 2003.

Bayón, Cristina, Teresa Manera, Gustavo Politis, and Silvia Aramayo. "Following the Tracks of the First South Americans." *Evolution: Education and Outreach* 4, no. 2 (2011): 205–17.

Beechey, F. W. *Narrative of a Voyage to the Pacific and Beering's Strait, to Co-Operate with the Polar Expeditions: Performed in His Majesty's Ship Blossom, under the Command of Captain F. W. Beechey, R.N., F.R.S. &C. In the Years 1825, 26, 27, 28.* London: H. Colburn and R. Bentley, 1831.

Benazzi, Stefano, Katerina Douka, Cinzia Fornai, Catherine C. Bauer, Ottmar Kullmer, Jiří Svoboda, Ildikó Pap, et al. "Early Dispersal of Modern Humans in Europe and Implications for Neanderthal Behaviour." *Nature* 479, no. 7374 (2011): 525–28.

Bergreen, Laurence. *Over the Edge of the World: Magellan's Terrifying Circumnavigation of the Globe.* first ed. New York: William Morrow, 2003.

Bethell, Leslie. "Britain and Latin America in Historical Perspective." In *Britain and Latin America: A Changing Relationship,* edited by Victor Bulmer-Thomas, 1–24. Cambridge, England: Cambridge University Press, 1989.

Blewitt, Mary. *Surveys of the Seas: A Brief History of British Hydrography.* London: Macgibbon & Kee, 1957.

Bolles, Edmund Blair. *The Ice Finders: How a Poet, a Professor, and a Politician Discovered the Ice Age.* Washington: Counterpoint, 1999.

Bosch, Marjolein D., Marcello A. Mannino, Amy L. Prendergast, Tamsin C. O'Connell, Beatrice Demarchi, Sheila M. Taylor, Laura Niven, Johannes van der Plicht, and Jean-Jacques Hublin. "New Chronology for Ksâr 'Akil (Lebanon) Supports Levantine Route of Modern Human Dispersal into Europe." *Proceedings of the National Academy of Sciences* 112, no. 25 (2015): 7683–88.

Bostrom, Nick. "Existential Risk Prevention as Global Priority." *Global Policy* 4, no. 1 (2013): 15–31.

Bowditch, Nathaniel. *The New American Practical Navigator Being an Epitome of Navigation: Containing All the Tables Necessary to Be Used with the Nautical Almanac, in Determining the Latitude, and the Longitude by Lunar Observations, and Keeping a Complete Reckoning at Sea: Illustrated by Proper Rules and Examples: The Whole Exemplified in a Journal, Kept from Boston to Madeira.* fourth ed. New York: E. M. Blunt and Samuel A. Burtus, 1817.

Brandoni, Diego, Brenda S. Ferrero, and Ernesto Brunetto. "Mylodon Darwini Owen (Xenarthra, Mylodontinae) from the Late Pleistocene of Mesopotamia, Argentina, with Remarks on Individual Variability, Paleobiology, Paleobiogeography, and Paleoenvironment." *Journal of Vertebrate Paleontology* 30, no. 5 (2010): 1547–58.

Brea, Mariana, Analía E. Artabe, and Luis A. Spalletti. "Darwin Forest at Agua De La Zorra: The First in Situ Forest Discovered in South America by Darwin in 1835." *Revista de la Asociación Geológica Argentina* 64, no. 1 (2009): 21–31.

Bricker, Mindy Kay. *Fukushima Daiichi Nuclear Power Station Disaster: Investigating the Myth and Reality.* London and New York: Routledge, 2014.

Browne, Janet. *Charles Darwin: A Biography, Vol. 1—Voyaging.* Princeton, N.J.: Princeton University Press, 1996.

———. *Charles Darwin: A Biography, Vol. 2—The Power of Place.* Princeton, N.J.: Princeton University Press, 2003.

———. *Darwin's Origin of Species: A Biography.* Books That Changed the World. New York: Atlantic Monthly Press, 2006.

Buckland, William. "On the Glacia-Diluvial Phaenomena in Snowdonia and the Adjacent Parts of North Wales." *Proceedings of the Geological Society of London* 2, no. 84 (1841): 579–84.

Burbank, Douglas West, and Robert S. Anderson. *Tectonic Geomorphology.* Malden, Mass.: Blackwell Science, 2001.

Burchfield, Joe D. *Lord Kelvin and the Age of the Earth.* Chicago: University of Chicago Press, 1990.

Burek, Cynthia V. "The First Female Fellows and the Status of Women in the Geological Society of London." *Geological Society, London, Special Publications* 317, no. 1 (2009): 373–407.

Burney, James. *A Chronological History of the Discoveries in the South Sea or Pacific Ocean, Part II.* London: G. & W. Nicol, 1806.

Caldcleugh, Alexander. *Travels in South America, During the Years 1819-20-21; Containing an Account of the Present State of Brazil, Buenos Ayres, and Chile. [with Plates and Maps].* London: John Murray, 1825.

———. "Some Observations on the Elevation of the Strata on the Coast of Chili." *Proceedings of the Geological Society of London* 2 no.48 (1837): 444–6.

Calder, Nigel. *How to Read a Nautical Chart: A Complete Guide to Understanding and Using Electronic and Paper Charts.* second ed. Camden, Me.: International Marine/McGraw-Hill, 2012.

California. State Earthquake Investigation Commission, Andrew C. Lawson, and Harry Fielding Reid. *The California Earthquake of April 18, 1906. Report of the State Earthquake Investigation Commission.* Carnegie Institution of Washington Publication. Washington: Carnegie Institution of Washington, 1908.

Callcott, Maria Graham. *Three Months Passed in the Mountains East of Rome, During the Year 1819.* London: Longmont, Hurst, Rees, Orme, and Brown, 1821.

———. *A Letter to the President and Members of the Geological Society, in Answer to Certain Observations Contained in Mr. Greenough's Anniversary Address of 1834.* London: T. Brettell, 1834.

———. *Journal of a Residence in Chile, During the Year 1822.* London: Longmont, Hurst, Rees, Orme, and Brown; John Murray, 1824.

Carlson, Anders E. "What Caused the Younger Dryas Cold Event?" *Geology* 38, no. 4 (2010): 383–84.

———. "The Younger Dryas Climate Event." In *The Encyclopedia of Quaternary Science*, edited by S. A. Silas, 126–34. Amsterdam: Elsevier, 2013.

Casadío, Silvio, and Miguel Griffin. "Sedimentology and Paleontology of a Miocene Marine Succession First Noticed by Darwin at Puerto Deseado (Port Desire)." *Revista de la Asociación Geológica Argentina* (Feb. 1, 2009): 83–89.

Chambers, Robert. *Vestiges of the Natural History of Creation.* London: J. Churchill, 1844.

Charpentier, J. de. "Notice Sur La Cause Probable Du Transport Des Blocs Erratiques De La Suisse." *Annales des mines* 8 (1835): 219–36.

Chatters, James C, Douglas J. Kennett, Yemane Asmerom, Brian M. Kemp, Victor Polyak, Alberto Nava Blank, Patricia A. Beddows, et al. "Late Pleistocene Human Skeleton and Mtdna Link Paleoamericans and Modern Native Americans." *Science* 344, no. 6185 (2014): 750–54.

Chatwin, Bruce. *In Patagonia.* New York: Penguin Classics, 2003.

Chester, Sharon R. *A Wildlife Guide to Chile: Continental Chile, Chilean Antarctica, Easter Island, Juan Fernandez Archipelago.* first ed. Princeton, N.J.: Princeton University Press, 2008.

Cione, Alberto L., Germán M. Gasparini, Esteban Soibelzon, Leopoldo H. Soibelzon, and Eduardo P. Tonni. *The Great American Biotic Interchange: A South American Perspective.* Amsterdam: Springer Netherlands, 2015.

Cisternas, M., B. F. Atwater, F. Torrejón, Y. Sawai, G. Machuca, M. Lagos, A. Eipert, et al. "Predecessors of the Giant 1960 Chile Earthquake." *Nature* 437, no. 7057 (2005): 404–07.

Clissold, Stephen. *Bernardo O'Higgins and the Independence of Chile.* New York: Praeger, 1969.

Cordingly, David. *Cochrane: The Real Master and Commander.* New York: Bloomsbury USA, 2007.

Crittenden, Max D. "Effective Viscosity of the Earth Derived from Isostatic Loading of Pleistocene Lake Bonneville." *Journal of Geophysical Research* 68, no. 19 (1963): 5517–30.

Cutler, Alan. *The Seashell on the Mountaintop: A Story of Science, Sainthood, and the Humble Genius Who Discovered a New History of the Earth.* New York: Dutton, 2003.

d'Orbigny, Alcide Dessalines, Paul Gervais, Gabriel Bibron, Valenciennes, H. Milne-Edwards, Pierre Hippolyte Lucas, Émile Blanchard, et al. *Voyage Dans L'amérique Méridionale: (Le Brésil, La République Orientale De L'uruguay, La République Argentine, La Patagonie, La République Du Chili, La République De Bolivia, La République du Pérou), Exécuté Pendant ees Annés 1826, 1827, 1828, 1829, 1830, 1831, 1832, et 1833.* nine vols. Paris and Strasbourg: Pitois-Levrault; Ve. Levrault, 1835.

Dalrymple, G. Brent. *The Age of the Earth.* Stanford, Calif.: Stanford University Press, 1991.

Darwin, Charles. "Observations of Proofs of Recent Elevation on the Coast of Chili, Made During the Survey of His Majesty's Ship *Beagle,* Commanded by Capt. FitzRoy, Rn." *Proceedings of the Geological Society of London* 2, no.48 (1837): 446–49.

———. "On Certain Areas of Elevation and Subsidence in the Pacific and Indian Oceans, as Deduced from the Study of Coral Formations." *Proceedings of the Geological Society of London* 2, no. 51 (1837): 552–54.

———. "On the Connexion of Certain Volcanic Phenomena, and on the Formation of Mountain-Chains and Volcanos, as the Effects of Continental Elevations." *Proceedings of the Geological Society of London* 2, no. 56 (1838): 654–60.

———. *Narrative of the Surveying Voyages of His Majesty's Ships* Adventure *and* Beagle, *between the Years 1826 and 1836: Journal and Remarks, 1832–1836.* Vol. 3, London: Henry Colburn, 1839.

———. "Note on a Rock Seen on an Iceberg in 61 South Latitude." *The Journal of the Royal Geographical Society of London* 9 (1839): 528–29.

———. "Observations on the Parallel Roads of Glen Roy, and of Other Parts of Lochaber in Scotland, with an Attempt to Prove That They Are of Marine Origin." *Philosophical Transactions of the Royal Society of London* 129 (1839): 39–81.

———. "On the Connexion of Certain Volcanic Phenomena in South America; and on the Formation of Mountain Chains and Volcanos, as the Effect of the Same Power by Which Continents Are Elevated." *Transactions of the Geological Society of London,* no. 3 (1840): 601–31.

———. ed. *The Zoology of the Voyage of H.M.S.* Beagle *under the Command of Captain FitzRoy, R.N., During the Years 1832 to 1836, Part 1. Fossil Mammalia.* London: Smith, Elder and Co., 1840.

———. *The Structure and Distribution of Coral Reefs. Being the First Part of the Geology of the Voyage of the* Beagle, *under the Command of Capt. FitzRoy, R.N. During the Years 1832 to 1836.* London: Smith Elder and Co., 1842.

———. "Notes on the Effects Produced by the Ancient Glaciers of Caernarvonshire, and on the Boulders Transported by Floating Ice." *The London and Edinburgh Philosophical Magazine and Journal of Science* 21, no. 137 (1842): 180–88.

———. "On the Distribution of the Erratic Boulders and on the Contemporaneous Unstratified Deposits of South America." *Transactions of the Geological Society of London,* no. 2 (1842): 415–31.

——. *Geological Observations on the Volcanic Islands Visited During the Voyage of H.M.S. Beagle, Together with Some Brief Notices of the Geology of Australia and the Cape of Good Hope. Being the Second Part of the Geology of the Voyage of the Beagle, under the Command of Capt. FitzRoy, R.N. During the Years 1832 to 1836.* London: Smith Elder and Co., 1844.

——. *Journal of Researches into the Geology and Natural History of the Various Countries Visited by H.M.S. Beagle, 1832–1836.* Second edition, corrected, with additions. ed. London: John Murray, 1845.

——. *Geological Observations on South America, Being the Third Part of the Geology of the Voyage of the Beagle, under the Command of Capt. FitzRoy, R.N., During the Years 1832 to 1836.* London: Smith, Elder and Co., 1846.

——. *A Monograph on the Sub-Class Cirripedia : With Figures of All the Species, Vol. 1, the Lepadidae; of, Pedunculated Cirripedes.* London: Ray Society, 1851.

——. *A Monograph of the Fossil Lepadidae, or, Pedunculated Cirripedes of Great Britain.* London: Palæontographical Society, 1851.

——. *A Monograph on the Sub-Class Cirripedia : With Figures of All the Species, Vol. 2, the Balanidae, (or Sessile Cirripedies); the Verrucidae.* London: Ray Society, 1854.

——. *A Monograph on the Fossil Balanidae and Verrucidae of Great Britain.* London: Palæontographical Society, 1854.

——. "On the Power of Icebergs to Make Rectilinear, Uniformly-Directed Grooves across a Submarine Undulatory Surface." *The London, Edinburgh, and Dublin Philosophical Magazine and Journal of Science* 10, no. 64 (1855): 96–98.

——. *On the Origin of Species by Means of Natural Selection, or the Preservation of Favoured Races in the Struggle for Life.* London: Murray, 1859.

——. "On the Thickness of the Pampean Formation, near Buenos Ayres." *Quarterly Journal of the Geological Society* 19, no. 1-2 (1863): 68–71.

Darwin, Charles, and Nora Barlow, ed. *The Autobiography of Charles Darwin, 1809–1882.* New York: W.W. Norton & Company, 1958.

Darwin, Charles, Paul H. Barrett, Peter J. Gautrey, Sandra Herbert, and Sydney Smith. *Charles Darwin's Notebooks, 1836–1844: Geology, Transmutation of Species, Metaphysical Enquiries.* London and Ithaca, N.Y.: British Museum (Natural History) and Cornell University Press, 1987.

Darwin, Charles, Frederick Burkhardt, and Conrad Martens. *The Beagle Letters.* Cambridge, England: Cambridge University Press, 2008.

Darwin, Charles, Frederick Burkhardt, and Sydney Smith. *The Correspondence of Charles Darwin: 1821–1860.* Cambridge, England: Cambridge University Press, 1985–2009.

Darwin, Charles, G. R. Chancellor, John van Wyhe, and K. Rookmaaker. *Charles Darwin's Notebooks from the Voyage of the Beagle.* Cambridge, England; New York: Cambridge University Press, 2009.

Darwin, Charles, and Francis Darwin. *The Life and Letters of Charles Darwin.* London: John Murray, 1887.

Darwin, Charles, and Sandra Herbert, ed. *The Red Notebook of Charles Darwin.* London and Ithaca, N.Y.: British Museum (Natural History) and Cornell University Press, 1980.

Darwin, Charles, and R. D. Keynes. *Charles Darwin's Beagle Diary.* Cambridge, England and New York: Cambridge University Press, 1988.

Darwin, Charles, and John van Wyhe. *Charles Darwin's Shorter Publications, 1829–1883.* Cambridge, England and New York: Cambridge University Press, 2009.

Darwin, George H. "On the Stresses Caused in the Interior of the Earth by the Weight of Continents and Mountains." *Philosophical Transactions of the Royal Society of London* 173 (1882): 187–230.

Daubeny, Charles. *A Description of Active and Extinct Volcanos: With Remarks on Their Origin, Their Chemical Phænomena, and the Character of Their Products: Being the Substance of Some Lectures Delivered before the University of Oxford, with Much Additional Matter.* London: W. Phillips; Oxford, England: Joseph Parker, 1826.

Davies, Gordon. L. *The Earth in Decay: A History of British Geomorphology, 1578–1878.* New York: American Elsevier Publishing, Inc., 1969.

Davison, Charles. *The Founders of Seismology.* Cambridge, England: The University Press, 1927.

De La Beche, Henry T., and Edward Hitchcock. *Researches in Theoretical Geology.* New York: F. J. Huntington & Co., 1837.

Delano, Amasa. *A Narrative of Voyages and Travels in the Northern and Southern Hemispheres: Comprising Three Voyages Round the World.* Boston: Printed by E. G. House, for the author, 1817.

Dengler, Lori, Sebastian Araya, Nicholas Graehl, Francisco Luna, and Troy Nicolini. "Factors That Exacerbated or Reduced Impacts of the 27 February 2010 Chile Tsunami." *Earthquake Spectra* 28, no. S1 (2012): S199–S213.

Dewey, James, and Perry Byerly. "The Early History of Seismometry (to 1900)." *Bulletin of the Seismological Society of America* 59, no. 1 (1969): 183–227.

DeYoung, Donald B. *Thousands, Not Billions: Challenging an Icon of Evolution; Questioning the Age of the Earth.* Green Forest, Ark.: Master Books, 2005.

Dillehay, Tom D. *Monte Verde: A Late Pleistocene Settlement in Chile.* Smithsonian Series in Archaeological Inquiry. two vols. Washington: Smithsonian Institution Press, 1989.

Dillehay, Tom D., Carlos Ramirez, Mario Pino, Michael B. Collins, Jack Rossen, and J. D. Pino-Navarro. "Monte Verde: Seaweed, Food, Medicine, and the Peopling of South America." *Science* 320, no. 5877 (2008): 784–86.

Dobbs, David. *Reef Madness: Charles Darwin, Alexander Agassiz, and the Meaning of Coral.* first ed. New York: Pantheon, 2005.

Dobrovolny, Ernest, and R. W. Lemke. "Engineering Geology and the Chilean Earthquakes of 1960." *Short papers in the geologic and hydrologic sciences: US Geol. Survey Prof. Paper* 424 (1961): C357–C59.

Dott Jr., Robert H., and Ian W. D. Dalziel. "Darwin the Geologist in Southern South America." *Earth Sciences History* 35, no. 2 (Feb. 2016): 303–45.

Du Toit, Alexander L. *Our Wandering Continents: An Hypothesis of Continental Drifting.* Edinburgh and London: Oliver and Boyd, 1937.

Duke, C. Martin. "The Chilean Earthquakes of May 1960." *Science* 132, no. 3442 (1960): 1797–802.

Dvorak, John J, and Giuseppe Mastrolorenzo. *The Mechanisms of Recent Vertical Crustal Movements in Campi Flegrei Caldera, Southern Italy.* Geological Society of America Special Papers. Vol. 263, Boulder, Colorado: Geological Society of America, 1991.

Ekman, Martin. "A Concise History of Postglacial Land Uplift Research (from Its Beginning to 1950)." *Terra Nova* 3, no. 4 (1991): 358–65.

———. *The Changing Level of the Baltic Sea During 300 Years: A Clue to Understanding the Earth.* Summer Institute for Historical Geophysics Åland Islands, 2009.

Elliott, J. M. *The Nautical Almanac and Astronomical Ephemeris, for the Year 1835: Published by Order of the Lords Commissioners of the Admiralty.* Published by Order of the Lords Commissioners of the Admiralty. 1834.

Ely, Lisa L., Marco Cisternas, Robert L. Wesson, and Tina Dura. "Five Centuries of Tsunamis and Land-Level Changes in the Overlapping Rupture Area of the 1960 and 2010 Chilean Earthquakes." *Geology* 42, no. 11 (2014): 995–98.

Evans, Arthur V., and C. L. Bellamy. *An Inordinate Fondness for Beetles.* A Henry Holt Reference Book. first ed. New York: Henry Holt and Company, 1996.

Evenson, Edward B., Patrick A. Burkhart, John C. Gosse, Gregory S. Baker, Dan Jackofsky, Andres Meglioli, Ian Dalziel, *et al.* "Enigmatic Boulder Trains, Supraglacial Rock Avalanches, and the Origin of 'Darwin's Boulders,' Tierra Del Fuego." *GSA Today* 19, no. 12 (2009): 4-10.

Everhart, Michael J. *Oceans of Kansas: A Natural History of the Western Interior Sea.* Life of the Past. Bloomington: Indiana University Press, 2005.

Fara, Patricia. *Erasmus Darwin: Sex, Science, and Serendipity.* Oxford, England: Oxford University Press, 2012.

Fariña, Richard A., Sergio F. Vizcaíno, and Gerardo De Iuliis. *Megafauna: Giant Beasts of Pleistocene South America.* Bloomington: Indiana University Press, 2013.

Farinati, Ester Amanda, Teresa Manera, and Rodrigo L Tomassini. "La Bahía Que Iluminó a Darwin." *Revista Española de Paleontología* 25, no. 1 (2010): 35–41.

Fernicola, Juan Carlos, José I. Cuitiño, Sergio F. Vizcaíno, M. Susana Bargo, and Richard F. Kay. "Fossil Localities of the Santa Cruz Formation (Early Miocene, Patagonia, Argentina) Prospected by Carlos Ameghino in 1887 Revisited and the Location of the Notohippidian." *Journal of South American Earth Sciences* 52 (2014): 94–107.

Fernicola, Juan Carlos, Sergio F. Vizcaíno, and Gerry de Iuliis. "The Fossil Mammals Collected by Charles Darwin in South America During His Travels on Board the HMS *Beagle.*" *Revista de la Asociación Geológica Argentina* 64, no. 1 (Feb. 1, 2009): 147–59.

Fernicola, Juan Carlos, Sergio F. Vizcaíno, and Richard A. Fariña. "The Evolution of Armored Xenarthrans and a Phylogeny of the Glyptodonts." In *The Biology of the Xenarthra,* edited by Sergio F. Vizcaíno and W. J. Loughry, 79–85. Gainesville: University Press of Florida, 2008.

Ferreiro, Larrie D. *Measure of the Earth: The Enlightenment Expedition That Reshaped Our World.* New York: Basic Books, 2011.

Feruglio, Egidio. *Descripción Geológica de la Patagonia.* three vols. Buenos Aires: Impr. y Casa Editora "Coni," 1949.

FitzRoy, Robert. *Narrative of the Surveying Voyages of His Majesty's Ships Adventure and Beagle, between the Years 1826 and 1836, Proceedings of the Second Expedition, 1831–1836.* Vol. 2 and Appendix: Henry Colburn, 1839.

Food and Agriculture Organization of the United Nations. *FAO Statistical Pocketbook.* Rome: FAO Statistical Yearbook, 2015.

Fox, Caroline. *Memories of Old Friends—Being Extracts from the Journals and Letters of Caroline Fox of Penjerrick, Cornwall from 1835 to 1871.* Horace N. Pym ed. Philadelphia: J. B. Lippincott & Co., 1884.

Frankel, Charles. *The End of the Dinosaurs: Chicxulub Crater and Mass Extinctions.* Cambridge, England and New York: Cambridge University Press, 1999.

Freeman, R. B. *Charles Darwin: A Companion.* Hamden, Conn.: Archon, 1978.

Fuis, Gary S., Peter J. Haeussler, and Brian F. Atwater. "A Tribute to George Plafker." *Quaternary Science Reviews* 113 (2015): 3–7.

Giambiagi, Laura B., Victor A. Ramos, Estanislao Godoy, P. Pamela Alvarez, and Sergio Orts. "Cenozoic Deformation and Tectonic Style of the Andes, between 33 and 34 South Latitude." *Tectonics* 22, no. 4 (2003).

Giambiagi, Laura B., Maisa Tunik, Victor Ramos, A., and Estanislao Godoy. "The High Andean Cordillera of Central Argentina and Chile Along the Piuquenes Pass-Cordón Del Portillo Transect: Darwin's Pioneering Observations Compared with Modern Geology." *Revista de la Asociación Geológica Argentina* 64, no. 1 (2009): 43–54.

Gilbert, G. K. *Lake Bonneville.* U.S. Geological Survey Monograph. Washington: U.S. Govt. Printing Office, 1890.

Goldewijk, Kees Klein, Arthur Beusen, and Peter Janssen. "Long-Term Dynamic Modeling of Global Population and Built-up Area in a Spatially Explicit Way: Hyde 3.1." *The Holocene* (2010).

Gould, Stephen Jay. *The Evolution of Gryphaea.* The History of Paleontology. New York: Arno Press, 1980.

——. *Punctuated Equilibrium.* first paperback ed. Cambridge, Mass.: Belknap Press of Harvard University Press, 2007.

Graham, Maria. "An Account of Some Effects of the Late Earthquakes in Chili. Extracted from a Letter to Henry Warburton, Esq. V.P.G.S.". *Transactions of the Geological Society of London, Second Series* 1, Part 2 (1824): 413–15.

Greene, Mott T. *Geology in the Nineteenth Century: Changing Views of a Changing World.* Cornell History of Science Series. Ithaca, N.Y.: Cornell University Press, 1982.

Greenough, George Bellas. "Address Delivered at the Anniversary Meeting of the Geological Society, on the 21st of February, 1834." *Proceedings of the Geological Society of London* 2, no. 35 (1838): 42–70.

Gribbin, John, and Mary Gribbin. *FitzRoy: The Remarkable Story of Darwin's Captain and the Invention of the Weather Forecast.* London: Review, 2003.

Grine, Frederick E., John G. Fleagle, and Richard E. Leakey. *The First Humans: Origin and Early Evolution of the Genus Homo: Contributions from the Third Stony Brook Human Evolution Symposium and Workshop, October 3–October 7, 2006.* Vertebrate Paleobiology and Paleoanthropology. Dordrecht, Netherlands: Springer, 2009.

Gutenberg, Beno, and C. F. Richter. *Seismicity of the Earth.* Geological Society of America Special Papers, 34. New York: The Society, 1941.

——. *Seismicity of the Earth and Associated Phenomena.* Princeton, N.J.: Princeton University Press, 1949.

——. *Seismicity of the Earth and Associated Phenomena.* second ed. Princeton, N.J.: Princeton University Press, 1954.

———. *Seismicity of the Earth and Associated Phenomena.* second ed. New York: Hafner, 1965.

Harper, C. G. *Holyhead Road: The Mail-Coach Road to Dublin, Vol. 2 Birmingham to Holyhead.* New York and Philadephia: Springer, 1902.

Haug, Gerald H., and Ralf Tiedemann. "Effect of the Formation of the Isthmus of Panama on Atlantic Ocean Thermohaline Circulation." *Nature* 393, no. 6686 (1998): 673–76.

Haynes, Gary. *American Megafaunal Extinctions at the End of the Pleistocene.* Vertebrate Paleobiology and Paleoanthropology Series. Dordrecht, Netherlands: Springer, 2009.

Head, Martin J., and Philip L. Gibbard. "Early-Middle Pleistocene Transitions: An Overview and Recommendation for the Defining Boundary." *Geological Society, London, Special Publications* 247, no. 1 (2005): 1–18.

Herbert, Sandra. *Charles Darwin, Geologist.* Ithaca: Cornell University Press, 2005.

Herbert, Sandra, and Michael B. Roberts. "Charles Darwin's Notes on His 1831 Geological Map of Shrewsbury." *Archives of Natural History* 29, no. 1 (2002): 27–29.

Herries Davies, G. L. *Whatever Is under the Earth: The Geological Society of London 1807–2007.* London: Geological Society, 2007.

Herschel, J. F. W. *A Preliminary Discourse on the Study of Natural Philosophy.* Longman, Rees, Orme, Brown and Green; and John Taylor, 1831.

Higham, Tom, Katerina Douka, Rachel Wood, Christopher Bronk Ramsey, Fiona Brock, Laura Basell, Marta Camps, et al. "The Timing and Spatiotemporal Patterning of Neanderthal Disappearance." *Nature* 512, no. 7514 (2014): 306–09.

Hoffecker, John F., Scott A. Elias, Dennis H. O'Rourke, G. Richard Scott, and Nancy H. Bigelow. "Beringia and the Global Dispersal of Modern Humans." *Evolutionary Anthropology: Issues, News, and Reviews* 25, no. 2 (2016): 64–78.

Holmes, Arthur. *Principles of Physical Geology.* Rev. printing. ed. New York: Ronald Press Co., 1945.

Holmes, Richard. *The Age of Wonder: How the Romantic Generation Discovered the Beauty and Terror of Science.* first Vintage Books ed. New York: Vintage Books, 2010.

Hooker, William Jackson, and John Smith. "Fitz-Roya Patagonica." *Curtis's Botanical Magazine* 77 (or Vol. 7 of the Third Series) (1851): Tab. 4616.

Humboldt, Alexander von, and Jason Wilson. *Personal Narrative.* London; New York: Penguin Books, 1995.

Hutton, James. *The Theory of the Earth.* Edinburgh, 1795.

Hyam, Ronald. *Britain's Imperial Century, 1815–1914: A Study of Empire and Expansion.* Cambridge Imperial and Post-Colonial Studies Series. third ed. Houndmills, Basingstoke, Hampshire; New York: Palgrave Macmillan, 2002.

Imbrie, John, and Katherine Palmer Imbrie. *Ice Ages: Solving the Mystery.* Cambridge, Mass.: Harvard University Press, 1986.

Iriondo, Martin, and Daniela Kröhling. "From Buenos Aires to Santa Fe: Darwin's Observations and Modern Knowledge." *Revista de la Asociación Geológica Argentina* 64, no. 1 (2009): 109–23.

Irving, Washington. *Tales of a Traveller.* London: Echo Library, 2007.

Isaacson, Walter. *Einstein: His Life and Universe.* New York: Simon & Schuster, 2007.

Jamieson, Thomas F. "On the Parallel Roads of Glen Roy, and Their Place in the History of the Glacial Period." *Quarterly Journal of the Geological Society* 19, no. 1-2 (1863): 235–59.

———. "On the Cause of the Depression and Re-Elevation of the Land During the Glacial Period." *Geological Magazine (Decade II)* 9, no. 10 (1882): 457–66.

Jefferson, Thomas F. "A Memoir on the Discovery of Certain Bones of a Quadruped of the Clawed Kind in the Western Parts of Virginia." *Transactions of the American Philosophical Society* 4 (1799): 246–60.

Jeffreys, Harold. *The Earth: Its Origin, History and Physical Constitution.* fourth ed. Cambridge, England: The University Press, 1962.

Kay, Suzanne Mahlburg, and Victor A. Ramos. *Field Trip Guides to the Backbone of the Americas in the Southern and Central Andes: Ridge Collision, Shallow Subduction, and Plateau Uplift.* Geological Society of America Field Guide 13. Boulder, Colo.: Geological Society of America, 2008.

Keynes, R. D. *The Beagle Record: Selections from the Original Pictorial Records and Written Accounts of the Voyage of H.M.S.* Beagle. Cambridge, England and New York: Cambridge University Press, 1979.

King, Philip Parker, and Robert FitzRoy. *Narrative of the Surveying Voyages of His Majesty's Ships* Adventure *and* Beagle: *Proceedings of the First Expedition, 1826–1830, under the Command of Captain P. Parker King.* Vol. 1, London: Henry Colburn, 1839.

Klein, Richard G., and Blake Edgar. *The Dawn of Human Culture.* New York: Wiley, 2002.

Koch, Paul L., and Anthony D. Barnosky. "Late Quaternary Extinctions: State of the Debate." *Annual Review of Ecology, Evolution, and Systematics* (2006): 215–50.

Kolbert, Elizabeth. *The Sixth Extinction: An Unnatural History.* First edition. ed. New York: Henry Holt and Company, 2014.

Kölbl-Ebert, M. "Observing Orogeny—Maria Graham's Account of the Earthquake in Chile in 1822." *Episodes* (1999).

Kurtén, Björn, and Elaine Anderson. *Pleistocene Mammals of North America.* New York: Columbia University Press, 1980.

La Pérouse, Jean-François de Galaup, and John Dunmore. *The Journal of Jean-François de Galaup de La Pérouse, 1785–1788.* Works Issued by the Hakluyt Society. two vols. London: Hakluyt Society, 1994.

Lange, Ian M. *Ice Age Mammals of North America: A Guide to the Big, the Hairy, and the Bizarre.* Missoula, Mont.: Mountain Press, 2002.

Lara, Luis E. "The 2008 Eruption of the Chaitén Volcano, Chile: A Preliminary Report." *Andean Geology* 36, no. 1 (2010): 125–29.

Lauder, Thomas Dick. "On the Parallel Roads of Lochaber." *Transactions of the Royal Society of Edinburgh* 9 (1821): 1–64.

Lewis, Cherry. *The Dating Game: One Man's Search for the Age of the Earth.* Cambridge, England and New York: Cambridge University Press, 2000.

Lewis, Cherry, and Simon J. Knell, eds. *The Making of the Geological Society of London.* London: Geological Society, 2009.

Litchfield, Henrietta Emma, and Emma Darwin. *Emma Darwin, Wife of Charles Darwin: A Century of Family Letters, by Her Daughter H. E. Litchfield.* Cambridge, England: Cambridge University Press, 1904.

Lomnitz, C. "Major Earthquakes of Chile: A Historical Survey, 1535–1960." *Seismological Research Letters* 75, no. 3 (May–June 2004): 368–78.

Long, Austin, Richard M. Hansen, and Paul S. Martin. "Extinction of the Shasta Ground Sloth." *Geological Society of America Bulletin* 85, no. 12 (1974): 1843–48.

Lowell, Wayne Russell, and Montis R. Klepper. "Beaverhead Formation, a Laramide Deposit in Beaverhead County, Montana." *Geological Society of America Bulletin* 64, no. 2 (1953): 235–44.

Lukefahr, S. D., and P. R. Cheeke. "Rabbit Project Development Strategies in Subsistence Farming Systems." *World Animal Review* 68 (1991): 60–70.

Lurie, Edward. *Louis Agassiz, a Life in Science.* Johns Hopkins paperback ed. Baltimore: Johns Hopkins University Press, 1988.

Lyell, Charles. *Principles of Geology, Being an Attempt to Explain the Former Changes of the Earth's Surface, by Reference to Causes Now in Operation.* Vol. 1, London: John Murray, 1830.

——. *Principles of Geology, Being an Attempt to Explain the Former Changes of the Earth's Surface, by Reference to Causes Now in Operation.* Vol. 2, London: John Murray, 1832.

——. *Principles of Geology, Being an Attempt to Explain the Former Changes of the Earth's Surface, by Reference to Causes Now in Operation.* Vol. 3, London: John Murray, 1833.

——. "The Bakerian Lecture: On the Proofs of a Gradual Rising of the Land in Certain Parts of Sweden." *Philosophical Transactions of the Royal Society of London* 125 (1835): 1–38.

——. *Life, Letters and Journals of Sir Charles Lyell, Bart: Edited by His Sister-in-Law Mrs. Lyell.* Volumes 1 and 2. Cambridge, England: Cambridge, University Press, 1881.

MacCulloch, John. "On the Parallel Roads of Glen Roy." *Transactions of the Geological Society of London* 4, no. 2 (1817): 314–92.

Mackenzie, Murdock, and James Horsburgh. *A Treatise on Marine Surveying.* London: Black, Kingsbury, Parbury, and Allen, 1819.

MacPhee, Ross D. E., M. A. Iturralde-Vinent, and Osvaldo Jiménez Vázquez. "Prehistoric Sloth Extinctions in Cuba: Implications of a New 'Last' Appearance Date." *Caribbean Journal of Science* 43, no. 1 (2007): 94–98.

Madariaga, Raul, Marianne Métois, Christophe Vigny, and Jaime Campos. "Central Chile Finally Breaks." *Science* 328, no. 5975 (2010): 181–82.

Major, Jon J., and Luis E. Lara. "Overview of Chaitén Volcano, Chile, and Its 2008–2009 Eruption." *Andean Geology* 40, no. 2 (2013): 196–215.

Manera de Bianco, Teresa, Sylvia A. Aramayo, and H. O. Ortiz. "Trazas de Pelaje en Icnitas de Megaterios en el Yacimiento Paleoicnológico de Pehuen Co (Pleistoceno Tardío), Provincia de Buenos Aires, Argentina." *Ameghiniana, Suplemento Resúmenes* 42 (2005): 73R.

Manera de Bianco, Teresa, Sylvia A. Aramayo, C. Zavala, and R. Caputo. "Yacimiento Paleoicnológico De Pehuen Co." *Un Patrimonio Natural en Peligro. Sitios de Interés Geológico, Comisión Sitios de Interés Geológico de la República Argentina, Instituto de Geología y Recursos Minerales, Servicio Geológico Minero Argentino (Eds). Buenos Aires: Artes Gráficas Papiros* (2008): 509–20.

Marquardt, Karl Heinz. *HMS Beagle: Survey Ship Extraordinary (Anatomy of the Ship).* London: Conway Maritime Press, 1998.

Martin, Fabiana, Manuel San Román, Flavia Morello, Dominique Todisco, Francisco J. Prevosti, and Luis A. Borrero. "Land of the Ground Sloths: Recent Research at Cueva Chica, Ultima Esperanza, Chile." *Quaternary International* 305 (2013): 56–66.

Martin, Paul S. *Twilight of the Mammoth: Ice Age Extinctions and the Rewilding of America (Organisms and Environments).* Berkeley: University of California Press, 2007.

Martínez, O., Jorge Rabassa, and Andrea Coronato. "Charles Darwin and the First Scientific Observations on the Patagonian Shingle Formation (Rodados Patagónicos)." *Revista de la Asociación Geológica Argentina* 64, no. 1 (2009): 90–100.

McCalman, Iain. *Darwin's Armada: Four Voyages and the Battle for the Theory of Evolution.* first American ed. New York: W. W. Norton & Co., 2009.

———. *The Reef: A Passionate History: The Great Barrier Reef from Captain Cook to Climate Change.* New York: Scientific American/Farrar, Straus and Giroux, 2015.

McHenry, Henry M. "Human Evolution." In *Evolution: The First Four Billion Years,* Michael Ruse and Joseph Travis eds., 256–80. Cambridge, Mass.: The Belknap Press of Harvard University Press, 2009.

McPhee, John. *Basin and Range.* New York: Farrar, Straus, Giroux, 1982.

Mellersh, H. E. L. *FitzRoy of the Beagle.* London: Hart-Davis, 1968.

Melnick, Daniel. "Rise of the Central Andean Coast by Earthquakes Straddling the Moho." *Nature Geoscience* 9 (2016): 401–7.

Melnick, Daniel, B. Bookhagen, H. P. Echtler, and M. R. Strecker. "Coastal Deformation and Great Subduction Earthquakes, Isla Santa Maria, Chile (37°S)." *Bulletin of the Geological Society of America* 118, no. 11-12 (2006): 1463–80.

Melnick, Daniel, Marco Cisternas, Marcos Moreno, and Ricardo Norambuena. "Estimating Coseismic Coastal Uplift with an Intertidal Mussel: Calibration for the 2010 Maule, Chile Earthquake (Mw=8.8)." *Quaternary Science Reviews* 42 (2011): 29–42.

Melnick, Daniel, Marco Cisternas, Robert L. Wesson, Marcos Moreno, and Marcelo Lagos. "Present and Past Land-Level Changes at Guafo Island, Chile (43.5°S): Transient Accumulation of Permanent Strain over the Seismic Cycle." Paper presented at the 2009 Portland GSA Annual Meeting, 2009.

Melnick, Daniel, Marcos Moreno, Shaoyang Li, Marco Cisternas, Juan Carlos Baez, Robert Wesson, Alan Nelson, and Michael Bevis. "Transient Evolution of Plate Coupling after the Giant 1960 Chile Earthquake at Guafo Island." Paper presented at the EGU General Assembly Conference Abstracts, 2015.

Meltzer, David J. *First Peoples in a New World: Colonizing Ice Age America.* Berkeley: University of California Press, 2009.

———. *The Great Paleolithic War: How Science Forged an Understanding of America's Ice Age Past.* Chicago: University of Chicago Press, 2015.

Meltzer, David J. "Pleistocene Overkill and North American Mammalian Extinctions." *Annual Review of Anthropology* 44 (2015): 33–53.

Melville, Herman. *Billy Budd and Other Stories.* New York: Penguin Classics, 1986.

———. *Moby-Dick; Or, the Whale.* New York: Penguin Classics, 2010.

Menard, Henry W. *The Ocean of Truth: A Personal History of Global Tectonics.* Princeton Series in Geology and Paleontology. Princeton, N.J.: Princeton University Press, 1986.

Mills, William. "Darwin and the Iceberg Theory." *Notes and Records of the Royal Society of London* 38, no. 1 (1983): 109–27.

Milne, David. "On the Parallel Roads of Lochaber, with Remarks on the Change of Relative Levels of Sea and Land in Scotland, and on the Detrital Deposits in That Country." *Transactions of the Royal Society of Edinburgh* 16, no. 03 (1847): 395–418.

Molnar, Peter. *Plate Tectonics: A Very Short Introduction*. Oxford, England and New York: Oxford University Press, 2015.

Monckeberg, Fernando, and C. O. Chichester. "Chilean Experience with Fortified Children's Formulas." In *Nutritional Improvement of Food and Feed Proteins*, edited by Mendel Friedman, 11–28. Boston: Springer, 1978.

Moreno, M., D. Melnick, M. Rosenau, J. Baez, J. Klotz, O. Oncken, A. Tassara, et al. "Toward Understanding Tectonic Control on the M W 8.8 2010 Maule Chile Earthquake." *Earth and Planetary Science Letters* 321 (2012): 152–65.

Moreno, Marcos, Matthias Rosenau, and Onno Oncken. "2010 Maule Earthquake Slip Correlates with Pre-Seismic Locking of Andean Subduction Zone." *Nature* 467, no. 7312 (2010): 198–202.

Muller, R., and Gordon J. MacDonald. *Ice Ages and Astronomical Causes: Data, Spectral Analysis and Mechanisms*. Springer-Praxis Books in Environmental Sciences. London and New York: Springer, 2000.

Naranjo, Jose A., and Charles R. Stern. "Holocene Tephrochronology of the Southernmost Part (42° 30'-45° S) of the Andean Southern Volcanic Zone." *Revista Geológica de Chile* 31, no. 2 (2004): 224–40.

National Research Council (U.S.). Committee on Abrupt Climate Change. *Abrupt Climate Change Inevitable Surprises*. Washington: National Academy Press, 2002.

Nelson, Richard W. *Darwin, Then and Now: The Most Amazing Story in the History of Science*. New York: iUniverse, Inc., 2009.

Newton, Isaac. *The Principia: Mathematical Principles of Natural Philosophy*. CreateSpace Independent Publishing Platform, 2013.

Nichols, Peter. *Evolution's Captain: The Dark Fate of the Man Who Sailed Charles Darwin Around the World*. first ed. New York: HarperCollins, 2003.

Nunn, Patrick D. *Oceanic Islands*. The Natural Environment. Oxford, England and Cambridge, Mass.: Blackwell, 1994.

O'Dea, Aaron, Harilaos A. Lessios, Anthony G. Coates, Ron I. Eytan, Sergio A. Restrepo-Moreno, Alberto L. Cione, Laurel S. Collins, et al. "Formation of the Isthmus of Panama." *Science Advances* 2, no. 8 (2016).

O'Leary, Michael J, Paul J. Hearty, William G. Thompson, Maureen E. Raymo, Jerry X. Mitrovica, and Jody M. Webster. "Ice Sheet Collapse Following a Prolonged Period of Stable Sea Level During the Last Interglacial." *Nature Geoscience* 6, no. 9 (2013): 796–800.

Officers of the HMS *Beagle*. *Santa Maria Island, South America, Coast of Chile*. Chart No. 1303. London: Hydrographic Office of the Admiralty, 1840.

Oldroyd, D. R. *Thinking About the Earth: A History of Ideas in Geology*. Studies in the History and Philosophy of the Earth Sciences. Cambridge, Mass.: Harvard University Press, 1996.

Oreskes, Naomi. *The Rejection of Continental Drift: Theory and Method in American Earth Science*. New York: Oxford University Press, 1999.

Oreskes, Naomi, and H. E. LeGrand. *Plate Tectonics: An Insider's History of the Modern Theory of the Earth*. Boulder, Colo.: Westview Press, 2001.

Oviatt, Charles G., Donald R. Currey, and Dorothy Sack. "Radiocarbon Chronology of Lake Bonneville, Eastern Great Basin, USA." *Palaeogeography, Palaeoclimatology, Palaeoecology* 99, no. 3-4 (1992): 225–41.

Owen, Richard. *Fossil Mammalia, Part 1. The Zoology of the Voyage of HMS Beagle, under the Command of Captain FitzRoy, R.N., During the Years 1832 to 1836.* Edited by Charles Darwin London: Smith Elder and Co., 1838–40.

———. *Report on British Fossil Reptiles. Part II.* [from the Report of the Eleventh Meeting of the British Association for the Advancement of Science; Held at Plymouth in July 1841]. London: Richard and John E. Taylor, 1841.

Owen, William Fitzwilliam, and Heaton Bowstead Robinson. *Narrative of Voyages to Explore the Shores of Africa, Arabia and Madagascar. [Edited by H. B. Robinson.].* London, 1833.

Palmer, Adrian, J. J. Lowe, and Jim Rose. *The Quaternary of Glen Roy and Vicinity.* London: Quaternary Research Association in association with Scottish Natural Heritage and Lochaber Geopark, 2008.

Parras, Ana, and Miguel Griffin. "Darwin's Great Patagonian Tertiary Formation at the Mouth of the Río Santa Cruz: A Reappraisal." *Revista de la Asociación Geológica Argentina* 64, no. 1 (2009): 70–82.

Pascual, Miguel, Paul Bentzen, Carla Riva Rossi, Greg Mackey, Michael T. Kinnison, and Robert Walker. "First Documented Case of Anadromy in a Population of Introduced Rainbow Trout in Patagonia, Argentina." *Transactions of the American Fisheries Society* 130, no. 1 (2001): 53–67.

Pearson, Paul Nicholas, and C. J .Nicholas. "'Marks of Extreme Violence': Charles Darwin's Geological Observations at St. Jago (São Tiago), Cape Verde Islands." *Geological Society, London, Special Publications* 287, no. 1 (2007): 239–53.

Pedoja, Kevin, Vincent Regard, Laurent Husson, Joseph Martinod, Benjamin Guillaume, Enrique Fucks, Maximiliano Iglesias, and Pierre Weill. "Uplift of Quaternary Shorelines in Eastern Patagonia: Darwin Revisited." *Geomorphology* 127, no. 3 (2011): 121–42.

Pennant, Thomas. A Tour in Scotland MDCCLXIX. Second Edition. London: B. White, 1772.

Perea, Daniel, ed. *Fósiles De Uruguay.* Montevideo: DIRAC, 2011.

Plafker, George. "Tectonic Deformation Associated with the 1964 Alaska Earthquake." *Science* 148, no. 3678 (1965): 1675–87.

———. "Possible Evidence for Downward-Directed Mantle Convection beneath the Eastern End on the Aleutian Arc." In *American Geophysical Union, 48th Annual Meeting,* 218–19. Washington, 1967.

———. *Tectonics of the March 27, 1964, Alaska Earthquake.* U.S. Government Printing Office, 1969.

———. "Alaskan Earthquake of 1964 and Chilean Earthquake of 1960: Implications for Arc Tectonics." *Journal of Geophysical Research* 77, no. 5 (1972): 901–25.

Plafker, G., and J. C. Savage. "Mechanism of the Chilean Earthquakes of May 21 and 22, 1960." *Geol. Soc. Am. Bull* 81, no. 4 (1970): 1001–30.

Playfair, John, and James Hutton. *Illustrations of the Huttonian Theory of the Earth.* Edinburgh: Cadell and Davies; William Creech, 1802.

Poma, Stella, Vanesa D. Litvak, Magdalena Koukharsky, E. Beatriz Maisonnave, and Sonia Quenardelle. "Darwin's Observation in South America: What Did He Find at Agua De La Zorra, Mendoza Province." *Revista de la Asociación Geológica Argentina* 64, no. 1 (2009): 13–20.

Potter, Ben A., Joel D. Irish, Joshua D. Reuther, Carol Gelvin-Reymiller, and Vance T. Holliday. "A Terminal Pleistocene Child Cremation and Residential Structure from Eastern Beringia." *Science* 331, no. 6020 (2011): 1058–62.

Press, Frank, and David Jackson. "Alaskan Earthquake, 27 March 1964: Vertical Extent of Faulting and Elastic Strain Energy Release." *Science* 147, no. 3660 (1965): 867–68.

Prichard, Hesketh Vernon Hesketh, Francisco P. Moreno, Arthur Smith Woodward, Oldfield Thomas, James Britten, A. B. Rendle, and John Guille Millais. *Through the Heart of Patagonia*. New York: D. Appleton and Company, 1902.

Puseman, Kathryn, Peter Kováčik, and R. A. Varney. "Identification and Ams Radiocarbon Dating of a Seed from Tirua, Chile." 6. Golden, Colo.: PaleoResearch Institute, 2013.

Pyne, Stephen J. *Grove Karl Gilbert, a Great Engine of Research*. Austin: University of Texas Press, 1980.

Quammen, David. *The Song of the Dodo: Island Biogeography in an Age of Extinction*. New York: Scribner, 1996.

———. *The Reluctant Mr. Darwin: An Intimate Portrait of Charles Darwin and the Making of His Theory of Evolution* (Great Discoveries). New York: W. W. Norton, 2006.

Quattrocchio, Mirta E., Cecilia M. Deschamps, Carlos A. Zavala, Sylvia C. Grill, and Ana M. Borromei. "Geology of the Area of Bahía Blanca, Darwin's View and the Present Knowledge: A Story of 10 Million Years." *Revista de la Asociación Geológica Argentina* 64, no. 1 (Feb. 1, 2009): 137–46.

Quoy, Jean René Constant, and Paul Gaimard. *Mémoire Sur L'accroissement Des Polypes Lithophytes Considéré Géoloiquement*. 1825.

Rackham, Oliver. *Woodlands*. new ed. London: William Collins, 2015.

Ramos, Victor A. "Darwin at Puente Del Inca: Observations on the Formation of the Inca's Bridge and Mountain Building." *Revista de la Asociación Geológica Argentina* 64, no. 1 (2009): 170–79.

Ramos, V. A., and B. Aguirre-Urreta. "Las Casuchas Del Rey: Un Patrimonio Temprano de la Integración Chileno-Argentina." XII Congreso Geológico Chileno Santiago, November 22–26, 2009, S5_021.

Rashid, Harunur, Leonid Polyak, Ellen Mosley-Thompson, and American Geophysical Union. *Abrupt Climate Change: Mechanisms, Patterns, and Impacts*. Geophysical Monograph 193. Washington: American Geophysical Union, 2011.

Rasmussen, Morten, Sarah L. Anzick, Michael R. Waters, Pontus Skoglund, Michael DeGiorgio, Thomas W. Stafford Jr., Simon Rasmussen, et al. "The Genome of a Late Pleistocene Human from a Clovis Burial Site in Western Montana." *Nature* 506, no. 7487 (2014): 225–29.

Raup, David M. *Extinction: Bad Genes or Bad Luck?* Introduction by Stephen Jay Gould. New York: W. W. Norton & Company, 1992.

———. "The Role of Extinction in Evolution." *Proceedings of the National Academy of Sciences* 91, no. 15 (1994): 6758–63.

Rector, John Lawrence. *The History of Chile*. The Greenwood Histories of the Modern Nations. Westport, Conn.: Greenwood Press, 2003.

Redmond, Brian G., H. Gregory McDonald, Haskel J. Greenfield, and Matthew L. Burr. "New Evidence for Late Pleistocene Human Exploitation of Jefferson's Ground Sloth

(Megalonyx Jeffersonii) from Northern Ohio, USA." *World Archaeology* 44, no. 1 (April 2012): 75–101.

Rees, Martin. *Our Final Hour: A Scientist's Warning.* New York: Basic Books, 2004.

Reid, Harry Fielding. *The California Earthquake of April 18, 1906: The Mechanics of the Earthquake/by Harry Fielding Reid.* Washington: Carnegie Inst., 1910.

Rhodes, Frank H. T. "Darwin's Search for a Theory of the Earth: Symmetry, Simplicity and Speculation." *The British Journal for the History of Science* 24, no. 2 (1991): 193–229.

Ricketts, Edward Flanders, and Jack Calvin. *Between Pacific Tides: An Account of the Habits and Habitats of Some Five Hundred of the Common, Conspicuous Seashore Invertebrates of the Pacific Coast between Sitka, Alaska, and Northern Mexico.* third ed. Stanford, Calif.: Stanford University Press, 1962.

Ritchie, G. S. *The Admiralty Chart: British Naval Hydrography in the Nineteenth Century.* New York: Elsevier, 1967.

Roberts, Michael B. "Darwin at Llanymynech: The Evolution of a Geologist." *The British Journal for the History of Science* 29, no. 04 (1996): 469–78.

———. "Darwin's Dog-Leg: The Last Stage of Darwin's Welsh Field Trip of 1831." *Archives of Natural History* 25, no. 1 (1998): 59–73.

———. "I Coloured a Map: Darwin's Attempts at Geological Mapping in 1831." *Archives of Natural History* 27, no. 1 (2000): 69–79.

———. "Just before the *Beagle*: Charles Darwin's Geological Fieldwork in Wales, Summer 1831." *Endeavour* 25, no. 1 (2001): 33–37.

———. "Genesis Chapter 1 and Geological Time from Hugo Grotius and Marin Mersenne to William Conybeare and Thomas Chalmers (1620–1825)." *Geological Society, London, Special Publications* 273, no. 1 (2007): 39–49.

———. "Adam Sedgwick (1785–1873): Geologist and Evangelical." In *Geological Society, London, Special Publications*, 155–70. London, 2009.

———. "Buckland, Darwin and the Attempted Recognition of an Ice Age in Wales, 1837–1842." *Proceedings of the Geologists' Association* 123, no. 4 (2012): 649–62.

Roberts, Neil. *The Holocene: An Environmental History.* third ed. Hoboken, N.J.: Wiley-Blackwell, 2014.

Robins, Nick S. *Coastal Passenger Liners of the British Isles.* Barnsley, England: Seaforth, 2011.

Rosen, Brian Roy. "Darwin, Coral Reefs, and Global Geology." *BioScience* 32, no. 6 (1982): 519–25.

Roussin, M. le Baron. *Navigation Aux Côtes Du Brésil.* Paris: Imprimerie Royal, 1821.

Ruddiman, William. F. *Plows, Plagues, and Petroleum: How Humans Took Control of Climate.* Princeton, N.J.: Princeton University Press, 2005.

Rudwick, Martin J. S. "Hutton and Werner Compared: George Greenough's Geological Tour of Scotland in 1805." *The British Journal for the History of Science* 1, no. 02 (1962): 117–35.

———. "Darwin and Glen Roy: A 'Great Failure' in Scientific Method?" *Studies in History and Philosophy of Science Part A* 5, no. 2 (1974): 97–185.

———. *The Great Devonian Controversy: The Shaping of Scientific Knowledge Among Gentlemanly Specialists.* Science and Its Conceptual Foundations. Chicago: University of Chicago Press, 1985.

——. *Bursting the Limits of Time: The Reconstruction of Geohistory in the Age of Revolution*. Chicago: University of Chicago Press, 2005.

——. *Worlds Before Adam: The Reconstruction of Geohistory in the Age of Reform*. Chicago: University of Chicago Press, 2008.

——. *Georges Cuvier, Fossil Bones, and Geological Catastrophes: New Translations and Interpretations of the Primary Texts*. Chicago: University of Chicago Press, 2008.

——. *The Parallel Roads of Glen Roy: In the Footsteps of Charles Darwin—a Field Guide*. London: Geological Society of London, History of Geology Group, 2009.

——. "The Origin of the Parallel Roads of Glen Roy: A Review of 19th Century Research." *Proceedings of the Geologists' Association* (2016).

Ruegg, J. C., A. Rudloff, C. Vigny, R. Madariaga, J. B. De Chabalier, J. Campos, E. Kausel, S. Barrientos, and D. Dimitrov. "Interseismic Strain Accumulation Measured by Gps in the Seismic Gap between Constitución and Concepción in Chile." *Physics of the Earth and Planetary Interiors* 175, no. 1 (2009): 78–85.

Satake, Kenji, and Brian F. Atwater. "Long-Term Perspectives on Giant Earthquakes and Tsunamis at Subduction Zones." *Annual Review of Earth and Planetary Sciences* 35 (2007): 349–74.

Sawai, Yuki, Yuichi Namegaya, Yukinobu Okamura, Kenji Satake, and Masanobu Shishikura. "Challenges of Anticipating the 2011 Tohoku Earthquake and Tsunami Using Coastal Geology." *Geophysical Research Letters* 39, no. 21 (2012).

Scholz, C. H. *The Mechanics of Earthquakes and Faulting*. second ed. Cambridge, England and New York: Cambridge University Press, 2002.

Secord, James A. *Controversy in Victorian Geology: The Cambrian-Silurian Dispute*. Princeton, N.J.: Princeton University Press, 1986.

Sedgwick, Adam. "Address to the Geological Society on the Evening of 18th February 1831 on Retiring from the President's Chair." *Proceedings of the Geological Society of London* 1 no. 20 (1834): 281–316.

——. "Address on Announcing the First Award of the Wollaston Prize." *Proceedings of the Geological Society of London* 1 no. 20 (1834): 270–80.

Sedgwick, Adam, John W. Clark, and Thomas M. Hughes. *Life and Letters of Adam Sedgwick, Vol. 1 and 2*. Cambridge, England: Cambridge University Press, 1890.

Shearer, Peter M. *Introduction to Seismology*. second ed. Cambridge, England: Cambridge University Press, 2009.

Shurtleff, William, and Akiko Aoyagi. *History of Soy Flour, Grits and Flakes (510 CE to 2013)*. Lafayette, Calif.: Soyinfo Center, 2013.

Sievers, Hellmuth A., Gullermo C. Villegas, Guillermo Barros, and Pierre Saint-Amand. "The Seismic Sea Wave of 22 May 1960 Along the Chilean Coast." *Bulletin of the Seismological Society of America* 53, no. 6 (July 29, 1963): 1125–90.

Silliman, Benjamin. "On the Reality of the Rise of the Coast of Chile, in 1822, as Stated by Mrs. Graham." *American Journal of Science and Arts* 28, no. 2 (1835): 236–47.

Simpson, George Gaylord. "The Beginning of the Age of Mammals in South America: Introduction. Systematics: Marsupialia, Edentata, Condylarthra, Litopterna and Notioprogonia." *Bulletin of the American Museum of Natural History* 91, no. 1 (1948).

————. *Splendid Isolation: The Curious History of South American Mammals.* New Haven, Conn.: Yale University Press, 1980.

Sissons, J. B. *The Evolution of Scotland's Scenery.* Edinburgh: Oliver & Boyd, 1967.

Smallwood, John. "Bouguer Redeemed: The Successful 1737–1740 Gravity Experiments on Pichincha and Chimborazo." *Earth Sciences History* 29, no. 1 (2010): 1–25.

Snyder, Laura J. *The Philosophical Breakfast Club: Four Remarkable Friends Who Transformed Science and Changed the World.* first ed. New York: Broadway Books, 2011.

Sobel, Dava. *Longitude: The True Story of a Lone Genius Who Solved the Greatest Scientific Problem of His Time.* New York: Walker, 1995.

Som, Sanjoy M., David C. Catling, Jelte P. Harnmeijer, Peter M. Polivka, and Roger Buick. "Air Density 2.7 Billion Years Ago Limited to Less Than Twice Modern Levels by Fossil Raindrop Imprints." *Nature* 484, no. 7394 (May 09 2013): 359–62.

Soulé, Bastien. "Post-Crisis Analysis of an Ineffective Tsunami Alert: The 2010 Earthquake in Maule, Chile." *Disasters* 38, no. 2 (2014): 375–97.

St. Amand, P. "South Chile Earthquake Swarm of 1960 [Abs.]." *Geological Society of America Bulletin* 71, no. 12, pt. 2 (1960): 1964.

Stauder, William, and G. A. Bollinger. "The Focal Mechanism of the Alaska Earthquake of March 28, 1964, and of Its Aftershock Sequence." *Journal of Geophysical Research* 71, no. 22 (1966): 5283–96.

Steadman, David W., Paul S. Martin, Ross D. E. MacPhee, A. J. Timothy Jull, H. Gregory McDonald, Charles A. Woods, Manuel Iturralde-Vinent, and Gregory W. L. Hodgins. "Asynchronous Extinction of Late Quaternary Sloths on Continents and Islands." *Proceedings of the National Academy of Sciences of the United States of America* 102, no. 33 (2005): 11763–68.

Stehli, Francis Greenough, and S. David Webb. *The Great American Biotic Interchange.* Topics in Geobiology. New York: Plenum Press, 1985.

Stoddart, David R. "Coral Islands, by Charles Darwin." *Atoll Research Bulletin* 88 (1962): 1–20.

————. "Ecology and Morphology of Recent Coral Reefs." *Biological Reviews* 44, no. 4 (1969): 433–98.

————. "Coral Reefs: The Last Two Million Years." *Geography* (1973): 313–23.

————. "Darwin, Lyell, and the Geological Significance of Coral Reefs." *The British Journal for the History of Science* 9, no. 2 (1976): 199–218.

Stokes, Pringle, and R. J. Campbell. "The Journal of the HMS *Beagle* in the Strait of Magellan." Chap. Part 2 In *Four Travel Journals: The Americas, Antarctica and Africa, 1775–1874* edited by Herbert K. Bealls. Works Issued by the Hakuyt Society, 141–251. London: Hakluyt Society, 2007.

Stott, Rebecca. *Darwin and the Barnacle: The Story of One Tiny Creature and History's Most Spectacular Scientific Breakthrough.* New York: W. W. Norton & Company, 2004.

Strelin, Jorge, and Eduardo Malagnino. "Charles Darwin and the Oldest Glacial Events in Patagonia: The Erratic Blocks of the Río Santa Cruz Valley." *Revista de la Asociación Geológica Argentina* 64, no. 1 (2009): 101–08.

Suess, Eduard, W. J. Sollas, and Hertha B. C. Sollas. *The Face of the Earth: (Das Antlitz Der Erde).* Oxford, England: Clarendon Press, 1904.

Sulivan, Bartholomew J. *Life and Letters of the Late Admiral Sir Bartholomew James Sulivan, KCB, 1810–1890*. The Organ of the Book Trade. Whitaker, 1896.

Swanson, Frederick J., Julia A. Jones, Charles M. Crisafulli, and Antonio Lara. "Effects of Volcanic and Hydrologic Processes on Forest Vegetation: Chaitén Volcano, Chile." (2013).

Swenson, Loyd S. *The Ethereal Aether: A History of the Michelson-Morley-Miller Aether-Drift Experiments, 1880–1930*. Austin: University of Texas Press, 1972.

Terry, James P., and James Goff. "One Hundred and Thirty Years since Darwin: 'Reshaping'the Theory of Atoll Formation." *The Holocene* 23, no. 4 (2013): 615–19.

Thackray, John C. *To See the Fellows Fight: Eye Witness Accounts of Meetings of the Geological Society of London and Its Club*. London: British Society for the History of Science, 2003.

Thomson, James. "The Seasons—Kindle Edition by James Thomson. Literature & Fiction Kindle Ebooks @ Amazon.Com." (2016).

Thomson, William. "On the Rigidity of the Earth." *Philosophical Transactions of the Royal Society of London* (1863).

Tonni, Eduardo Pedro, Alberto Luis Cione, and Leopoldo Héctor Soibelzon. "The Broken Zig-Zag: Late Cenozoic Large Mammal and Tortoise Extinction in South America." *Revista del Museo Argentino de Ciencias Naturales* 5 (2003).

Toomey, M., A. D. Ashton, and J. T. Perron. "Profiles of Ocean Island Coral Reefs Controlled by Sea-Level History and Carbonate Accumulation Rates." *Geology* 41, no. 7 (July 28, 2013): 731–34.

Turner, Gerard L'Estrange. *Scientific Instruments, 1500–1900: An Introduction*. London and Berkeley: Philip Wilson, University of California Press, 1998.

Venter, Oscar, Eric W. Sanderson, Ainhoa Magrach, James R. Allan, Jutta Beher, Kendall R. Jones, Hugh P. Possingham, et al. "Sixteen Years of Change in the Global Terrestrial Human Footprint and Implications for Biodiversity Conservation." *Nature Communications* 7 (2016).

Vigny, C., A. Socquet, S. Peyrat, J. C. Ruegg, M. Metois, R. Madariaga, S. Morvan, et al. "The 2010 M_w 8.8 Maule Megathrust Earthquake of Central Chile, Monitored by GPS." *Science* 332, no. 6036 (June 17, 2011): 1417–21.

Vine, Frederick John, and Drummond Hoyle Matthews. "Magnetic Anomalies over Oceanic Ridges." *Nature* 199, no. 4897 (1963): 947–49.

Vizcaíno, Sergio F., Richard A. Fariña, and Juan C. Fernicola. "Young Darwin and the Ecology and Extinction of Pleistocene South American Fossil Mammals." *Revista de la Asociación Geológica Argentina* 64, no. 1 (Feb. 1, 2009): 160–69.

Vizcaíno, Sergio F., Teresa Manera, and Juan C. Fernicola. "Viaje Al Sepulcro De Los Gigantes: Darwin Y Los Mamiferos Fósiles De América Del Sur." [In Spanish]. *Ciencia hoy* 19, no. 113 (2009): 68–73.

Wahlross, Sven. *Mutiny and Romance in the South Seas: A Companion to the Bounty Adventure*. Topsfield, Mass.: Salem House, 1989.

Walker, Cyril, and David Ward. *Smithsonian Handbooks: Fossils*. London and New York: DK, 2002.

Walker, Malcolm. *History of the Meteorological Office*. Cambridge, England and New York: Cambridge University Press, 2012.

Watts, A. B. *Isostasy and Flexure of the Lithosphere.* Cambridge, England and New York: Cambridge University Press, 2001.

Webster, William Henry Bayley, and Henry Foster. *Narrative of a Voyage to the Southern Atlantic Ocean, In the Years 1828, 29, 30, Performed in H. M. Sloop Chanticleer, Etc.* two vols. London: Richard Bentley, 1834.

Wegener, Alfred. *The Origin of Continents and Oceans.* New York: Dover Publications, 1966.

Wesson, Robert L., Daniel Melnick, Marco Cisternas, Marcos Moreno, and Lisa L. Ely. "Vertical Deformation through a Complete Seismic Cycle at Isla Santa Maria, Chile." *Nature Geoscience* 8, no. 7 (2015): 547–51.

Whewell, William. *The Philosophy of the Inductive Sciences, Founded Upon Their History.* London: Parker, 1840.

Wigley, P., P. Dolan, T. Sharpe, and H. S. Torrens, *Strata Smith: His 200 Year Legacy, Digitally Enhanced Maps & Sections by William Smith, George Bellas Greenough, John Cary & Richard Thomas 1796–1840.* London: Geological Society Of London, 2007. https://www.amazon.com/Strata-Smith-Digitally-Greenough-1796-1840/dp/1862392447.

Willacy, Mark. *Fukushima: Japan's Tsunami and the Inside Story of the Nuclear Meltdowns.* Sydney: Pan Macmillan Australia, 2013.

Williams, J.E.D. *From Sails to Satellites: The Origin and Development of Navigational Science.* Oxford, England : Oxford University Press, 1992.

Winchester, Simon. *The Map That Changed the World: William Smith and the Birth of Modern Geology.* first ed. New York: HarperCollins, 2001.

Woodward, Horace B. *The History of the Geological Society of London.* London: Geological Society of London, 1907.

Zappettini, Eduardo O., and José Mendía. "The First Geological Map of Patagonia." *Darwin in Argentina. Revista de la Asociación Geológica Argentina* 64, no. 1 (2009): 55–59.

Zárate, Marcelo, and Alicia Folguera. "On the Formations of the Pampas in the Footsteps of Darwin: South of the Salado." *Revista de la Asociación Geológica Argentina* 64, no. 1 (2009): 124–36.

Photo Insert Credits

Figure 1.
Portrait of Charles Darwin by George Richmond (1809-1896). Source: Wikimedia Commons from *Origins*, Richard Leakey and Roger Lewin. https://commons.wikimedia.org/wiki/File:Charles_Darwin_by_G._Richmond.png

Figure 2.
From Schmidt, Herman John, 1872-1959: Portrait and landscape negatives, Auckland district. Ref: 1/1-001318-G. Alexander Turnbull Library, Wellington, New Zealand.

Figure 3.
Drawing by Conrad Martens, engraving by Thomas Landseer. From FitzRoy, Robert, *Narrative of the surveying voyages of His Majesty's ships* Adventure *and* Beagle *between the years 1826 and 1836: describing their examination of the southern shores of South America and the Beagle's circumnavigation of the globe: Volume 2.* London: Henry Colburn, 1839. Source: Biodiversity Heritage Library. http://www.biodiversitylibrary.org/bibliography/84933#/summary

Figure 4.
Drawing by Joseph Smit from Hutchinson, H. N., *Extinct monsters: A popular account of some of the larger forms of ancient animal life.* Fourth and Cheaper Edition. London: Chapman & Hall, 1896. Source: Biodiversity Heritage Library. http://www.biodiversitylibrary.org/bibliography/14948#/summary

Figure 5.
Drawing by DiBrd. Source: Wikimedia Commons. https://commons.wikimedia.org/wiki/File:Megatherum_DB.jpg

Figure 6.
From Lankester, E. Ray, *Extinct Animals*, New York: Henry Holt, 1905. Source: Wikimedia Commons. https://commons.wikimedia.org/wiki/File:Toxodon_skeleton.jpg

Figures 7-8.
Photos by the author.

Figure 9.
From Owen, Richard. *Fossil Mammalia, Part 1. The Zoology of the Voyage of HMS Beagle, under the Command of Captain FitzRoy, R.N., During the Years 1832 to 1836.* Edited by Charles Darwin London: Smith Elder and Co., 1838–40. Source: Biodiversity Heritage Library. http://www.biodiversitylibrary.org/bibliography/14216#/summary

Figure 10.
Photo: © Rolex Awards/Marc Latzel.

Figure 11.
Photo: © Rolex Awards/Quentin Deville.

Figure 12.
Drawing by Conrad Martens, engraving by S. Bull. From FitzRoy, Robert, *Narrative of the surveying voyages of His Majesty's ships* Adventure *and* Beagle *between the years 1826 and 1836: describing their examination of the southern shores of South America and the Beagle's circumnavigation of the globe: Volume 2.* London: Henry Colburn, 1839. Source: Biodiversity Heritage Library. http://www.biodiversitylibrary.org/bibliography/84933#/summary

Figures 13-14.
Photos by the author.

Figure 15.
Drawing by Conrad Martens, engraving by Thomas Landseer. From FitzRoy, Robert, *Narrative of the surveying voyages of His Majesty's ships* Adventure *and* Beagle *between the years 1826 and 1836: describing their examination of the southern shores of South America and the Beagle's circumnavigation of the globe: Volume 2.* London: Henry Colburn, 1839. Source: Biodiversity Heritage Library. http://www.biodiversitylibrary.org/bibliography/84933#/summary

Figures 16-19.
Photos by the author.

Figure 20.
From Darwin, Charles R. *Geological observations on the volcanic islands and parts of South America visited during the voyage of H.M.S. Beagle,* New York: D. Appleton and Company,1891. Source: Biodiversity Heritage Library. http://www.biodiversitylibrary.org/bibliography/61452#/summary

Figures 21-22.
Photos by Gayle Gordon.

Figure 23.
Portion of cross section from Darwin, Charles R. *Geological observations on the volcanic islands and parts of South America visited during the voyage of H.M.S. Beagle,* New York: D. Appleton and Company, 1891. Source: Biodiversity Heritage Library. http://www.biodiversitylibrary.org/bibliography/61452#/summary

Figures 24-28.
Photos by the author.

Figure 29.
Drawing by John C. Wickham, engraving by S. Bull. From FitzRoy, Robert, *Narrative of the surveying voyages of His Majesty's ships* Adventure *and* Beagle *between the years 1826 and 1836: describing their examination of the southern shores of South America and the Beagle's circumnavigation of the globe: Volume 2.* London: Henry Colburn, 1839. Source: Biodiversity Heritage Library. http://www.biodiversitylibrary.org/bibliography/84933#/summary

Figures 30-31.
From Ramos, Victor A. "Darwin at Puente Del Inca: Observations on the Formation of the Inca's Bridge and Mountain Building." *Revista de la Asociación Geológica Argentina* 64, no. 1 (2009): 170–79.

Figure 32.
From Beechey, Frederick William, Narrative of a voyage to the Pacific and Beering's strait, to co-operate with the polar expeditions: performed in His Majesty's ship Blossom, under the command of Captain F.W. Beechey . . . in the years 1825, 26, 27, 28 . . . London: H. Colburn and R. Bentley, 1831. Biodiversity Heritage Library. http://www.biodiversitylibrary.org/bibliography/6235#/summary

Figure 33.
Modified from NASA Earth Observatory image created by Jesse Allen and Robert Simmon, using EO-1 ALI data from NASA EO-1 team. http://earthobservatory.nasa.gov/IOTD/view.php?id=76791

Figure 34.
From Darwin, Charles R. "Observations on the Parallel Roads of Glen Roy, and of Other Parts of Lochaber in Scotland, with an Attempt to Prove That They Are of Marine Origin." *Philosophical Transactions of the Royal Society of London* 129 (1839): 39–81. Source: Philosophical Transactions of the Royal Society of London. http://rstl.royalsocietypublishing.org/content/129/39

Figure 35.
Photo by Richard Crowest. Source: Wikimedia Commons, licensed under Creative Commons Attribution-Share Alike 3.0 Unported license. https://commons.wikimedia.org/wiki/File:Parallel_Roads.JPG

Figure 36.
Photo by Gayle Gordon.

Figures 37.
Photo by the author.

Figures 38-40.
Photos by Daniel Melnick.

Figures 41-43.
Photos by the author.

Figure 44.
Officers of the HMS *Beagle. Santa Maria Island, South America, Coast of Chile.* Chart No. 1303. London: Hydrographic Office of the Admiralty, 1840. Source: Scanned from original at U.S. Library of Congress.

Figures 45-47.
Photos by the author.

Figures 48-51.
Modified after Atwater, Brian F., Geological Survey (U.S.), National Tsunami Hazard Mitigation Program (U.S.), and Universidad Austral de Chile. *Surviving a Tsunami—Lessons from Chile, Hawaii, and Japan.* Circular 1187. Revised and reprinted 2005. ed., Reston, Va. and Denver, Co.: U.S. Dept. of the Interior, U.S. Geological Survey, 2005. https://pubs.usgs.gov/circ/c1187/

Figure 52.
© AP Photo/ Natacha Pisarenko.

Figure 53.
© Ric Francis/ZUMA Press/Almay Stock Photo

Figure 54.
Photo by the author.

Figure 55.
Graphic by Marco Cisternas. Base image: Google, Digital Globe.

Figure 56.
Photo by Lisa Ely.

Figures 57-58.
Photos by the author.

Figure 59.
Photo by Daniel Melnick

Figure 60-61.
Photo by the author.

Figure 62.
Photo by Julia Jones.

Acknowledgments

This book is a tribute to the genius of Charles Darwin. It rests on the foundation built by historians who have studied his contributions, and geologists who have advanced their science. My first debt is to Darwin scholars and historians of geology, most notably Janet Browne, Martin J. S. Rudwick, and Sandra Herbert. As I was trying to sharpen the focus of the story presented here, and began to zero in on Darwin's ideas about tectonics, I was particularly encouraged to proceed by a paper by Frank H. T. Rhodes.

The book would not likely have been possible without the resources available from the Darwin Online and Darwin Correspondence projects at Cambridge University, as well as the Jefferson County, Colorado, Public Library and its participation in the Prospector system of the Colorado Alliance of Research Libraries.

My own fieldwork got off to a terrific start when I approached Alan Nelson and he suggested that I contact Brian Atwater. Through Brian's kindness and support, I got off the ground and later met Marco Cisternas and Lisa Ely. Then through Marco, I met Daniel Melnick. All five have become good friends and colleagues. I especially want to thank Marco's wife Patricia Martínez for her patience and hospitality.

The research in Chile by Lisa and Marco has been supported by grants over the last several years from the National Geographic Society, the U.S. National Science Foundation (NSF), and the Chilean equivalent of NSF, FONDECYT. Daniel's work has been supported by the German Science Foundation. I was a formal investigator on only one of the NSF grants, but I have benefitted directly or indirectly from all of these.

Beyond Brian, Marco, Lisa, Daniel, and Alan, I have been privileged to work in Chile together with (in something like chronological order) Marcelo Lagos, Yuki Sawai, Kako Rodriguez, Caitlin Orem, Daniel Ramirez, James Goff, Catherine Chagué-Goff, Zamara Fuentes, Ed Garrett, Tina Dura, Julius Jara, Simon Engelhart, Diego Muñoz, Jessica Pilarczyk, Matías Carvajal, Marcos Moreno, Isabel Hong, Ben Horton, Bre MacInnes, Daria Nikitina, Cristian Araya, Alexandra Ruiz, Cyntia Mizobe, and Hermann Fritz. I was received with kindness by the SERNAGEOMIN-USGS team at Chaitén in 2010 including Álvaro Amigo, John Pallister, John Eichelberger, Jake Lowenstern, Rick Hoblitt, Tom Pierson, and Jon Major. Meeting Julia Jones, Charlie Crisafulli, and Fred Swanson at Chaitén was a special bonus. In Chile I also owe special thanks to Paul Delaveau, Carlos Vergara (and to his fellow Carabinero whose name I unfortunately did not record), Señora Ruth Nuñez, and Javier Basque for help on Isla Santa María, and to Señora Rosa Vergara who cooked marvelous *curantos* for us in Maullín.

I am extremely grateful to so many people in Chile who let us dig holes on their land or were kind enough to share their stories with us including José Luis Ruíz, Gabriel Gutíerrez, José Luis Díaz, Doña Elena Riquelme, Joel and Constanza Riquelme, Doña Josefina Astorga, José Luis Montero, and Luis Bravo.

Greg McDonald was very kind to get me started on the ground sloths and to put me in touch with paleontologists in Argentina and Uruguay. Daniel Perea and Valeria Mesa were extremely gracious to take me into the field with them in Uruguay, as were Teresa Manera and Rodrigo Tomassini in Argentina. Juan Carlos Fernicola, Sergio

Vizcaíno, and Marcelo Zárate were all generous with their time for discussions. Special thanks also to Javier Defferrari and his wife Claudia of the Estancia La Porteña in Uruguay.

Alejandro Aranda and Horacio Osorio led our trek across the Andes, which we enjoyed with a group including eight delightful people from Mendoza. Laura Giambiagi was kind to discuss Darwin's geology of the Andes with me. David Catling was a terrific field companion in Patagonia and on the Río Santa Cruz. I owe special thanks to Milthon Rischmann and Ruben (Flaco) Alfonso for making our kayak trip possible. I am also grateful to Steve Austin for sharing his experience on the Río Santa Cruz with me.

Jonathan Hodge helped me get started in the UK. Michael Roberts was extremely helpful with Darwin in Wales and with his perspective on Sedgwick. He and his wife, Andrea, have become good friends and are wonderful host and hostess. Peter Worsley told me that Darwin was "besotted with icebergs." Special thanks also to Joy Hendry and Mrs. Hilda Padel, and to Guy Hannaford of the UK Hydrographic Office and to Caroline Lam, Archivist at the Geological Society of London.

I am grateful to the organizers and participants in writers' workshops at the Lighthouse Writers Workshop of Denver, Aspen Summer Words, and the Taos Summer Writers' Conference for their support and critiques. Very special thanks to Shari Caudron, Harrison Candelaria Fletcher, Bill Loizeaux, Greg Campbell, and Jane von Mehren who all played key roles in helping me shape this narrative.

I am especially grateful to friends, colleagues from the USGS and elsewhere, and family who plowed through one or more chapters giving me helpful feedback including Richard Leonard, John Unger, Alan Feiger, Tom Neel, Sumner Brown, Jill McCarthy, Rich Briggs, Dan McNamara, Joy Hendry, Michael Roberts, Marco Cisternas, Daniel Melnick, Bill Ellsworth, Ray Watts, Pradeep Talwani, Alan Lindh, Sarah Gillingham, and Alex Wesson. Lisa Ely, John Sumner, Elizabeth Wesson, and especially Gayle Gordon, earned the title of super-readers.

Other colleagues at the USGS have been extremely supportive and helpful, notably Peter Haeussler, Mark Petersen, Buddy Schweig, and Bob Thompson. Thanks to Bob for his introduction to the ground sloths of the southwestern U.S.

The encouragement of Phil Smith and Ruth Wesson was especially important in the early stages of this project. Sadly neither is around to see the final product.

I cannot overemphasize my gratitude to Marco Cisternas and Lisa Ely for their support at every stage from fieldwork to writing.

I am exceptionally grateful to Jane von Mehren, my agent, who worked with me and helped me finally get across the line with a publishable concept, and to Jessica Case for her confidence in that concept, her improvements through editing, and her overall encouragement. Drew Wheeler caught more spelling errors than I care to admit.

Most important of all was my wife, Gayle Gordon, who not only accompanied me on important parts of the journey and read every word multiple times, but who also tolerated my time away from home and in front of my computer, and paid the opportunity cost that involved. She suffered not so much from the paroxysms of the planet, as from the obsession of her husband.

Thank you to all, and to anyone I have inadvertently omitted.

Notwithstanding my gratitude for all this help, I accept full responsibility for the opinions and interpretations expressed here, as well as for any and all errors and omissions.

Index

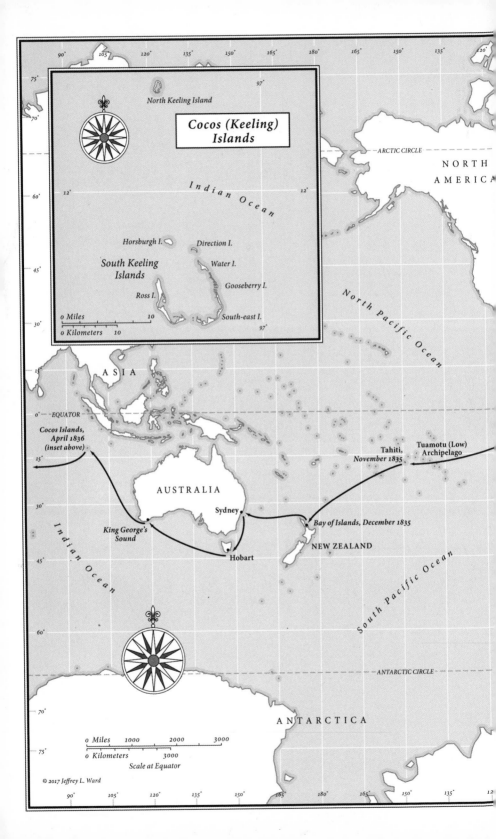

Cocos (Keeling) Islands

North Keeling Island

Indian Ocean

Horsburgh I.
Direction I.
Water I.
South Keeling
Islands
Gooseberry I.
Ross I.
South-east I.

0 Miles 10
0 Kilometers 10

ARCTIC CIRCLE

NORTH
AMERICA

North Pacific Ocean

ASIA

EQUATOR

Cocos Islands,
April 1836
(inset above)

Tahiti,
November 1835

Tuamotu (Low)
Archipelago

AUSTRALIA

Sydney

King George's
Sound

Bay of Islands, December 1835

NEW ZEALAND

Hobart

Indian Ocean

South Pacific Ocean

ANTARCTIC CIRCLE

ANTARCTICA

0 Miles 1000 2000 3000
0 Kilometers 3000
Scale at Equator

© 2017 Jeffrey L. Ward